GENETICS OF NEUROLOGICAL AND PSYCHIATRIC DISORDERS

Research Publications:
Association for Research in Nervous and Mental Disease

Volume 60

Genetics of Neurological and Psychiatric Disorders

*Research Publications:
Association for Research in Nervous and
Mental Disease*
Volume 60

Editors

Seymour S. Kety, M.D.
*Mailman Research Center
Belmont, Massachusetts and
Harvard Medical School
Boston, Massachusetts*

Lewis P. Rowland, M.D.
*H. Houston Merritt Clinical
Research Center for Muscular Dystrophy
and Related Diseases
Columbia-Presbyterian Medical Center
New York, New York*

Richard L. Sidman, M.D.
*Children's Hospital Medical Center
Boston, Massachusetts and
Harvard Medical School
Boston, Massachusetts*

Steven W. Matthysse, Ph.D.
*Mailman Research Center
McLean Hospital
Belmont, Massachusetts and
Harvard Medical School
Boston, Massachusetts*

Raven Press ■ New York

Raven Press, 1140 Avenue of the Americas, New York, New York 10036

© 1983 by Raven Press Books, Ltd. All rights reserved. This book is protected by copyright. No part of it may be reproduced, stored in a retrieval system, or transmitted, in any form or by any means, electronic, mechanical, photocopying, recording, or otherwise, without the prior written permission of the publisher.

Library of Congress Cataloging in Publication Data
Main entry under title:

Genetics of neurological and psychiatric disorders.

(Research publications / Association for Research in Nervous and Mental Disease; v. 60)
Includes bibliographies and index.
1. Nervous system—Diseases—Genetic aspects.
2. Mental illness—Genetic aspects.   3. Neurogenetics.
I. Kety, Seymour S. II. Series: Research publications (Association for Research in Nervous and Mental Disease; v. 60. [DNLM: 1. Nervous system diseases—Familial and genetic. 2. Mental disorders—Familial and genetic. W1 RE233p v. 60 / WL 100]
RC346.G46 1983   616.8'0442   82-16672
ISBN 0-89004-626-3

Great care has been taken to maintain the accuracy of the information contained in the volume. However, Raven Press cannot be held responsible for errors or for any consequences arising from the use of the information contained herein.

Materials appearing in this book prepared by individuals as part of their official duties as U.S. Government employees are not covered by the above-mentioned copyright.

ASSOCIATION FOR RESEARCH IN NERVOUS AND MENTAL DISEASE

OFFICERS 1980

Seymour S. Kety, M.D.
President
Mailman Research Center
McLean Hospital
Belmont, MA. 02178

Lewis P. Rowland, M.D.
Vice-President
Neurological Institute
710 West 168th Street
New York, N.Y. 10036

Richard L. Sidman, M.D.
Vice-President
Children's Hospital Medical Center
300 Longwood Avenue
Boston, MA. 02115

Steven W. Matthysse, Ph.D.
Mailman Research Center
McLean Hospital
Belmont, MA. 02178

Bernard Cohen, M.D.
Secretary-Treasurer
Mt. Sinai School of Medicine
1 Gustave Levy Place
New York, N.Y. 10029

Richard Friedman, M.D.
Assistant Secretary Treasurer
New York Cornell-
Westchester Division
White Plains, N.Y. 10605

TRUSTEES

Melvin D. Yahr, M.D.
Chairman
Shervert Frazier, M.D.
Robert Katzman, M.D.
Seymour S. Kety, M.D.
Lawrence C. Kolb, M.D.
Robert Michels, M.D.
Fred Plum, M.D.
Dominick Purpura, M.D.
Clark T. Randt, M.D.
Lewis P. Rowland, M.D.
Edward Sachar, M.D.
Albert J. Stunkard, M.D.

HONORARY TRUSTEES

Francis J. Braceland, M.D.
Clarence C. Hare, M.D.

COMMISSION—1980

Roscoe O. Brady, M.D.
Bethesda, Maryland

James F. Crow, Ph.D.
Madison, Wisconsin

Park S. Gerald, M.D.
Boston, Massachusetts

L. Erlenmeyer-Kimling, Ph.D.
New York, New York

Seymour S. Kety, M.D.
Boston, Massachusetts

Stephen Matthysse, Ph.D.
Boston, Massachusetts

Dominick Purpura, M.D.
Bronx, New York

Lewis P. Rowland, M.D.
New York, New York

Edward Sacher, M.D.
New York, New York

Richard L. Sidman, M.D.
Boston, Massachusetts

Albert J. Stunkard, M.D.
Philadelphia, Pennsylvania

COMMITTEE ON NOMINATIONS

Shervert H. Frazier, M.D. Chairman
Clark T. Randt, M.D.
Dominick Purpura, M.D.

COMMITTEE ON ADMISSIONS

Sidney Malitz, M.D. Chairman
Margaret Hoehn, M.D.
Arthur Schwartz, M.D.

COMMITTEE ON PUBLIC RELATIONS

Arthur Zitrin, M.D., Chairman
Robert Barrett, M.D.
Ivan Bodis-Wollner, M.D.

Publisher's Note

Titles marked with an asterisk (*) are out of print in the original edition. Some out-of-print volumes are available in reprint editions from Hafner Publishing Company, 866 Third Avenue, New York, N.Y. 10022.

  I. (1920) *Acute Epidemic Encephalitis (Lethargic Encephalitis)
  II. (1921) *Multiple Sclerosis (Disseminated Sclerosis)
  III. (1923) *Heredity in Nervous and Mental Disease
  IV. (1924) The Human Cerebrospinal Fluid
  V. (1925) *Schizophrenia (Dementia Praecox)
  VI. (1926) *The Cerebellum
  VII. (1922) *Epilepsy and the Convulsive State (Part I)
    (1929) *Epilepsy and the Convulsive State (Part II)
  VIII. (1927) *The Intracranial Pressure in Health and Disease
  IX. (1928) *The Vegetative Nervous System
  X. (1929) *Schizophrenia (Dementia Praecox) (Communication of Vol. V)
  XI. (1930) *Manic-Depressive Psychosis
  XII. (1931) *Infections of the Central Nervous System
  XIII. (1932) *Localization of Function in the Cerebral Cortex
  XIV. (1933) *The Biology of the Individual
  XV. (1934) *Sensation: Its Mechanisms and Disturbances
  XVI. (1935) *Tumors of the Nervous System
  XVII. (1936) *The Pituitary Gland
  XVIII. (1937) *The Circulation of the Brain and Spinal Cord
  XIX. (1938) *The Inter-relationship of Mind and Body
  XX. (1939) *Hypothalamus and Central Levels of Autonomic Function
  XXI. (1940) *The Disease of the Basal Ganglia
  XXII. (1941) *The Role of Nutritional Deficiency in Nervous and Mental Disease
  XXIII. (1942) *Pain
  XXIV. (1943) *Trauma of the Central Nervous System
  XXV. (1944) *Military Neuropsychiatry
  XXVI. (1946) *Epilepsy
  XXVII. (1947) *The Frontal Lobes
  XXVIII. (1948) *Multiple Sclerosis and the Demyelinating Diseases
  XXIX. (1949) *Life Stress and Bodily Disease
  XXX. (1950) *Patterns of Organization in the Central Nervous System
  XXXI. (1951) Psychiatric Treatment
  XXXII. (1952) *Metabolic and Toxic Diseases of the Nervous System
  XXXIII. (1953) *Genetics and the Inheritance of Integrated Neurological Psychiatric Patterns
  XXXIV. (1954) *Neurology and Psychiatry in Childhood
  XXXV. (1955) Neurologic and Psychiatric Aspects of Disorders of Aging
  XXXVI. (1956) *The Brain and Human Behavior
  XXXVII. (1957) *The Effect of Pharmacologic Agents on the Nervous System
  XXXVIII. (1958) *Neuromuscular Disorders
  XXXIX. (1959) *Mental Retardation
  XL. (1960) Ultrastructure and Metabolism of the Nervous System
  XLI. (1961) Cerebrovascular Disease
  XLII. (1962) *Disorders of Communication
  XLIII. (1963) *Endocrines and the Central Nervous System
  XLIV. (1964) Infections of the Nervous System
  XLV. (1965) *Sleep and Altered States of Consciousness
  XLVI. (1968) Addictive States
  XLVII. (1969) Social Psychiatry
  XLVIII. (1970) Perception and Its Disorders
  XLIX. (1971) Immunological Disorders of the Nervous System
  50. (1972) Neurotransmitters
  51. (1973) Biological and Environmental Determinants of Early Development
  52. (1974) Aggression
  53. (1974) Brain Dysfunction in Metabolic Disorders
  54. (1975) Biology of the Major Psychoses: A Comparative Analysis
  55. (1976) The Basal Ganglia
  56. (1978) The Hypothalamus
  57. (1979) Congenital and Acquired Cognitive Disorders
  58. (1980) Pain
  59. (1981) Brain, Behavior, and Bodily Disease

Preface

This volume summarizes current knowledge concerning the genetics of nervous and mental disorders and the neural substrates upon which they develop. As in the case of other research publications of the Association for Research in Nervous and Mental Disease, the topic was carefully selected for its timeliness and significance, and because of the numerous advances in knowledge that have occurred over the past several years. The topics that comprise the eighteen chapters were selected to represent the most significant and recent foci of research interest at both the basic and clinical levels and in each area, an outstanding individual was invited to prepare a succinct and comprehensive review.

The first chapters deal with the genetic aspects of the development of the nervous system and some of its particular components. A thoughtful and provocative contribution by a distinguished population geneticist is the keynote for the remaining chapters that focus on clinical disorders and the genetic and environmental contributions to their etiology and pathogenesis. These chapters describe new strategies and techniques and the resulting knowledge that pertains to particular neurological and psychiatric disorders.

The application of new knowledge, strategies, and powerful techniques from molecular biology, embryology, biochemistry, virology, and epidemiology have brought new progress in our understanding of the genetic contributions to nervous and mental diseases. There are imaginative and promising contributions to the diagnosis of genetic disorders and the prospect of treating metabolic errors of genetic origin from outstanding laboratories of molecular biology. New insights into the etiology at the environmental as well as genetic levels also have emerged, and a wealth of knowledge regarding the processes interposed between etiology and clinical disorder has developed in a number of disciplines.

The meetings of the Association for Research in Nervous and Mental Disease and the volumes that subsequently emerge are designed to fulfill the expectations of basic neuroscientists and practitioners of neurology and psychiatry who desire sound and authoritative reviews of fields of current interest and significance. It is the hope of the editors that this volume will be found useful by them and a valuable contribution to the archival literature.

The Editors

Acknowledgments

The Editors wish to acknowledge the support of the Trustees of the Association for Research in Nervous and Mental Disease who chose the topic and gave valuable advice on the selection of the contributors. The organization of the meeting and the editing of this volume was greatly facilitated by Mrs. Ruth Warren and Miss Sandra Cole, respectively. The staff of Raven Press provided invaluable assistance in the final preparation and publication of the volume.

Contents

1   On the Pathways of Neural Development
    Cyrus Levinthal and Françoise Levinthal

19 Experimental Neurogenetics
    Richard L. Sidman

47 Genetic Control of Developmental Antigens
    Dorothea Bennett

55 Genetic Control of the Number of Dopamine Neurons in the Brain: Relationship to Behavior and Responses to Psychoactive Drugs
    Donald J. Reis, J. Stephen Fink, and Harriet Baker

77 Genetics of Neuronal Form
    Steven Matthysse and Roger Williams

93 Some Perspectives from Population Genetics
    James F. Crow

105 Observations on Genetic and Environmental Influences in the Etiology of Mental Disorder from Studies on Adoptees and Their Relatives
    Seymour S. Kety

115 Use of the Danish Adoption Register for the Study of Obesity and Thinness
    Albert J. Stunkard, Thorkild Sørensen, and Fini Schulsinger

121 Genetics of the Major Psychoses
    Elliot S. Gershon

145 Genetic Heterogeneity in Alcoholism and Sociopathy
    C. Robert Cloninger and Theodore Reich

167 Application of Recombinant DNA Techniques to Neurogenetic Disorders
    David Housman and James F. Gusella

173 Recombinant DNA and the Analysis of Cytogenetic Disorders Associated with Mental Retardation
    P.S. Gerald and G.A. Bruns

- 181 Genetic Errors and Enzyme Replacement Strategies
  Roscoe O. Brady
- 195 Dominant Ataxias
  Roger N. Rosenberg
- 215 Genetic Heterogeneity of the Hexoaminidase Deficiency Diseases
  William G. Johnson
- 239 Glycogen-Storage Diseases of Muscle: Genetic Problems
  Lewis P. Rowland and Salvatore DiMauro
- 255 Genetic Predisposition to Environmental Factors
  Jack P. Antel and Barry G. W. Arnason
- 273 Familial Spongiform Encephalopathies
  David M. Asher, Colin L. Masters, D. Carleton Gajdusek, and Clarence J. Gibbs, Jr.
- 293 Subject Index

Contributors

Jack P. Antel
Department of Neurology
University of Chicago Pritzker School of Medicine and
Division of Biological Sciences
Chicago, Illinois 60637

Barry G. W. Arnason
Department of Neurology
University of Chicago Pritzker School of Medicine and
Division of Biological Sciences
Chicago, Illinois 60637

David M. Asher
Laboratory of Central Nervous System Studies
National Institute of Neurological and Communicative Disorders and Stroke
National Institutes of Health
Bethesda, Maryland 20205

Harriet Baker
Laboratory of Neurobiology
Department of Neurology
Cornell University Medical College
New York, New York 10021

Dorothea Bennett
Sloan-Kettering Institute for Cancer Research
New York, New York 10021

Roscoe O. Brady
Developmental and Metabolic Neurology Branch
National Institute of Neurological and Communicative Disorders and Stroke
National Institutes of Health
Bethesda, Maryland 20205

G. A. Bruns
Department of Pediatrics
Harvard Medical School
Boston, Massachusetts 02115

C. Robert Cloninger
Department of Psychiatry
Washington University School of Medicine and
Jewish Hospital of St. Louis
St. Louis, Missouri 63110

James F. Crow
Department of Medical Genetics
University of Wisconsin
Madison, Wisconsin 53706

Salvatore DiMauro
Department of Neurology and the
H. Houston Merritt Clinical Research Center for Muscular Dystrophy and Related Diseases
Columbia-Presbyterian Medical Center
New York, New York 10032

J. Stephen Fink
Laboratory of Neurobiology
Department of Neurology
Cornell University Medical College
New York, New York 10021

D. Carleton Gajdusek
Laboratory of Central Nervous System Studies
National Institute of Neurological and Communicative Disorders and Stroke
National Institutes of Health
Bethesda, Maryland 20205

P. S. Gerald
Clinical Genetics Division
Children's Hospital Medical Center
Boston, Massachusetts 02115

Elliot S. Gershon
Section on Psychogenetics
Biological Psychiatry Branch
National Institute of Mental Health
Bethesda, Maryland 20205

CONTRIBUTORS

Clarence J. Gibbs, Jr.
Laboratory of Central Nervous System
  Studies
National Institute of Neurological and
  Communicative Disorders and Stroke
National Institutes of Health
Bethesda, Maryland 20205

James F. Gusella
Department of Neurology
Massachusetts General Hospital
Boston, Massachusetts 02114

David Housman
Center for Cancer Research
Massachusetts Institute of Technology
Cambridge, Massachusetts 02139

William G. Johnson
Columbia University
College of Physicians and Surgeons
New York, New York 10032

Seymour S. Kety
Laboratories for Psychiatric Research
Mailman Research Center
Belmont, Massachusetts 02178 and
Department of Psychiatry
Harvard Medical School
Boston, Massachusetts 02115

Cyrus Levinthal
Department of Biological Sciences
Columbia University
New York, New York 10027

Françoise Levinthal
Department of Biological Sciences
Columbia University
New York, New York 10027

Colin L. Masters
Laboratory of Central Nervous System
  Studies
National Institute of Neurological and
  Communicative Disorders and Stroke
National Institutes of Health
Bethesda, Maryland 20205

Steven Matthysse
Mailman Research Center
McLean Hospital
Belmont, Massachusetts 02178

Theodore Reich
Department of Psychiatry
Washington University School of
  Medicine and
Jewish Hospital of St. Louis
St. Louis, Missouri 63110

Donald J. Reis
Laboratory of Neurobiology
Department of Neurology
Cornell University Medical College
New York, New York 10021

Roger N. Rosenberg
Laboratory of Cellular Neurobiology
Department of Neurology
University of Texas Health Science
  Center
Southwestern Medical School
Dallas, Texas 75235

Lewis P. Rowland
Department of Neurology and the
H. Houston Merritt Clinical Research
  Center for Muscular Dystrophy and
  Related Diseases
Columbia-Presbyterian Medical Center
New York, New York 10032

Fini Schulsinger
Psykologisk Institut
Kommunehospitalet
Copenhagen, Denmark

Richard L. Sidman
Department of Neuroscience
Children's Hospital Medical Center
Boston, Massachusetts 02115

Thorkild Sørensen
Psykologisk Institut
Kommunehospitalet
Copenhagen, Denmark

Albert J. Stunkard
Department of Psychiatry
University of Pennsylvania
Philadelphia, Pennsylvania 19104

Roger Williams
Mailman Research Center
McLean Hospital
Belmont, Massachusetts 02178

Genetics of Neurological and Psychiatric Disorders, edited by Seymour S. Kety, Lewis P. Rowland, Richard L. Sidman, and Steven W. Matthysse. Raven Press, New York © 1983.

On the Pathways of Neural Development

Cyrus Levinthal and Françoise Levinthal

Department of Biological Sciences, Columbia University, New York, New York 10027

It is frequently convenient to divide the problems of basic neurobiology into three groups, related to how a nervous system is formed, how it works, and how it is modified. Our laboratory has been concerned with the first two sets of problems: what are the rules governing the orderly assembly of the large array of neurons formed by the visual system, and how are some of its simple functions carried out through the connections whose formation we have studied.

The central problems of developmental neurobiology are, for the most part, the same as those that have concerned developmental biologists since the early decades of this century, after biologists in 1905 rediscovered the seminal work of Mendel on heredity published in 1866. The development of formal genetics by Morgan, his collaborators, and contemporaries had as one of its main objectives the understanding of how genes function and how they are controlled in such a way that cells differentiate from each other and interact in an organized fashion.

The inheritance of familial traits such as nose, ears, and hands makes it clear that, in some way, recognizable morphological features are inherited, and from the apparently strict inheritance of certain features it seems that the genes involved must be either very few or in very closely linked clusters. This does not, of course, imply that all the genes that determine the shapes of noses, for example, are in the same cluster, but only that those which determine the characteristics that cause us to see resemblance are closely linked. However, likeness of facial features remains largely anecdotal, and detailed quantitative studies of the morphological similarity of nerve cells has been more extensive.

When we speak of understanding morphology, we mean both the external morphology of organs and organisms and the morphology of the individual cells that constitute the organ. We would like to find out as much as we can about the control of this morphology. To what extent is it determined by the genes of the organism, and to what extent is it determined by the environment either of the animal or of the individual cells? With some understanding of these processes, we then want to ask what are the gene products that result in the particular structure observed, how are the genetic interactions expressed, and how is the expression of these genes controlled by the cellular environment?

By and large, we expect the gene products important in controlling the cellular morphology to be proteins, and many of them will probably be membrane proteins because cells that must recognize and be altered by other cells are likely to be affected by proteins on their membranes. We also assume that the way in which an individual cell grows is dependent on its history in addition to what it recognizes in its environment.

There is rather good experimental evidence that the number of genes that control the morphology of brains is not enormous. Investigators who have tried to isolate mutant organisms affecting the nervous system have estimated that the number of mutations is between a few hundred and a few thousand. However, in higher organisms such as humans, the total number of nerve cells is hundreds of billions—many orders of magnitude more than the maximum possible number of genes. Thus it seems clear that one cannot consider the hypothesis that individual genes control the shapes of individual cells. The genetic instructions given to most nerve cells must be in the form of some kind of *class instructions*. The question then should be: what are the class instructions that individual genes give to a group of cells that lead them to grow into a particular pattern?

In trying to answer this question, our laboratory has taken a strategy that differs somewhat from that used by many other investigators. We do not look for mutations that affect the nervous system. Instead, we have studied organisms in which the genetic background can be precisely controlled so that we can examine different organisms that have the same genes. We try to determine how similar the details of the anatomy of individual cells in these organisms are and, if we make changes (deletions, alterations, or additions of individual cells), the consequences of the modifications and the rules regulating them.

STUDIES AT EARLY DEVELOPMENT

The general logic of our approach requires several attributes for the organisms under study. First, we want to work with the simplest organism having the required characteristics. Several years ago, we started studying the small water flea, *Daphnia magna*. *Daphnia* has only a few thousand cells in its nervous system and we restricted our analysis to nerves directly involved in the primary connections of the visual system. There are only a few hundred such cells. To be specific, there are 176 photoreceptor cells and 110 first-order nerve cells in the optic ganglion, and they make precisely ordered connections that are the same from one organism to another. Can we understand how those connections are made and why they are made in the order and pattern observed?

As a second attribute, these organisms should have as little genetic variation as possible. For *Daphnia* and several other organisms used in our laboratory, this is possible because these organisms reproduce, as a natural part of their life cycle, by parthenogenesis, whereby a female makes offspring identical to herself without any genetic intervention of a male. This type of reproduction is found naturally in several simple systems, and in these systems we can obtain genetically identical organisms with no difficulty. In more complex systems, for example, in the case of certain lower vertebrates such as frogs and fishes, there are ways by which such genetic identity can be induced in the laboratory.

Another general aspect of our studies is one that is extraordinarily old-fashioned. We started with the notion that it is necessary to make detailed observations of the shape of individual cells and how they change during development. This kind of biology was very actively pursued from about 1885 to 1920, when it tended to be bypassed by biochemistry and was almost forgotten until a few years ago. The methods we have used are conceptually similar to those used at the turn of the century. If one wants to study the detailed anatomy of a three-dimensional structure, there is no present alternative to cutting the piece of tissue into many serial thin sections and looking at each section with a light or electron microscope. Then, from the mass of photographic data, one reconstructs the three-dimensional structure and tries to find out how it is put together. Among the main methodological differences between what we are doing in the laboratory and what was done in the 1890s are the use of electron microscopes instead of relying exclusively on light microscopes and the extensive use of computer graphics to handle the large amount of three-dimensional information we are trying to extract from the set of photographs.

In order to study the details of the interactions, we have developed several procedures for handling large numbers of serial sections, both optically and computationally. These procedures and the devices that support them are called the CARTOS system, for Computer Aided Reconstruction by Tracing Of Serial Sections (8,9,18). This system permits visualization of the three-dimensional reconstruction of individual neurons or collections of neurons as well as their interactions.

Our objective was to study the normal development of these small animals and see if the rules by which they develop could be inferred. We try to make models, mathematical, computational, or otherwise, and ask what predictions the model would allow about the ability to correct various defects or induced alterations in the system. We then carry out experiments to determine if the results confirm or refute the model.

THE WATER FLEA AS A MODEL SYSTEM

Daphnia magna is a few millimeters long with a compound eye containing 22 lenses, each surrounded by eight photoreceptor cells. The 176 receptor cells make first-order connections with 22 sets of five cells in the optic ganglion. Several years ago, we analyzed the detailed anatomy of these 286 (176 receptor and 110 laminar cells in the optic ganglion) cells as they grow and make connections with each other in the embryonic animal and proposed a model that seems to explain some aspects of their orderly development (10,13,16).

In this model, we assume that the embryonic cells that are to become the photoreceptors and those that are to become ganglion cells are specified early in development as to their general type but not as to their specific identity. Thus, the 176 receptor cells have the necessary genes activated to produce receptor cells but they are not necessarily specified as to which one of the 176 they are to become. One of these presumptive receptor cells, which happens to be close to the midline of the animal, starts to differentiate and send out an axon that grows towards the midplane of the head and then backwards in the general direction of the presumptive ganglion cells. When the first axon reaches the group of presumptive laminar cells, it makes contact with five of them in a reproducible manner. In this process of "making contacts," the laminar cell wraps around the growing axon and establishes a very close membrane-to-membrane contact in which transient gap junctions are observed (Fig. 1). These junctions last for only approximately 20 min, a short time compared to the 50 hr required for the developmental time of the embryo. We postulated that the gap junctions provide some form of trigger for further developmental events.

Shortly after the first receptor cell starts to differentiate, others that will be part of the same group of eight also differentiate and send out axons that follow along the path taken by the first fibers. In electron micrographs, the following axon gives the appearance of growing along the surface of the lead fiber. As the follower fiber reaches the region where the leader has been wrapped by ganglion cells, the close contact of the wrapping process is loosened. At approximately the same time, the laminar cells themselves differentiate (Fig. 2) and send out fibers that travel in the same direction as those from the receptor cells. We suggest that the laminar cells receive a signal from the lead fiber while they are tightly wrapped, and this signal triggers the further differentiation process of the ganglion cells.

Several additional features of the proposed model are necessary to account for even these limited aspects of the differentiation and growth of the 286 nerve cells. However, its important general features can be summarized briefly. We make the following assumptions: First, all the presumptive receptor cells are operating under the same set of genetic instructions and all the laminar cells are operating under a different set of instructions. Second, the specific identity and the subsequent growth and development of an individual nerve cell is determined by the place of its cell body at the time it receives a signal to differentiate. Third, growing nerves have, under certain conditions, the ability to follow closely along the surface of another nerve that grew at an earlier time.

Obviously, any model that relies on such a simple set of assumptions cannot claim to represent a complete description of neural development. At a minimum, several additional features reflecting the history of the embryo at the time of the differentiation of the first nerve cells would have to be included. For example, even if this model were completely correct, signals from tissue not made of nerves would still be required to provide the guiding pathway that the first nerve fibers could follow.

After we had formulated this limited type of model based on detailed anatomical observations during early development, Professor Eduardo Macagno, our colleague at Columbia, initiated a series of experimental tests designed to determine the validity of various aspects of it (17). Specifically, he investigated the ways in which destroying or delaying the development of a few receptor nerve cells would affect the subsequent development of the entire set of 286 cells. The results obtained in these experiments confirmed the general features of the model as we proposed it.

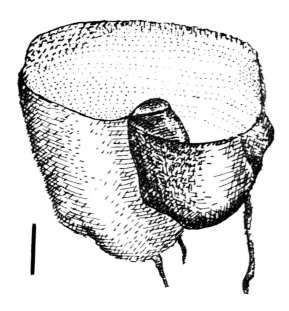

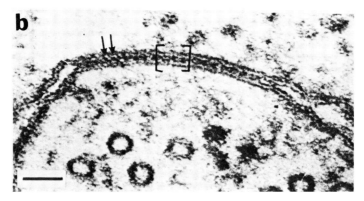

FIG. 1. A: A drawing made from a computer reconstruction showing the relationship between the lead axon and the recently contacted neuroblasts. In order to show the groove where the lead fiber rests, only half of the midplane neuroblast is illustrated here. This neuroblast will start to differentiate into a lamina cell shortly after its contact with the lead fiber. (Bar = 1 μm.) **B:** High-power micrograph of the labile contact established between the lead fiber and the neuroblast. The seven-layered structure of the gap junction is best seen between the brackets. The arrow indicates the area of periodic cross-sections where the gap junction has been sectioned more obliquely (Bar = 400 Å.) (14).

THE SHAPES OF IDENTIFIED CELLS

Most of the morphological studies of individually identified neurons designed to determine their similarities or differences have been carried out in invertebrates

FIG. 2. Computer reconstruction of one lamina cell at various times during development. **Right:** cell of an adult. Proceeding towards the left, the same cell is shown from a 47-, 44-, 37.5-, and 34-hr-old embryo. Because the adult cell was traced from low-power micrographs, some of the fine branches were not identified or recorded.

using three-dimensional reconstruction of groups of cells in defined volumes of tissue (16) or of single identified cells injected with a dye (5). These studies show that identified cells in invertebrates have general shapes that are characteristic and recognizable but exhibit anatomical variations, thought of as "developmental noise." A few systems in vertebrate organisms have uniquely identifiable cells amenable to this type of detailed anatomical study. One of these is the Mauthner cell system in fishes. These cells are a pair of large motor neurons located in the medulla, mediating the escape response.

The morphology of these cells was determined in two species of small tropical fish, each of which can be maintained so that their genetic background is constant. One of these species, *Poecilia formosa*, has a parthenogenetic pattern of reproduction as a normal part of its life cycle. The other species, the common Zebra fish *(Brachydanio rerio)*, can be made to reproduce parthenogenetically. Clones of animals are produced with individuals having identical genes except for the rare spontaneous mutations. These procedures were developed by Dr. George Steissinger and his colleagues at the University of Oregon (21). When identified cells of these two species were analyzed, two rather unexpected observations were made. It was found that in addition to the two Mauthner motor neurons long known as uniquely identifiable neurons, there are many other motor neurons that can be classified as unique by virtue of their position relative to the Mauthner and their general shape and orientation. In a small region of the medulla, approximately 300 μm thick, more than 20 pairs of neurons are uniquely identifiable and are reproducibly present in all animals examined.

The other unexpected result of this analysis came from work with *Poecilia formosa*, which was the first species we studied. As expected, there was considerable variability in the shape and branching pattern from cell to cell and for the same cell from animal to animal when the three-dimensional reconstructions were examined in detail. However, surprisingly, there appeared to be a uniformity in the branching complexity of *all* of the cells from a single animal, and this variation showed itself as an animal-to-animal variation. Thus, in some animals most of the identified cells were more complex, and in others most of the cells were less complex (Fig. 3). This was true even though the animals examined had been from the same brood, were grown under as nearly identical conditions as we could maintain, and had the same weight, size, and external appearance at 6 months of age when they were sacrificed.

The *Poecilia formosa* is a "natural" species that is the result of the semi-fertile mating between two related but well defined species. Each such mating produces a female fish that can then produce offspring with no genetic contribution from a male. Thus, each of the initial matings can produce a different clone of genetically identical heterozygous, diploid animals as similar genetically as human identical twins. In addition, the *Poecilia* are live-bearers: the embryos develop in the body cavity of the mother. From these characteristics, two hypotheses present themselves to account for the animal-to-animal variation of the individuals analyzed. First, the fact that the animals are *heterozygous* diploid means that chromosome masking could possibly produce individual differences in certain cell lines, as does the expression of one or the other of the X chromosomes in female primates (15). Second, the individual differences might be the result of different environments for the embryos growing in the body cavity of the mother. Such differences are not reflected in any gross features of the adults because when the animals were sacrificed they all seemed to be identical; but they still might involve factors that affect the early development of neurons and their branching patterns, differences of afferent inputs during development (1,7) or differences in the chemical environment (18). It is certainly true that there were great differences in the extent to which the embryos were crowded in the body cavity of gravid females.

In order to evaluate these possibilities, the same analyses were repeated with the Zebra fish *(Brachydanio rerio)*, which is an egg-laying animal, so that the embryonic environment could be held constant in a Petri dish. In addition, techniques are available to produce either heterozygous or homozygous clones. The analysis of neurons of individual animals from such clones provides quite clear answers. Regardless of whether one used homozygous or heterozygous animals, the Zebra fish did not show the animal-to-animal variation, although they still have the same degree of "developmental noise" corresponding to variations between individual neurons. Our tentative conclusion is that the crowding in the body cavity of the mother produces these rather dramatic animal-to-animal variations in the complexity of the nerves observed in *Poecilia formosa*. The implications that a major effect on neural development can arise from differences in the embryonic environments

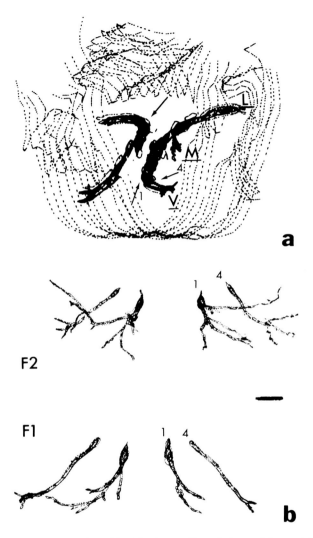

FIG. 3. A: Computer reconstruction of the Mauthner cells and of part of the medulla in *Poecilia formosa*. The contour of the medulla has been traced each 35 μm *(dotted lines)*. The two Mauthner cells are the large neurons on each side of the brain. They have three dendrites: ventral (v), lateral (l), and medial (m). The axons of the Mauthner cells leave the soma *(arrow)*, going towards the midline of the medulla. After crossing the midline, they turn towards the spinal cord where they run until its caudal end. A number of large motoneurons lie on each side of the ventral dendrite. We have studied two of them, Muller 1 (1) and Muller 4 (4), illustrated in part B. **B:** The axons of the Muller cells run uncrossed in the spinal cord, parallel to the Mauthner axons and are not shown in these computer drawings of two pairs of Muller cells in two different brains of isogenic fishes (F1) and (F2). The dendritic branching and the total volume occupied by these cells are strikingly different. The branching pattern is very simple in F1 and rather elaborate in F2. (Bar = 100 μm).

within the body cavity of the mother are obviously important. We have not as yet investigated the phenomenon further.

VERTEBRATE OPTIC NERVE FORMATION

More recently, we have been studying with Dr. Neil Bodick (when he was a graduate student at Columbia), the early development of the visual system of the Zebra fish (2,11). In this small vertebrate, the optic nerve of the adult contains approximately 50,000 nerve fibers. As in larger vertebrates with many more fibers, the optic nerve carries the visual information from the eye to the brain of the animal. An ordered map of the visual world is produced at the arborization of the optic fibers on the surface of the optic tectum. A question that has concerned developmental neurobiologists for many years relates to the mechanism used by this large number of nerve fibers that permits each fiber from the retina to reach its correct position in the brain. In an adult primate there are approximately 2 million fibers, and the optic nerve may be several centimeters long. However, by observing the developmental process in this small fish, one is immediately struck with an important, although obvious, generalization that applies to all animals from *Daphnia* to higher vertebrates. The distance that a nerve fiber must travel in order to make its connections in the embryo is much shorter than the distance that the nerve spans in the adult. In addition, the optic tecta in the embryos are immediately behind the eyes, and the path of the optic nerve is straight and direct compared with the path in the adult (Fig. 4).

The three-dimensional reconstructions made of optic nerves in embryo and newborn Zebra fish show some similarities with the observations made in the *Daphnia*. In very young embryos, the position that a particular fiber takes in the nerve is directly related to the position occupied by its cell body when the fiber first started its growth (Fig. 5). Furthermore, the position of the fiber within the nerve is, at least in part, determined by the fact that the growth cones on the tips of growing fibers are in intimate contact with more mature fibers from neighboring cell bodies. In addition, the most recently differentiated fibers always lie on the same side of the nerve as it leaves the eye (Fig. 6).

Many reconstructions of embryonic and newborn optic nerves have allowed us to formulate several rules to account for the organization of optic nerve fibers during their differentiation. These rules are sufficient to account for the orderly arrangement of fibers in the optic nerve. Furthermore, they are simple enough so that we can begin to suggest models as to what proteins would have to be produced by the genes in order to express such a process. The rules, or generalizations, that we derive from the anatomy of the optic nerves in early development are as follows:

(a) There is a radial pattern of fiber outgrowth in the retina recognized by several investigators that can be understood as a reflection of the mitotic pattern alone or in combination with a radial triggering of differentiation. In either case, one would consider the formation of such a pattern as the result of signaling between cells:

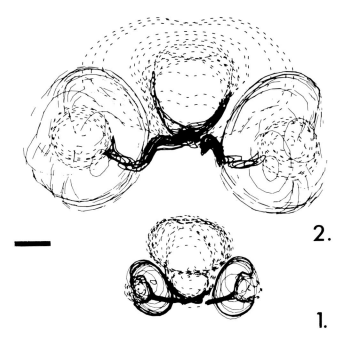

FIG. 4. Computer reconstruction of the eye, optic nerve, and brain of a newborn (1) and a 1-month-old Zebra fish (2). The length of the optic nerve from the exit of the eye to the optic chiasma is 55 μm in a newborn compared with 1,500 μm in the adult fish (7 months old). (Bar = 100 μm.)

those that have recently initiated axonal growth would signal their neighbor to do the same a short time later.

(b) The growing tips of newly differentiating fibers follow the axons of neighboring cells. To account for this neighbor-following, we assume that the growth cones have a tendency to follow more mature axons and that axons from neighboring cells are more readily available than others simply because of their proximity to the fiber starting to grow.

(c) The overall pattern of growth from the eye to the brain can be accounted for by these two rules plus the guidance provided by two nonneuronal structures: the choroid fissure remaining after the stalk connecting the eye to the brain is closed and the major retinal artery adjacent to the optic nerve. In the newborn animal all the growth cones are found on the side of the nerve adjacent to the artery (Levinthal and Levinthal, *in preparation*).

The position of the individual fibers in the optic nerve of an adult animal does not show the same orderly pattern observed in the newborn. After the fibers have made connection with their termination points in the brain, individual fibers or groups can change their relative position along the intervening region in an apparently random manner (Fig. 7). This relative motion of the fibers in the nerve of

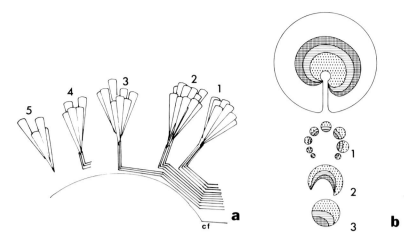

FIG. 5. A: Illustration of the order of ganglion cell differentiation inside a retina, drawn from a computer 3-D reconstruction of retina from a very early embryo. On the right, near the choroid fissure (cf), is the group of cells that differentiated first (1). Then a second group (2) differentiated with its axons growing near the ventral surface, displacing the first fasciles from the choroid fissure and the ventral surface. The group at the extreme left is still at the neuroblast stage. **B:** Diagrammatic representation of retinal ganglion axons in the developing nerve. The upper drawing represents the position of the ganglion cell bodies in the retina. The lower three drawings correspond to cross-sections through the optic nerve, indicating the oldest axons from the first cells to differentiate *(dots)*, younger axons *(hatched)*, the youngest axons *(cross-hatched)*, which still have their growth cones in or near the eye. **1:** A section near the optic nerve head, but still in the eye, where the small groups of axons have formed. Large fasciles are distributed along the vitreal surface. **2:** Further along the nerve, the fascicles have merged and form a crescent. **3:** Still further, the crescent has become more rounded, with the youngest axons, which are still growing, found on the ventral surface.

the mature animal does not affect the functional connectivity established in the embryo, but it does mean that important features of the developmental anatomy cannot necessarily be inferred by examining the anatomy of the adult.

Although a contrary view has been taken by many neurobiologists, we assume that the mechanisms involved in nerve regeneration in an adult animal are probably not the same as those used in establishing the initial connection in an embryo. The distances traveled by a nerve fiber to reach a specific target are much shorter in the embryo than for a regenerating nerve in the adult. In addition, there is a most important difference between the two processes: when a nerve is crushed or cut, the residual fibers can remain in place for a long period (6). The possibility that these residual fibers are used as guiding structures for ingrowing nerves is difficult to eliminate. On the other hand, to the extent that the growth of nerves in a developing embryo must rely on guiding structures, the guidance can be produced only by nonneuronal tissue or nerve fibers from recently differentiated cells.

FIRST-ORDER CONNECTIONS IN THE FISH TECTUM

A related set of investigations carried out in our laboratory with Peter Sajovic concerns the physiology and therefore the specific connections made by the optic

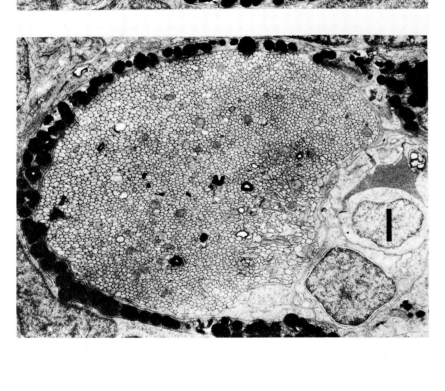

FIG. 6. Electron micrographs of cross-sections of optic nerve from a newborn Zebra fish. Both are taken from sections near the optic nerve head, inside the eye. They are separated by 3 μm. On each of them, irregular profiles of growth cones can be seen near the ventral part of the nerve. They correspond to the youngest axons, the oldest ones being assembled at the dorsal part of the nerve, shown at the top of the figure. (Bar = 1 μm.)

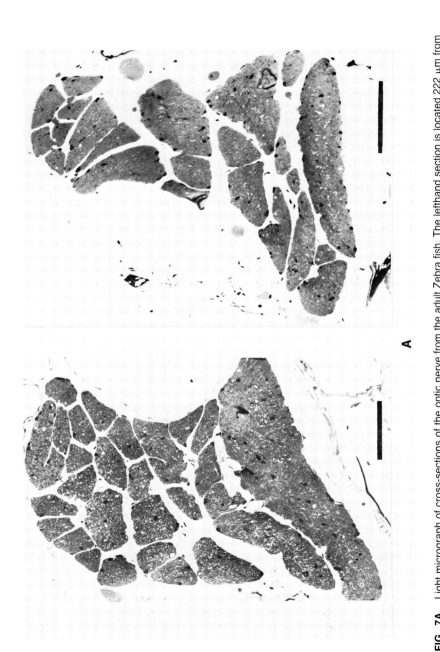

FIG. 7A. Light micrograph of cross-sections of the optic nerve from the adult Zebra fish. The lefthand section is located 222 μm from the exit of the optic nerve to the retina. The section on the right is further away at 280 μm. Note the differences between the bundles in these two sections. (Bar = 100 μm.)

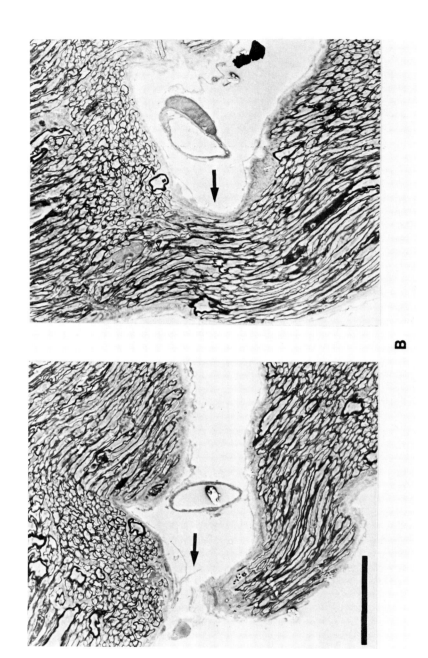

FIG. 7B. Electromicrographs of fibers passing from one bundle to an adjacent one. The sections are separated from each other by 1 μm. The merging of the two bundles is indicated (*arrow*). (Bar = 10 μm.)

nerve fibers when they reach the visual tectum of the Zebra fish. Although there has been a great deal of work on the electrophysiological mapping of the optic fibers onto the surface of the tectum where their initial arborization takes place, there is little information on the wiring of the neurons in the tectum that process the visual imputs. Most of the single-unit recordings obtained by Sajovic arose from cells in the stratum periventricularum (SPV) region of the tectum, which contains the vast majority of the tectal cells. The results suggested that many of the physiological responses can be understood if one assumes that a large set of cells in the SPV receives primary input from the retinal ganglion cells as excitatory signals and also contributes and receives inhibitory signals from neighboring SPV cells (Fig. 8). This model provides a satisfactory explanation for a rather large number of physiological responses to various visual stimuli (20).

It is important for this discussion to note that the observed physiological properties of the tectum can be understood in a context in which a large fraction of the half-million SVP cells in the tectum of the Zebra fish are carrying out the same function. They have different positions in the tectum by virtue of the growth pattern, which results in a wave of differentiation across the structure. However, we have no reason to think that different members of the set of SVP cells express different genes or

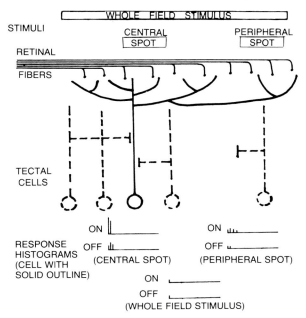

FIG. 8. Proposed model of the local inhibition of type I tectal cells in response to retinal stimuli. The model assumes (i) that tectal type I cells receive excitatory input from retinal fibers and delayed inhibitory input from neighboring tectal cells; and (ii) that the excitatory process is more readily saturated than the inhibitory process. These two hypotheses are sufficient to explain the responses, illustrated in the lower part of the diagram, to a central-spot, a peripheral, or a whole-field stimulus.

even the same genes to different extents. It seems more likely that, as in the case of the retinal ganglion cells, the same rules are expressed on all the cells, or that at most there are a few sets of cells expressing different rules.

SOME SIMPLE CONCLUSIONS AND FUTURE PROSPECTS

As a general model for the genetic control of the development of the nervous system we suggest the following.

At any time during embryonic development, we can observe the set of cells already present and ask what rules determine their subsequent growth, migration, and differentiation. In the case of developing neurons we believe that, to a large extent, the primary factors guiding their growth are the nature of the surrounding tissues. For large arrays of neurons, like the ganglion cells, the guidance is provided primarily by other nerves that are more mature. The growth of the initial fibers must, of course, be guided by nonneuronal structures. Both initially and to a considerable extent during later development, the neighboring blood vessels seem to play an essential guiding role in providing the general pathway for the fibers to follow. If we push the question further back to ask what determines the position of blood vessels, we obviously get to the general problem of understanding the sequential unfolding of the developmental program in any multicellular organism.

In addition to the types of guidance rules suggested here, there must also be many different phenomena that are important for determining the final connections made by neurons. There is extensive plasticity in virtually all vertebrate systems examined carefully (3), and it can also be demonstrated in invertebrate systems (4). Furthermore, there is clear evidence of selective cell death as a significant factor in obtaining the "correct" connections. Which of the changes in neural connectivity after the general positions have been attained reflect the influence of neuronal function or saturation of synaptic receptors or any of a multitude of other factors that have been suggested is impossible to determine at this time. However, it does seem clear that in some simple systems one can understand the observed growth patterns with particularly simple rules.

In general terms, the ideas that have been developed from observations of neuroanatomy during early development support the notion that rather simple rules of cellular growth and interactions can account for the behavior of growing neurons, even very large arrays of them. However, the question that must now be asked is how can these apparently simple rules result from the sequential expression of genes that govern the growth of neurons in general and of individual neural types in particular? We know that single gene mutations in mice can result in the loss of individual cell type. Thus, among many examples, Sidman and his collaborators have shown that the staggerer mutation leads ultimately to the loss of granular cells in the mouse cerebellum. However, we have at present no idea as to the number of genes involved in producing the growth pattern specific for granular cells.

The recent advances in immunology and molecular biology should make it possible to start investigations of neuronal differentiation at a molecular level. Monoclonal antibodies and nucleic acid hybridization techniques, combined with detailed anatomical observations in developing systems, can be expected to lead to more specific models of neuronal differentiation specificity than has been possible by anatomical and physiological studies alone.

ACKNOWLEDGMENT

This work was supported by grants from the National Institutes of Health (NS-09821 and RR-00442).

REFERENCES

1. Berry, M., and Bradley, P. M. (1976): The growth of the dendritic trees of Purkinje cells in irradiated agranular cerebellar cortex. *Brain Res.*, 116:361–388.
2. Bodick, N., and Levinthal, C. (1980): Growing optic nerve fibers follow neighbors during embryogenesis. *Proc. Natl. Acad. Sci. U.S.A.*, 77:4374–4378.
3. Cotman, C. W. (1978): *Neuronal Plasticity*. Raven Press, New York.
4. Friedlander, D., and Levinthal, C. (1982): Anomalous identified neurons in the prawn. *J. Neurosci.* (2:2).
5. Goodman, C.S., Pearson, K.G., and Heitler, W. J. (1979): Variability of identified neurons in grasshoppers. *Comp. Biochem. Physiol.*, 64A:455–462.
6. Gyula-Lazar (1980): Long-term persistence, after eye-removal, on unmyelinated fibers in the frog visual pathway. *Brain Res.*, 199:219–224.
7. Kimmel, C. B., Schabtach, E., and Kimmel, R. J. (1977): Developmental interactions in the growth and branching of the lateral dendrite of Mauthner's cell *(Ambystoma mexicanum)*. *Dev. Biol.*, 55:244–259.
8. Levinthal, C., and Ware, R. (1972): Three dimensional reconstruction from serial sections. *Nature*, 236:207–210.
9. Levinthal, C., Macagno, E. R., and Tountas, C. (1974): Computer-aided reconstruction from serial sections. *Fed. Proc.*, 33:2336–2340.
10. Levinthal, F., Macagno, E. R., and Levinthal, C. (1976): Anatomy and development of identified cells in isogenic organisms. *Cold Spring Harbor Symp. Quant. Biol.*, 60:321–331.
11. Levinthal, F., and Levinthal, C. (1980): Development of retinotectal connections. *Soc. Neurosci. Abstr.*, 6:293.
12. Levinthal, F., and Levinthal, C. (1982): Development and organization of the retinotectal connections *(in preparation)*.
13. Lopresti, V., Macagno, E. R., and Levinthal, C. (1973): Structure and development of neuronal connections in isogenic organisms: Cellular interactions in the development of optic lamina of *Daphnia*. *Proc. Natl. Acad. Sci. U.S.A.*, 70:433–437.
14. Lopresti, V., Macagno, E. R., and Levinthal, C. (1974): Structure and development of neuronal connections in isogenic organisms: Transient gap junctions between growing optic axons and lamina neuroblasts. *Proc. Natl. Acad. Sci. U.S.A.*, 71:1098–1102.
15. Lyon, M. F. (1962): Sex chromatin and gene action in the mammalian x-chromosome. *Am. J. Hum. Genet.*, 14:135–148.
16. Macagno, E. R., Lopresti, V., and Levinthal, C. (1973): Structure and development of neuronal connections of isogenic organisms: variations and similarities in the optic system of *Daphnia magna*. *Proc. Natl. Acad. Sci. U.S.A.*, 70:57–61.
17. Macagno, E. R. (1978): Mechanism for the formation of synaptic projections in the arthropod visual system. *Nature*, 275:318–320.
18. Macagno, E. R., Levinthal, C., and Sobel, I. (1979): Three-dimensional computer reconstruction of neurons and neuronal assemblies. *Annu. Rev. Biophys. Bioeng.*, 8:323–351.
19. McConnell, P., and Berry, M. (1978): The effects of undernutrition on Purkinje cell dentritic growth in the rat. *J. Comp. Neurol.*, 177:159–172.
20. Sajovic, P., and Levinthal, C. (1981): Visual system of zebrafish optic tectum. *J. Neuroscience*, *(in press)*.
21. Streissinger, G., Walker, C., Dower, N., and Knauber, D. (1981): Production of clones of homozygous diploid zebrafish. *Nature*, 291:293–296.

Genetics of Neurological and Psychiatric Disorders, edited by Seymour S. Kety, Lewis P. Rowland, Richard L. Sidman, and Steven W. Matthysse. Raven Press, New York © 1983.

Experimental Neurogenetics

Richard L. Sidman

Department of Neuroscience, Children's Hospital Medical Center, Boston, Massachusetts 02115

The term "experimental" in the title of this chapter implies the question, "Given our preponderant interest in the neurogenetics of our own species, what experimentally accessible material is available that may allow us to probe beyond what is feasible with human organisms, cells, or cell products?" I shall focus on the mouse, since my experience is greater with this species than with any other, the number of pertinent mutations available is relatively large, formal genetic analysis is more advanced than in any other mammalian species except man, and neural organization is reasonably familiar to students of the human nervous system. A far broader survey of organisms and problems has appeared recently (30), as have several briefer and more specialized accounts (4,9,60,69,104,109).

Only a few of the mutations affecting the nervous system in mice appear to correspond directly to particular inherited human diseases. Our interest in them does not lie in how closely they reproduce the specific features of human disorders, for even if the genetic locus is the same, the exact perturbations in the gene are unlikely to match, and the two species are sufficiently different in organization and rate of development that phenotypic differences are to be expected. Correspondence must be sought on the deeper, and fundamentally more interesting, levels of formal genetics (chromosomal map position, neighboring genetic loci, and intragenic structure), homologous gene products and target cells, and equivalence of the developmental or physiological event that is perturbed. Many other mutations in the mouse, not recognized today to match identified human diseases, are important because they illuminate normal developmental and organizational principles or give insights into nervous system pathophysiology. The following examples serve as illustrations.

MUTATIONS AFFECTING BEHAVIOR

Tottering: Absence and Focal Motor Epilepsy

One of the early uses of mutations affecting locomotor behavior in the mouse was for elaboration of the formal genetic map. Such behavioral abnormalities are virtually as easy to observe and score among segregating progeny in a linkage test as are coat color phenotypes. Tottering (gene symbol, *tg*), an autosomal recessive mutation, appeared "spontaneously" at the Jackson Laboratory in 1957, and when it was recognized 5 years later to be closely linked to a skeletal mutant, oligosyndactylism (26), it served to establish the new linkage group XVIII (27), now known to be on chromosome 8 (90).

In the original report (26), a complex episodic behavioral abnormality was described in which affected mice at about 2 to 3 weeks of age display frequent brief seizures featuring an arrest of general body movement, often with single myoclonic jerks or twitches. By about a week later, these attacks have evolved into an elaborate motor seizure pattern. A seizure is commonly heralded by increased running activity with abrupt changes of direction or by dystonic or spastic postures, sometimes interspersed with moments of immobility. This phase is followed by focal tonic–clonic seizures that spread slowly and variably from one part of the body to another. The final stage involves bilateral clonic movements of forelimbs alone. A total episode averages 20 to 30 min and may last a full hour.

More recently (75), electrocorticographic recordings from unanesthetized, free-moving affected mice have shown abnormal bursts of bilaterally synchronous and symmetrical spike-and-wave sequences, six to seven per second, over the cerebral hemispheres (Fig. 1). These spike–wave paroxysms (see also 51) are always accompanied behaviorally by a sudden arrest of movement and a fixed staring posture, often with twitching of vibrissae or jaw. The more complex motor seizure patterns are accompanied by low-voltage desynchronized activity interspersed with long runs of generalized theta (4–7 Hz) waves, the intermittent tonic–clonic movements correlating with bursts of high-voltage spikes, sharp waves, and slow waves (Fig. 1). The murine spike–wave electrical pattern accompanied by behavioral arrest resembles the common recurrent absence seizures of human childhood. The resemblance is reinforced by autoradiographic data visualizing increased uptake of ^{14}C-labeled 2-deoxy-D-glucose symmetrically into various rostral brainstem structures (75), observations consistent with the proposed "centrencephalic" origin of human absence attacks (81). The point of theoretical interest, and of potential practical value in the long run, is that a single gene mutation, with a single (possibly modifiable) molecular disturbance at its core, can generate a complex but rather stereotyped functional disturbance involving extensive parts of the nervous system (74).

Hypopigmentation: Effects on Organization of the Visual System

Human albinos have abnormalities of visual function beyond what can be accounted for simply on the basis of excess entry of ambient light secondary to deficiencies in the amount of screening pigment in eye tissues (17). The structure of the visual pathways is abnormal in albino mutants of all mammalian species studied, including mouse, rat, rabbit, guinea pig, ferret, mink, cat, tiger, and human (summarized in 59). Many questions come to mind. Is the disorder referable to the albino genetic locus or to hypopigmentation per se? What might be the primary

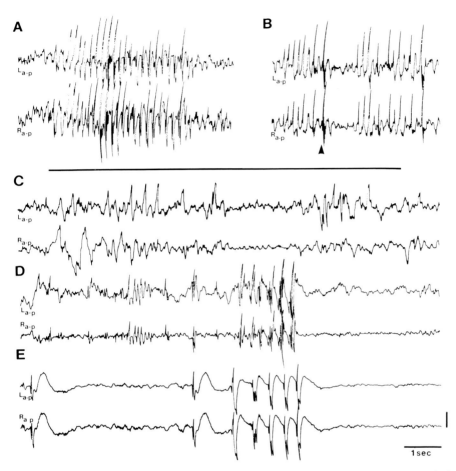

FIG. 1. Bipolar electrocorticogram (ECOG) recordings (left and right, anterior to posterior) from cerebral cortex of unanesthetized, unrestrained *tg/tg* mice **A:** Bilaterally synchronous, spike–wave bursts (six to seven per second) in a 32-day-old *tg/tg* mouse. The animal displayed sudden behavioral arrest and maintained fixed, staring posture throughout the spike–wave discharge. Larger-amplitude spikes were accompanied by myoclonic head jerks. **B:** Transient bilateral spike-burst suppression induced by startle (air puff, *arrow*) in a 12-week-old *tg/tg* mutant. The burst sequence demonstrates subharmonic spiking at a rate of three spikes per second superimposed on waves of six per second. **C,D:** An ECOG recording during stage 2 of spontaneous, partial tonic–clonic seizure in an adult *tg/tg* mouse. **C:** Generalized and lateralized spike-and-slow-wave dysrhythmias correspond with a behavioral sequence of bilateral and unilateral limb clonus. **D:** Spiking at a rate of six spikes per second and a rare burst of spike waves, three per second, appear briefly during spontaneous partial motor seizure. **E:** Polyspike–wave complexes during pentylenetetrazole-induced activation of the stereotyped motor seizure pattern. Calibration: **A–C,** 100 μV; **D,E,** 150 μV. (From Noebels and Sidman, ref. 75, with permission.)

cellular target and primary molecular mechanism? What is the relationship between structural and behavioral defects?

An advantage of the mouse is that many alleles at the albino, *c*, locus are available as well as mutations at several independent genetic loci affecting pigmentation, the majority of them on the same inbred C57BL/6J strain background (56). Many pigmentation mutations have also been conserved, for primarily economic reasons, in mink (94). Not all of these mouse and mink mutations affect the visual system, but clearly more do so than just mutations at the c locus. The common feature, then, is not likely to be referable to the enzyme tyrosinase, for which the c locus is either the structural or a regulative gene (32). Tyrosine levels correlate poorly with visual pathway disturbances in mice (56) and humans (18,28).

The primarily affected cell in the nervous system appears to be the retinal ganglion cells, although this is not necessarily the direct cellular target of the hypopigmentation mutations. The constant feature in all affected mammals is an increased ratio of crossed to uncrossed fibers in the optic tracts (59). This apparently straightforward observation, however, does not answer whether the affected cells comprise the whole retinal ganglion cell population or some subset, or whether the problem is in cell genesis, cell viability, rate of cell maturation, timing of axonal growth to and through the optic chiasm, axonal branching, or dying back of axons. The problem is that to date, all measurements have been of ratios, rather than absolute numbers, of contralateral versus ipsilateral axons, for lack of adequate quantitative methods for nerve fibers or their parent neuron somas. Siamese cats (homozygous for a mutant allele at the c locus) show disproportionate disturbance of one of the three subsets of retinal ganglion cells, the large and mainly ipsilaterally projecting Y cells (13). In the pigmented mouse, about 1/100 of the conventionally positioned retinal ganglion cells project ipsilaterally, compared to about one-fifth of those cells whose somas are "displaced" into the outer part of the inner nuclear layer; in albino mice, relatively few of those displaced, ipsilaterally projecting neurons have been demonstrated (22).

The evidence is compelling that the primary action of hypopigmentation mutations is extrinsic to the retinal ganglion cell or its axon. First, these cells lack melanin pigment. Second, mosaic "flecked" mice were examined by Guillary et al. (29). These mice carry a translocation in which the segment of chromosome 7 carrying the wild-type allele at the c locus has been translocated to the X chromosome, where it is subject to random inactivation in cells of female mice. The intact autosome carries the mutant c allele, so that X-inactivated pigment cells show the albino phenotype, and pigmented tissues show the "flecked" configuration. If the c mutation acted intrinsically in retinal ganglion cells, Guillary et al. (29) predicted that flecked mice should have about half the normal number of ipsilaterally projecting axons in the optic tracts, but, in fact, they found the number to be greater than that in the pigmented control mice and concluded that the mechanism of gene action is extrinsic to the retinal ganglion cell. Third, a comparison of various independent hypopigmentation genes in mink and mice indicates that the common factor in all animals with abnormal visual systems in hypopigmentation of the retinal pigment epithelium and not other pigmented tissues, including the choroid layer of the eye (56,94). Further, the degree of increase in the contralateral/ipsilateral axon ratio was graded and was more or less proportional to the degree of hypopigmentation in the pigment epithelial layer (56). A possible developmental mechanism has been suggested (110) in which a layer of transient pigmented epithelial cells along the embryonic optic stalk may influence the trajectory of early growing retinal ganglion cell axons.

The initial laboratory clue relating albinism and the visual system was behavioral, obtained in rats (98). With recognition of both behavioral and optic tract abnormalities in Siamese cats, correlative studies came to focus on cats rather than rodents. Organization of the visual system may be abnormal at many levels, and the simplest interpretation of the total range of data is that functional appropriateness, not simply genetic blueprinting, is somehow the organizing principle (96; summarized in 30).

The abnormal decussation pattern at the optic chiasm in Siamese cats leads to a systematic anomaly in layering and in visual field topographic representation in the lateral geniculate body, with a given lateral geniculate receiving information about the total temporal half of the contralateral eye's visual field (in the correct geniculate layer A), up to 20° of the nasal half of its visual field (in the incorrect geniculate layer A_1), and correspondingly less than half of the ipsilateral eye's nasal field. The abnormal geniculate organization appears to revertebrate downstream in the primary visual cortex in one of two ways. If the proportion of misrouted axons at the chiasm is relatively large (the "Boston" pattern), the entire geniculocortical radiation becomes rearranged so that both the appropriate and inappropriate fields are topographically represented in area 17, and the vertical meridian, which ordinarily is represented at the border between areas 17 and 18, is displaced about 20° to either side. Further reverberations are seen with references to transcallosal connections. Normally, axons in the corpus callosum connect the vertical meridian of the visual field that is represented in each hemisphere so as to eliminate a potential "seam" in perceived visual space.

In "Boston" Siamese cats, the callosal connections are anatomically abnormal but functionally appropriate, connecting visually equivalent points in the two hemispheres (96). Even the projections from the 17/18 border back to the lateral geniculate (different neurons from those projecting across the corpus callosum) are functionally appropriate and, correspondingly, anatomically abnormal (97). A different solution is achieved in Siamese cats with a lesser proportion of misrouted axons at the chiasm (the "Midwestern" pattern). Here, the geniculocortical projection is visually inappropriate and somehow is functionally suppressed (50). No higher-order abnormalities have been described. Perhaps the general rule is that a main influence in development of connections in the visual system is visual appropriateness. It is noteworthy that visual evoked potential data in human albinos suggest that there may be two distinct patterns in anatomical pathways and connectivity between retina and visual cortex in this species also (8).

The tottering and hypopigmentation mutants strongly suggest that a relatively simple initial event in the nervous system may reverberate spatially through a wide-ranging chain of connected neurons and temporally through a span of stages from early embryonic development to functional maturation. This statement expresses, in a particular context, the fairly general belief among geneticists that most genes affect most cells. When we begin our analysis with the complex full-blown phenotype, as is usually the case, the tasks of dissecting back to primary events and of assembling developmental sequences (both tasks of fundamental importance) may indeed be formidable. The following two behavioral mutants are at an early stage in this process of analysis and can be summarized briefly.

Diabetes: Hypothalamus-Mediated Behavior

The autosomal recessive mutation diabetes, *db*, is usually considered to cause a "metabolic" rather than a "behavioral" disorder, but either term simply expresses investigator bias; both the primary target cell and the pathophysiological mechanism are unknown (117). Affected mice become extraordinarily obese, eat and drink excessively, regulate temperature poorly, and are sterile (44). The brain has been suggested as a site of action of the *db* genetic locus (12,134), although a genetically different mutation, obese, with a virtually identical phenotype, may act at a site extrinsic to the central nervous system (12). By focusing on the sterility problem in affected females, Johnson and Sidman (49) adduced evidence that the hypothalamus is functionally abnormal. Ovaries grafted from *db/db* homozygotes to histocompatible wild-type females were fertile (44,49). The hypoplastic vaginal epithelium, uterus, and ovaries of mutant females responded quantitatively normally to exogenous hormonal stimulation. Although gonadotropin release from the pituitary was reduced in female mutant mice, the response to exogenous gonadotropin-releasing hormone (GnRH) was normal. Endogenous hypothalamic GnRH detected by radioimmunoassay was significantly higher in castrated affected females than in castrated controls. The conclusion from these data was that infertility of *db/db* females results from inadequate gonadotropic stimulation and not from an unresponsive reproduction system, probably on the basis of inadequate GnRH release from the hypothalamus. The primary problem could lie in the hypothalamus, further upstream in the CNS, or elsewhere (11,49).

Mocha: Episodic Aggressive Behavior

Mocha, *mh*, was described as an autosomal recessive disorder affecting coat color and inner ear but with additional behavioral features suggestive of a wider neurological syndrome (55). Brief mention was made that affected females cared poorly for their newborn offspring. A second, independently occurring mutation at the Jackson Laboratory was named mocha-2J (mh^{2J}) (105). It was found to have a mild axonal degenerative disorder in PNS and CNS but, more interesting, to show an unusual behavioral syndrome. Affected females became remarkably aggressive toward babies or other adults at around the time of parturition. Specifically, they would attack and kill their own or other babies by making gross puncturing bites and might inflict similar puncturing or slashing wounds on rumps and tails of male or female adults added to the home cage.

Mice of normal strains will sometimes kill their young, especially their first litters. In our C57BL/6J colony, all pups in 10 of 163 consecutive first litters delivered by isolated females were killed by the second postnatal day, with signs of cannibalism or neglect but without puncture wounds. Most wild-type females that destroy their first litters do raise subsequent litters. In approximately the same time period, by contrast, each of five homozygous *mh* or *mh^{2J}* females killed all progeny in eight of nine litters within 24 hr of delivery, and at least one pup in each of the destroyed litters bore several tooth marks and/or punctures (Victor, W. M. Cowan, and R. L. Sidman, *unpublished observations*). The one surviving litter was the second born to a *mh^{2J}/mh^{2J}* female housed in a cage with two wild-type foster mothers who took main roles in rearing the litter to weaning age. The same *mh^{2J}/mh^{2J}* female destroyed all pups in her first and third litters, born in the absence of other adults. Two *mh* and four *mh^{2J}* homozygous males sired litters with heterozygous or wild-type females. Each male was present at the time of delivery and did not attack the young.

Reactions to foreign pups also were different for wild-type and mutant females. Wild-type adults of either sex, when isolated in a fresh cage to which foreign newborn pups are then added, usually react to the pups by sniffing them briefly, engaging in exploratory activity about the cage, and then gather the pups and sit or lie near them, often in a nursing posture. Most *mh* or *mh^{2J}* homozygotes of either sex are likely to react in a comparable setting by sniffing the pups and then running wildly about the cage for up to 10 min, commonly scattering the pups as they do so. The adult is then likely, with some variations in intensity and exact pattern, to seize, vigorously puncture, and kill the pups, one after another. We have seen a mild version of this behavior at 5 weeks of age and the fullblown behavior, in those mice destined to react severely, by 9 weeks.

Comparison of the *mh* and *mh^{2J}* alleles is inadequate because the mutations are available on different genetic backgrounds, *mh* on the MG/Ln inbred strain (carrying the linked mocha and grizzled mutations in repulsion, ref. 55), and *mh^{2J}* on the strain of origin, C3H/HeJ (105). We are transferring *mh^{2J}* onto the C57BL/6J background, on which it breeds better than on the coisogenic C3H/HeJ background, but have been unable to sustain the *mh* allele congenic with the C57BL/6J strain.

Allowing for the interpretive complications of different genetic backgrounds, a few additional phenotypic features deserve mention. The two alleles have somewhat different effects on coat color, probably not accounted for entirely by differences in other pigmentation genes. Even when each mutation is partially inbred onto nonagouti C57BL/6J stock, the *mh* homozygote is gray, and the *mh^{2J}* homozygote is brownish in coat colar. The *mh/mh^{2J}* hybrid "double heterozygote" is intermediate in coat color. Among the features of greater interest is our immediate context, many *mh^{2J}* homozygotes have occasional staring attack ("absence") seizures and spike–wave abnormalities in their electrocorticograms (J. L. Noebels, Victor, J. Cowen, and R. L. Sidman, *unpublished observations*), rather like those described above for tottering. The seizure frequency and intensity may vary with genetic background. Mice homozygous for the *mh* allele have somewhat different electrocorticographic abnormalities. In parallel with seizure activity, affected mice emit curious "chirping" sounds, easily audible across a normal-sized laboratory room. It is our impression that neither the seizures nor the chirping are coordinated temporally with the infanticidal behavior.

MUTATIONS AFFECTING CEREBELLAR STRUCTURE

The attractiveness of the cerebellum for the investigator stems from its relatively simple and stereotyped cellular structure, its continued development well beyond birth in rodents and man, and the availability of many mutations affecting the same cell types. The latter feature is probably an outcome of the ease in screening for the locomotor phenotypes that commonly accompany cerebellar disorders and the nonlethal quality of many of these affections. The list of available cerebellar mutations in the mouse has been enlarging rapidly (compare the lists in 9,69,100,107) an now totals about 18 independent genetic loci (Table 1).

Homologous Cerebellar Mutations in Mouse and Man

Three diseases that apparently affect the same genetic locus in each species have been recognized.

Mottled

Several alleles at this locus on the X-chromosome of the mouse are known. The most intensively studied, because of its similarity to human Menkes' kinky hair syndrome, has been brindled, Mo^{br}. Affected Mo^{br}/Y mice cease to grow by postnatal day 12, develop a mild tremor and uncoordination, and usually die by day 20, although a few have lived and sired progeny (107). Other alleles have greater or lesser effects in a number of organs, all of the pleiotropic changes being thought referable to defective intracellular copper transport and compartmentalization and to secondary reduction in activity of various copper-requiring enzymes in pigment cells, hair, fibroblasts, elastic tissue, and perhaps other cells. The original mottled, Mo, mutation, for example, is an embryonic lethal. Blotchy, Mo^{blo}, males often survive beyond weaning, whereas the very mildly affected pewter, Mo^{pew}, males reach adulthood and reproduce.

Human cases show extensive granule cell neuronal loss in the cerebellar cortex as well as in layer IV of many parts of cerebral cortex and in relay nuclei of thalamus (128). Purkinje cells show somatic filopodia and spines, presumably persisting beyond the usual developmental period, and underdeveloped dendritic trees with increased concentrations of thorns on primary dendrites, swellings at many dendritic branch points, and a paucity of tertiary branchlets (41,84,128). Somewhat similar observations have been made in preweaning Mo^{br}/Y (132,133) and Mo^{blo}/Y (R. L. Sidman and K. W. T. Caddy, *unpublished observations*) mice, except that granule cell loss appears less than in human cases; adult Mo^{pew} males show only a mild reduction in cerebellar cortical volume. Heterozygous females, mouse or human, are considered to be normal.

Brindled male mice can be rescued, and the neuropathological changes lessened, by injection of cupric chloride postnatally (46,62,71). No corresponding therapeutic

TABLE 1. Cerebellar mutants of the mouse

Name	Gene symbol	Chromosome	Genetic background[a]	Characteristics[b]
Direct homologs of human mutations				
Brindled	Mo^{br}	X	C3H/HeJ-$Mo^{br-J}/+$	One of several alleles at the Mottled locus; closest match to human Menkes' kinky hair disease; in mouse, a juvenile lethal affecting cerebellar Purkinje more than granule cell neurons and also affecting selected cerebral cortical and thalamic neurons.
Blotchy	Mo^{blo}	X	C57BL/6J-Mo^{blo}	Less severe than Mo^{br} but with a similar range of target cells within and outside the nervous system
Pewter	Mo^{pew}	X	CBA/J-Mo^{pew} (M)	Mildest allele in the Mo series, with fertile adult males showing slight reduction in cerebellar size
Beige	bg	13	C57BL/6J-$bg/+$ SB/Le-A^W/A^w-bg sa/bg sa SJL/J-bg/bg $c^+/c^+/p^+/p^+$ C57BL/6J-bg^J (M) C3H/HeJ-$bg^{2J}/+$ (M)	Homologous with human Chediak–Higashi disease; affects cerebellar Purkinje cells at 5 months and later
Wasted	wst	—	B6C3-a/a-wst	Similar to human ataxia telangectasia, with abnormalities evident from about postnatal day 20 to death by day 30; cerebellar Purkinje cells small and reduced in number by terminal stage
Mutation affecting predominantly the Purkinje neuron				
Reeler	rl	5	C57BL/6J-rl/+ B6C3-a/a-rl	Malpositioning of early cerebellar and cerebral neurons migrating to cortices
Staggerer	sg	9	C57BL-$+d$ se/sg $++$	Neonatal deficit of Purkinje cells; failure of viable ones to develop tertiary (branchlet) spines; secondary loss of granule cells
Lurcher	Lc	6	B6CBA-A^{w-J} A-Lc	Loss of Purkinje cells beginning in second postnatal week; secondary loss of granule cell and inferior olivary neurons
Purkinje cell degeneration	pcd	13	C57BL/6J-pcd C3H/HeJ-pcd	Loss of Purkinje cells beginning in third postnatal week and of selected retinal, olfactory, and thalamic neurons on other temporal schedules
Nervous	nr	8	BALB/cGr-nr/+ (M) C3HeB/FeJ-nr	Loss of Purkinje cells beginning in third postnatal week, but more slowly and less completely than in pcd

TABLE 1. (contd.)

				Mutations affecting predominantly the granule cell neuron
Weaver	wv	16	C57BL/6J-wv/+ (M) B6CBA-A^w-J/A-wv	Failure of early postnatal granule cells to differentiate an axon, translocate the cell body, and survive in vivo and in vitro; similar but less extreme effects in heterozygote
Tortured	tor	—	Noninbred	Granule cell loss beginning near end of first postnatal month and progressing for more than a year; degeneration of many other CNS and PNS neuronal classes
				Other mutations affecting cerebellar cortex
Leaner	tg^{la}	8	C57BL/6J-Os+/+ tg^{la}	Loss, mainly but not exclusively in the anterior lobe, of granule cells in first postnatal month and of Purkinje and Golgi II neurons in second month; mildly progressive loss of all three types thereafter
Rolling mouse Nagoya	tg^{rol}	8	C3H-tg^{rol}	Perinatal disordering of cerebellar cortical cells mainly, but not exclusively, in anterior lobe
Meander tail	mea	4	Noninbred	Embryonic malformation of rostral part of fourth ventricle and adjacent structures, with major impact on anterior lobe of cerebellum
Swaying	sw	15	B6C3-a/a-sw	Early embryonic inductive disorder of rhombencephalon and otocyst, with variable malformation of cerebellar paraflocculus
Kreisler	kr	2	B6CBA-A^w/A-we a kr	Probable underproduction of Purkinje and granule cells; Purkinje cell degeneration late in first postnatal month
Stumbler	stu	—	C3H/HeJ-stu/+ (M)	Persistence or development of somatic and primary dendritic spines on Purkinje cells
Hyperspiny Purkinje cell	hpc	—	Noninbred	
				Mutations affecting cerebellar nuclei
Vibrator	vb	11	B6D2-vb/+ B6C3-a/a-vb	Progressive degeneration of neurons in all cerebellar nuclei in first postnatal month as well as in sensory ganglia and selected sites in spinal cord and brainstem; a spinocerebellar system degeneration
Cerebellar outflow degeneration	cod	—	BALB/cGr-cod/+	Neuron degeneration by middle of first postnatal month in cerebellar, vestibular, and red nuclei; selective axonal degeneration in many other tracts of spinal cord and brain

[a] Genetic backgrounds are listed for Jackson Laboratory (33) and our own stocks. Additional inbred strains may be available elsewhere. The symbol (M) after a stock indicates maintenance coisogenic with the strain of origin.
[b] See text for further details and for key references.

result has been achieved in humans, perhaps because copper has not been given on the appropriate schedule. The forms of abnormal Purkinje and other neurons suggest perturbation of development at late fetal stages (128), and administration of copper postnatally in affected humans may be relatively ineffective. The site or mechanism of action of exogenous copper is unclear (61,83). Deficient activity of dopamine-β-hydroxylase, a copper enzyme, as well as reduced norepinephrine and normal dopamine concentrations, has been described in brains of male mice with several of the *Mo* alleles (45,47), but these facts are difficult to correlate with either the behavioral signs or the neuropathology. A new lead may be available through the establishment of an SV40-transformed, HPRT-deficient tissue culture cell line from a Mo^{blo}/Y mouse; the cells express an abnormal phenotype, a fourfold increase in incorporation of ^{64}Cu, and are suitable for man–mouse somatic cell hybridization (43).

Beige

The autosomal recessive beige, *bg*, mutation has occurred several times independently at the Jackson Laboratory and is available on several different inbred backgrounds (33,79). It is homologous with Chediak–Higashi disease in children, a disorder featuring abnormally enlarged lysosomes in several types of cells (including Purkinje cells of the cerebellum and Schwann cells of peripheral nerve), with hypopigmentation, marked susceptibility to infection, and reduced life-span, often with terminal lymphoreticular malignancy (58,131). Beginning at about 5 years of age, a progressive peripheral neuropathy is common, and some patients show mental retardation, seizures, and/or cerebellar ataxia. In affected mice, the lysosomal abnormality in cerebellar Purkinje cells is an expression of intrinsic gene action, as shown in *bg* ↔ + chimeras (67). Affected mice live longer, relative to the normal life-span, than do affected humans and, beginning by about 5 months of age, show a selective and progressive loss of cerebellar Purkinje neurons (E. Murphy, *personal communication*). This is the oldest age at which a genetic degeneration of Purkinje cells has been described in mice (see below).

Wasted

This important autosomal recessive mutation was recognized at the Jackson Laboratory in 1972 and was named wasted, *wst*, in 1981 (118). Homozygotes develop a fine tremor at about 20 days of age, followed by progressive flaccidity of the hind limbs and death at about 28 days of age. It closely resembles the remarkably pleiotropic human disease, ataxia telangectasia (80). Affected mice show no weight gain after day 20, marked lymphoid hypoplasia in thymus-dependent and thymus-independent areas, leucopenia, decreased responsiveness on delayed hypersensitivity tests with sheep erythrocytes, a fourfold increase in incidence of chromosomal damage in colchicine-treated bone marrow smears, and probably an increased susceptibility to radiation-induced chromosomal injury (L.C. Shultz, *personal com-*

munication). Purkinje cells in the cerebellar cortex were shrunken and reduced in number in terminal *wst/wst* mice.

Mutations Affecting Predominantly the Purkinje Neuron

Although these diseases are, for the present, peculiar to the mouse, they do illustrate general developmental principles. None of these genetic loci affects Purkinje cells exclusively, and in some, this cell may not even be the direct target or the sole target of gene action, but a good case can be made that the Purkinje cell alteration dominates the phenotypic expression.

Reeler

The reeler locus wild-type allele, as well as mutant allele, acts extrinsically on early generated young neurons migrating to cortices so as to control laminar position in the radial vector between periventricular germinal sites and ultimate cortical resting sites (9,69). Purkinje neurons of cerebellar cortex, as well as neurons in most parts of cerebral cortex, are positionally controlled very exactly by the *rl* locus, with very little modulating influence by the rest of the genome (10). Control is not exerted in terms of proliferative behavior of Purkinje cell precursors or in terms of the vector or timing of cell migration but seems to involve some aspect of local environmental signaling close to, or within, the target cortical area.

The reeler mutation, then, perturbs cortical development in a unique manner, making neuron position the experimental variable. Cell form, as opposed to cell position, is not fundamentally at issue, for even the most aberrant Purkinje cells in reeler homozygotes are recognizable by experienced morphologists as members of the Purkinje class. An interesting challenge will be to specify, say for a computer, what are the denominating features that define the Purkinje class despite distortions of position, nearest-neighbor relationships, orientation, size, and dendritic mass and branching pattern. One of the most stable features is the formation of characteristic postsynaptic spines, even in the face of a gross deficit of presynaptic axons (85). Another is the relative paucity of abnormal synaptic inputs onto Purkinje cells, although some do develop (63).

We may ask, then, "What is the effect of cell position on cell form?" Most of the abnormalities in Purkinje cell form in reeler homozygotes appear to reflect local environmental properties (9). If the Purkinje cell soma lies in the granular layer or at the border of granular and molecular layers, and some of its dendrites gain access to a zone packed with relatively normal parallel fibers (granule cell axons) in the molecular layer, the dendrites will locally take on a relatively normal volume and branching pattern and will be oriented at a 90° angle to the parallel fibers. If the Purkinje cell is entirely confined to one zone (granular layer or deep white matter), it assumes a multipolar form with reference to its dendrites. If the dendrites cross the boundary between granular layer and deep white matter, the cell again is likely to display a unipolar dendritic pattern, as it does in normal position. Presumably, in this last case, the cell may have formed primary dendrites, or at least somatic filopodia, in several directions from the soma and then resorbed all but the one that crossed the interface between disparate cortical zones.

Staggerer

The obvious phenotypic abnormality, recognized at the time of the initial description 20 years ago (108), is a very small cerebellum based mainly on a dramatic reduction in the net number of granule cell neurons. However, some of the earliest and most important developmental problems involve the Purkinje neurons (100,101). For example, uniquely among mutant or exogenous developmental disorders, staggerer Purkinje neurons generate few tertiary (branchlet) spines (39,101,114) and thus virtually lack the normal synaptic target sites of parallel fibers. Cell position and cell size are sufficiently abnormal in the mature mutant cerebellum that classification of cell types is inexact by light microscopic criteria. However, recent studies have established that the number of Purkinje neurons is reduced to approximately 25% of normal, with consistent regional variation in severity along the mediolateral axis (34). Analysis of chimeras proves conclusively that the Purkinje neuron is a direct target of the sg gene and that the gene directly influences cell size and position as well as dendritic development (35). A numerical analysis of two $sg \leftrightarrow$ t chimeras suggests further that the reduced number of Purkinje neurons in staggerer homozygotes is also an intrinsic effect but does not answer whether cell genesis or some aspect of cell maintenance is so controlled (36).

Other features of the mutant phenotype include altered physiological reactivity of Purkinje cells (20), multiple innervation of Purkinje cells by climbing fibers (19), reduction in size of external granular layer (the source of granule cell neurons and probably of interneurons of the molecular layer) from the day of birth onward (52,136,137), reduced mass of cerebellar nuclear complex (92,93), and arrested or altered differentiation of cerebellar glycoproteins (31,119). It is not yet clear which of these, if any, are expressions of the primary alteration of Purkinje cells. Granule cell neurons from staggerer neonates survive *in vitro* even better than cells from controls (65,66). The reduced proliferative activity in the external granular layer, for example, may be secondary to the Purkinje cell defect, and this in turn implies a normal integration of Purkinje and granule cell neuron numbers by action of Purkinje cells on the granule cell precursor population (101). Likewise, the degeneration in postnatal weeks 3 to 5 of those granule cells that do form may be secondary to the failure of granule cells to effect durable synaptic connections with the abnormal Purkinje cells (52,102,114).

Lurcher

This autosomal semidominant mutation causes cytological abnormalities by postnatal day 10 (P10) and death, beginning by about 2 weeks of age, of almost 100% of cerebellar Purkinje neurons (5). In addition, about 90% of cerebellar granule cells and 80% of inferior olivary neurons degenerate later (the latter group in two phases, one at about P35–60 and the other at P90–120) (5). Analysis of $Lc \leftrightarrow +$

chimeras indicates that the *Lc* genetic locus acts intrinsically in Purkinje cells, whereas the death of granule and olivary neurons is extrinsic, probably a consequence of loss of their major synaptic target, the Purkinje cell (127). No clues are available as to the basis for the selective (and, so far, unique) action of the *Lc* mutation on Purkinje cells or as to the role of the wild-type allele at this locus.

Purkinje Cell Degeneration

The *pcd* mutation, an autosomal recessive (68), also acts intrinsically in the cerebellar Purkinje cell (67). The first cytological abnormality that consistently distinguishes *pcd* homozygotes from littermates is seen by electron microscopy at P15, when normal Purkinje cells have matured sufficiently to have lost the mass of free polysomes between nucleus and initial segment of axon, but affected cells still retain it (54). Other abnormalities in the configuration of strands of rough endoplasmic reticulum may already be present, as well as apparently immature intracisternal A particles (doughnut-shaped RNA virus-like particles about 90 nm in diameter) and nematosomes (cytoplasmic organelles previously found mainly in normal mature sympathetic ganglion neurons) in cell body cytoplasm, axon hillock, and initial axonal segment. These changes are unique among the mutant phenotypes examined to date. Synaptogenesis appears to be qualitatively normal until P18, when the tertiary dendrites of Purkinje cells in the mutant appear stunted, and many of the branchlet spines are normal in shape and covered with glial processes rather than parallel fiber synapses. Some parallel fiber synapses are made directly onto Purkinje dendritic shafts, an arrangement absent in controls. The first evidence of overt degeneration of Purkinje cells was obtained by light microscopy at P18 and involved about 50% of the cells by P24 and almost all by P30. At P24, most Purkinje axon terminals on cell bodies of lateral nucleus neurons have degenerated (1% of the target cell body surface is occupied at this time with Purkinje terminals, compared to 35% in controls), although only half of the Purkinje cell bodies are gone (91).

Both a possible clue and a complication to the eventual unraveling of the mode of action of the *pcd* locus is the curious range of target cells affected. Retinal photoreceptor cells begin to degenerate almost as early as do the Purkinje cells and are lost slowly and progressively for more than 1 year (68,70). Mitral neurons of the olfactory bulb have begun to degenerate by 13 weeks, with a peak perhaps as late as 38 to 45 weeks (68). Additional cell losses occur sharply between P50 and P60 in discrete parts of the thalamus, all neurons degenerating in the ventral medial geniculate nucleus and many neurons in the mediodorsal, submedial, ventrolateral, and posterior nuclei (78). No evidence has been adduced in support of an extrinsic transsynaptic or vascular action of the *pcd* gene on these targets. The same question is at the core of many human degenerative diseases; namely, what does the disease mechanism recognize in common among the apparently disparate target cell populations?

Nervous

Although the Purkinje cell appears to be the major cerebellar neuronal target also of this autosomal recessive disease (106), chimerism studies have not yet yielded a clear answer as to whether the *nr* locus acts intrinsically or extrinsically (R. J. Mullen, *personal communication*). The initial cytological abnormality on record is a change in the shape of mitochondria in the cell body and dendrites from elongated to spherical in some Purkinje cells at P9 and in all of them at P15 (53). Cell death begins a few days later than in *pcd* homozygotes, proceeds slightly more slowly, but continues probably for the life of the animals (116) and comes ultimately to involve 50 to 90% or more of Purkinje cells, depending on cerebellar region and general genetic background. The surviving cells regain a normal mitochondrial morphology. At about 10 weeks of age, cerebellar benzodiazepine receptors are reduced 25 to 29% (111). Retinal photoreceptor cells degenerate progressively and also show rounded mitochondria before they die. The mechanism of photoreceptor cell loss is probably different in *pcd* and *nr* mice (70) and probably is not secondary in some nonspecific way to Purkinje cell degeneration, since lurcher mice lose Purkinje cells but have intact retinas. Neurons in some other CNS regions in young *nr* homozygotes may show rounded mitochondria, but these later revert to normal without much accompanying cell death.

These five cerebellar mutants displace or lose Purkinje neurons at different ontogenetic stages, and the consequences for the granule cell population are somewhat different in each case. In reeler, the early postnatal external granule layer of the cerebellum is focally reduced in proportion to the local deficit of Purkinje cells, and granule cell genesis is correspondingly decreased, to give a subnormal cerebellar size at all postnatal ages (85). Later death of granule cells has not been described. Also, in staggerer, the external granule layer is reduced in thickness and probably in area virtually from birth onward. Since the time of occurrence of the Purkinje cell deficit is unknown, one cannot say whether the underdevelopment of the external granule layer relates to the reduced number or altered quality of Purkinje cells or to some other aspect of the disease. Those granule cells that are generated come to translocate normally to the granular layer, receive qualitatively appropriate mossy fiber inputs, and form all normal classes of output synapses except the parallel fiber–Purkinje dendritic spine synapse. The subsequent rather abrupt degeneration of almost all of these granule cells is likely to be a consequence of their failure to make contact with their major synaptic target (101,102,114). This in turn implies some trophic influence passing in retrograde direction from Purkinje to granule cell, a mechanism analogous to that which underlies "programmed cell death" or "target organ effects" in some other parts of the nervous system (14,15,82).

The more modest disturbance of inferior olivary neurons in staggerer compared to lurcher may relate to survival in staggerer of about 25% of the target Purkinje cells and/or to the early timing of the Purkinje disorder, with polyinnervation of the surviving Purkinje cells by climbing fibers. In lurcher, Purkinje cell loss is extreme but occurs somewhat later, mainly in the third and fourth postnatal weeks.

Most granule cell and inferior olivary neurons will degenerate, but several weeks later, well after the time of the granule cell loss in staggerer. In the Purkinje cell degeneration and nervous mutants, the Purkinje neuron loss occurs some days later than in lurcher, and there is no deficit of olivary neurons and a milder and still more delayed loss of granule cells—as though the putative retrograde trophic influence had a duration of effect somewhat proportional to the duration of contact between pre- and postsynaptic neurons prior to degeneration of the Purkinje cell. This hypothesis of long-term transsynaptic trophic relationships merits exploration on a time scale of months or years in various parts of the CNS.

Mutations Affecting Predominantly the Granule Cell Neuron

We have seen that the granule cell neuron may bear the brunt of the effect in examples such as the reeler and staggerer disorders in which the primary target of the mutation is the immature Purkinje cell. Several human case reports describe predominant dysfunction of cerebellar granule cells (3,40,48,76,123,124), but the limited sample of time points in the course of the diseases and the limited histopathological methods preclude much insight into the primary cellular target. This may be a very difficult issue even in the experimental animal cases, as exemplified by the following mutants.

Weaver

As described above for reeler and staggerer, the weaver homozygote has a small cerebellum containing very few granule cell neurons (100). These cells are generated in normal concentration in the external granular layer but fail to translocate inward, fail to generate the bipolar cytoplasmic processes destined normally to differentiate into the parallel fibers, and die within several days after their genesis (87). The important idea that impaired migration is fundamental, not merely a consequence of impaired cell viability, came from the evidence of Razai and Yoon (89) that weaver heterozygotes show impaired rates of granule cell body translocation from the external granular layer, across the molecular layer, to the granular layer. Rakic and Sidman (87) ascribed this abnormal granule cell behavior to an abnormality of the radially disposed Bergmann glial fibers of the molecular layer, but Sotelo and Changeux (115) reasoned that the granule cell neuron was a primary target. Granule cell history was difficult to follow in $wv \leftrightarrow +$ chimeras because β-glucuronidase activity (23), the independent genetic marker that had served conveniently for Purkinje cells (69), failed to serve for classifying the source of granule cells. More suitable for this purpose but still not ideal is the nuclear marker available in ichthyosis mouse mutants (25), which has given preliminary evidence that the wv locus does act intrinsically in cerebellar granule cells (24). Whether or not it also acts in glial cells remains unresolved.

All studies agree that the Purkinje cell is basically not at risk from the mv mutation but responds secondarily to the absence of a parallel fiber milieu (86,112). Of the many genetic and nongenetic causes of selective granule cell loss during development (9), weaver presents the purest example. Purkinje cell dendritic branches that never at any stage are contacted by parallel fibers nonetheless generate spines with prominent "postsynaptic" densities (38,87) and maintain them for at least 2 years in contact with glial cells rather than axonal boutons (88). Such Purkinje cells differentiate so as to acquire smooth somatic surfaces and, with few exceptions (112), qualitatively correct synaptic inputs from other neurons than granule cells (86,88). They do, however, show the reduced dendritic mass and "weeping willow" configuration typical of all *in vivo* and *in vitro* situations in which Purkinje cell dendrite formation proceeds in a granuloprival milieu (9).

Descriptive studies *in vivo* are unlikely in themselves to answer what is the basic problem with weaver granule cells, and a beginning has been made at *in vitro* analysis. Messer and Smith (66) were the first to use this approach and were unable to distinguish cerebellar cells from early postnatal, +/+, +/wv, and wv/wv mice in monolayer culture. In a microwell reaggregation paradigm (121), +/+ granule cells migrated along flattened cells attached to the culture substratum and along the abundant cables composed mainly of neurites and some glial processes that came to interconnect the reaggregates. Comparable cells from wv/wv mice formed reaggregates at a normal rate and of similar size and number as in the controls but showed reduced cell migration and cable formation (120). Prior extraction of the horse serum component of the medium with ethanol–ether led to a doubling of the number of cables per culture well in the controls and an increase from about 4% to 50% of control values on the part of wv/wv cells.

More recently, a *wv* gene dose effect has been uncovered on neuron survival and neurite formation in monolayer cultures of dissociated cerebellar cells from 7- and 8-day-old mice (129). Of the cells present on day 1 *in vitro*, fewer than 50% survived to day 3 in wv/wv cultures, more than 60% survived in +/wv cultures, and more than 90% in +/+ cultures. Survival at day 6 *in vitro* was 20%, 40%, and 90%, respectively (Fig. 2, middle panel). By labeling the DNA-synthesizing cells with ^{3}H-thymidine *in vivo* 6 hr before setting up the cultures (almost all labeled cells residing at this time in the external granule layer) and following the labeled cells in autoradiograms at 1, 2, and 3 days *in vitro*, the fate of subpopulations of external granule cells could be followed. There was little further division *in vitro*. Of the labeled cells present at day 1 *in vitro*, 88% were gone from wv/wv cultures by day 3, compared with 20% and 10% from +/wv and +/+ cultures, respectively. When the losses from the total-cell and labeled-cell populations were compared, it became evident that recently generated (labeled) cells are more at risk in wv/wv than in +/wv cultures. This is consistent with the shorter life-span of the wv/wv granule cells *in vivo* (89).

The impairment of neurite formation *in vitro* appears also to correspond to the *in vivo* abnormality (87) and to show a *wv* allele dose effect (129). Whether results were expressed in terms of percentage of cells bearing neurites of various length classes, or, as shown in Fig. 2, in terms of total length of all neurites per cell (left panel), and the product of mean total length times number of cells per field (right panel), the quantitative differences among the three genotypes at 6 days *in vitro*

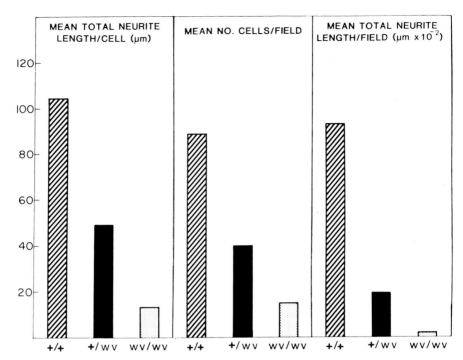

FIG. 2. Neurite growth at 6 days *in vitro*. The mean total neurite length/cell **(left panel)** was obtained by computer-associated digitization of all fibers originating from a total of 20 neurons in triplicate cultures of +/+, +/*wv*, and *wv/wv* cerebellar cultures. This value was multiplied by the number of cells/field **(center panel**, data averaged from six contiguous fields in each of two cultures of each genotype) to obtain the mean total neurite length/field **(right panel)**. (From Willinger et al., ref. 129, with permission.)

are striking. Further, when the length of the longest neurite was measured for each of 20 cells in each of the three geneotypes in 6-day cultures, the mean lengths were 9.0, 26.8, and 41.8 μm for *wv/wv*, +/*wv*, and +/+, respectively. In *wv/wv* cultures, 90% of the cells had neurites less than 20 μm in length, and the longest ones were 40 to 50 μm, as compared to 110 to 120 μm in +/*wv* and 140 to 150 μm in +/+ cultures. Neurite-generating behavior was followed more closely for selected individual living cells by time-lapse cinematography (130). Cells from homozygous weaver mice showed unusually active somatic and neuritic surfaces, with frequent formation and withdrawal of small cytoplasmic blebs and irregular activity of growth cones. Initiation of cytoplasmic processes suggestive of neurites was frequent, actually more so than in +/*wv* or +/+ cells. However, these newly initiated processes were commonly withdrawn soon afterward and, even when sustained and elongated for some hours, were relatively unstable along the shaft and at the growth cone. The behavior suggests that the *wv* genetic locus may be affecting cell surface membrane stability or cytoskeletal organization or both.

Tortured

This newly discovered autosomal recessive mutation has severe effects on behavior and on neuron survival in several parts of the central and peripheral nervous systems (16). The behavioral phenotype, featuring grossly abnormal postures and movements of all limbs, is recognized at about 17 days after birth and reaches maximum severity by 28 days. Although severely handicapped, affected mice may live for 1 year or longer. The change of particular note in the cerebellum is a progressive degeneration of granule cell neurons, beginning before the end of the first postnatal month and becoming very intense in the third month. By 1 year of age, relatively few granule cells remain except in the nodulus and uvula, where degeneration is still active. Purkinje cell loss is milder, so that in the moderately small cerebellar cortex at 1 year of age the residual Purkinje cell bodies are highly concentrated and prominent in their usual position at the inner boundary of the now reduced molecular layer. We have not recognized any major early effect on mossy fibers that might account transsynaptically for the granule cell degeneration, nor is it obvious what might be the property in common between cerebellar granule cells and the other degenerating neuronal populations in sensory ganglia, dorsal and ventral horns of spinal cord, several brainstem components, and basal ganglia.

Other Mutations Affecting Cerebellar Cortex

Mutants of this group have been analyzed less than most of those reviewed above and are grouped on the basis of ignorance into this "waste-basket" category (Table 1).

Leaner

This autosomal recessive was initially considered to be an independent cerebellar mutation (107,135) and only later was found to be allelic with tottering, tg (122). A third mutant allele, rolling mouse Nagoya, tg^{rol}, has also been described (77). Mice homozygous for tg^{la} or tg^{rol} show loss of granule and Purkinje cells in the cerebellar cortex (64,73,107), whereas a structural lesion has not been detected in tg homozygotes (75), and instead, a spike–wave seizure disorder is the major feature, as reviewed above. Neither tg^{la}/tg^{la} nor tg^{rol}/tg^{rol} mice have overt seizures. Rolling (1) and tottering (57) mice both have an increased noradrenergic axonal input, as visualized by fluorescence microscopy, in the cerebellar cortex, and tg^{rol} homozygotes have increased tyrosine hydroxylase and dopamine-β-hydroxylase activities in cerebellum and brainstem (72).

Other recent studies have defined the cell loss and its distribution in the cerebellar cortex more exactly in leaner homozygotes (37). Granule cells appear to be affected first, degenerating between P10 and P40, then more slowly until about P120, and probably continuing at a mild pace for life. Purkinje cell loss is just discernable at P26, reaches nearly 50% by P60, and then continues at a slower rate. Golgi II cells of the granule layer also become reduced about 50%. Granule and Purkinje cell loss is more extensive anteriorly than posteriorly, superficially resembling the pathology of certain anterior lobe syndromes in man (2), whereas reduction in the Golgi II population is regionally uniform.

Meander Tail

This autosomal recessive mutation causes, as the name implies, both neurological and skeletal abnormalities (42). Both are variable in severity. As a first approximation, the more severe the tail deformity, the more severe the neurological abnormality. The behavioral ataxia reflects an anatomical disorder that is already very prominent in the anterior lobe of the cerebellum (vermis and medial parts of hemispheres) by 10 days of age, the earliest age studied (R. L. Sidman, *unpublished observations*) and is probably present by birth or earlier. Both Purkinje and granule cell populations are severely reduced in number in the anterior lobe and are affected, although less prominently, elsewhere. Computer-aided three-dimensional reconstruction has revealed a more subtle volumetric reduction in parts of the posterior cerebellum and in the topographically related deep nuclear components (Ware and R. L. Sidman, *unpublished observations*). Efforts are in progress to stabilize the expression of the disorder by placing the *mea* mutation on an inbred genetic background.

Swaying

This disorder affects approximately the same parts of the cerebellum that bear the brunt of the tg^{la} and *mea* disorders summarized above, except that the disease mechanism is different. Swaying homozygotes have a malformation in which the fourth ventricle displays an abnormal recess extending dorsally between cerebellum and inferior colliculus; these colliculi are displaced laterally, the cerebellum is divided sagittally in the midline, and its anterior lobe is reduced in volume (100). Within the anterior vermis, the tissue is distorted into multiple small islands and folds of cortical tissue instead of forming one smoothly curving sheet; the cellular arrangement into cortical layers is essentially preserved within the individual little islands (see Figs. 1 to 4 in ref. 100). The malformation is already prominent at embryonic day 16 and bears some resemblance to human Dandy–Walker malformations.

Kreisler

This is another early embryonic malformation in which inductive relationships between the rhombencephalon and otocyst are disturbed (21). Although the consequences in the CNS are somewhat variable from individual to individual, many of the affected mice show a relatively severe and selective reduction in one lobule of the cerebellum, the paraflocculus, and thus provide another "experiment" involving regional disorder of the cerebellum.

Stumbler

Homozygous stumblers have smaller than normal cerebella by postnatal day 9, the earliest age studied, and probably earlier (6,7). The disorder is quantitatively subtle. Although the concentrations and distributions of Purkinje and granule cells appear normal, the absolute numbers of both must be subnormal from P10 onward, since the volume of the cerebellum and the surface area of its cortical layers are reduced and actually never increase beyond the size attained at P10. No degenerating cells, beyond the numbers seen in control specimens, have been detected up to P20, and the disorder is therefore considered tentatively to involve underproduction of granule and Purkinje cells. Further, all Purkinje cells are abnormal at P10 and later. They show somatic spines, multiple primary dendrites, and small bushy dendritic trees like those of younger normal cells as well as increased mitochondrial profiles and disorganized endoplasmic reticulum, features never encountered in normal Purkinje cells at any age. From P20 to P35, the oldest age studied, degenerating Purkinje cells are encountered, but those cells not actually degenerating appear to be reverting to normal form, a phenomenon already met in the case of the nervous mutant (see above). The disordered Purkinje cells share certain features with those in the *Mo* mutants (see above) and with *hpc* (see below) and may help in time to unravel the detailed controlling factors in Purkinje cell differentiation.

Hyperspiny Purkinje Cell

The *hpc/hpc* homozygote has been described only briefly (113). Affected animals are recognized by postnatal day 8 by their shuffling, hesitant gait and die by weaning age. In the third week, the cerebellum as a whole is slightly reduced in size but is normally organized. The external half of the granular layer is vacuolated, and the white matter contains many degenerating fibers, but the Purkinje cells show the most marked deviations from normal. The dendritic trees are small, with few tertiary branchlets or branchlet spines, whereas the cell body and thick dendritic trunks are studded with spines. Climbing fibers innervate somatic spines and spines on the main dendritic stem but extend no further distally. The Purkinje cell–climbing fiber system may be the cellular targets of the mutation.

Mutations Affecting Cerebellar Nuclei

Although any mutation affecting cerebellar cortex must almost necessarily reverberate within the deep nuclei of the cerebellum, we are concerned here with mutations that affect the nuclei disproportionately (Table 1). They cause somewhat different behavioral disorders than do the "cortical"mutations, but since many neuronal systems are affected in each of these "nuclear" diseases, a precise anatomico–behavioral correlation is not yet possible.

Vibrator

Although this mutation has been available for many years at the Jackson Laboratory, where it originated in 1961 (107), it is only now beginning to receive serious attention. A recessive on chromosome 11, the *vb* mutation, is being studied on a DBA/2J X C57BL/6J recombinant inbred strain and on an outcrossed background with and without the linked marker mutation, vestigial tail (125). From about P10 to P12, affected mice show a generalized fine rapid tremor of the head, trunk, proximal limbs, and tail which causes them to "vibrate" about the long axis of the body and which distinguishes them from all other mice. By about P22, the tremor becomes coarser, perhaps in part because of a desynchronization of coordination between muscle groups, and locomotion becomes impaired as hindlimbs develop enfeebled extensor tone, gait becomes increasingly unstable, and placing reactions become jerky and inaccurate. Inbred mice then lose the ability to stand, hold the hindlimbs extended but hypotonic on passive manipulation, develop fasciculations in limbs and face, breathe irregularly, rapidly lose weight, and die. Many go in and out of coma for a few days before they succumb. Noninbred homozygous affected animals may live for 6 months or more in the moderately uncoordinated stage.

The pathology is that of a spinocerebellar system degeneration, i.e., a progressive neuronal degeneration that is already marked at P13 to P24 and affects dorsal root sensory ganglion cells and their axons, neuron somas in Rexed's laminae VI and VII in the intermediate gray matter of the spinal cord at all segmental levels, and neurons in the medullary and pontine reticular formation, the lateral vestibular nucleus, red nucleus, and the cerebellar nuclei (125,126). At the cellular level, the outstanding changes are central chromatolysis, nuclear eccentricity, swelling of the neuron soma, and, in many cells, appearance of multiple intracytoplasmic and intraaxonal vacuoles which sometimes appear to exclude all other organelles locally. One older animal (P164) showed, in addition, atrophy of the posterior vermis and hemispheres of the cerebellum.

Cerebellar Outflow Degeneration

The complex neurological phenotype in this autosomal recessive mutation has been presented only in preliminary form (99,103). The mutation occurred in W. F. Hollander's colony *(personal communication)* and has been placed by us onto a BALB/c inbred background. At P12 to P15, the affected animals show slight impairment of balance when pushed about in the cage. Within a few more days, they develop a tremor of trunk, ataxia, and a consant nodding of the head punctuated occasionally by myoclonic jerking. By about P45, they appear to be somewhat lethargic and are unable to remain afloat or swim when placed in water, although they manage to eat and drink with little difficulty in their home cages. They live a normal life-span but, like most of the neurological mutant animals, do not breed.

The pathological process consists of remarkably widespread, but not random, degeneration of axons with preservation of their parent cell bodies and of severe degeneration of cell bodies in the cerebellar nuclei and in brainstem and spinal cord centers synaptically connected with the cerebellum. The earliest changes we have seen were present at P11 and consisted of swelling of a few cell bodies in each of several of the vestibular nuclei and degeneration of a few axons, especially of axon terminals in a few brainstem sites. By P15, cellular changes were more advanced in lateral and superior vestibular, red, and cerebellar nuclei. Changes proceeded from swelling to severe chromatolysis to cell death and came to involve progressively more neurons in these sites and in additional ones. By P270, approximately 90% of the large neurons had disappeared from the red nucleus and from all of the cerebellar nuclei, slightly fewer from the lateral and superior vestibular nuclei, and about 50% from nucleus pontis and nucleus reticularis tegmenti pontis. The pattern of cell body and axonal changes is incompatible with transneuronal degeneration as a major causal event. The early cell swelling and chromatolysis may be a response to the concurrent axonal disorder, but neither the fundamental pathological process nor its distribution has been explained.

FUTURE PROSPECTS

Both human and mouse neurogenetics are at the threshold of major technical advances that are likely to render many of today's questions and concerns obsolete (95). One can discern the general shape of an approach to intragenic structure, molecular control mechanisms, and direct progression from gene to gene products without proceeding laboriously from complex final phenotype back towards primary molecules. The big problem that may remain, perhaps, will be the analysis of developmental sequences between direct gene product and final phenotype. This will be a general issue in genetics and developmental biology, but nowhere will it be more compelling than with reference to the mammalian nervous system, the most complex of all biological organs and probably the one most responsive to its environment throughout the entire span of life.

Several actions must be taken with mouse stocks today if we wish to gain the experimental material necessary to exploit the new technical prospects. First, it becomes mandatory to save as many mutations as possible on their initial coisogenic backgrounds, if they happen to have occurred in inbred strains, and to make all other mutations of interest congenic with an inbred strain, say C57BL/6J. Preferably, each mutation should be rendered congenic with two genetically disparate strains so that F_1 hybrid affected, as well as inbred affected, animals will be available. These steps are essential for the quantitative comparison of mutant and wild-type alleles, for the comparison of independent mutations with one another, for the effective use of genetic markers to study intrinsic versus extrinsic gene action. Such strains are crucial also for the extremely important analysis of environmental effects, which can only be done reliably if the genome is held as constant as possible. Further, all "repeat" mutations should be saved, since it is statistically very unlikely that two mutations will cause the same change within a gene. That is, recurrences are not boring; they are the prime tools for future fine-structural genetic analysis.

REFERENCES

1. Adachi, K., Sobue, I., Tohyama, M., and Shimizu, N. (1975): Changes in the cerebellar noradrenaline nerve terminals of the neurological murine mutant rolling mouse Nagoya: A histofluorescence analysis. *IRCS Med. Sci.*, 3:329–330.

2. Adams, R. D., and Sidman, R. L. (1968): *Introduction to Neuropathology.* McGraw-Hill, New York.
3. Ambler, M., Pogacar, S., and Sidman, R. (1969): Lhermitte–Duclos disease (granule cell hypertrophy of the cerebellum). Pathological analysis of the first familial cases. *J. Neuropathol. Exp. Neurol.*, 28:622–642.
4. Baumann, N. (1980): *INSERM Symposium No. 14, Neurological Mutations Affecting Myelination.* Elsevier/North Holland, Amsterdam.
5. Caddy, K. W. T., and Biscoe, T. J. (1979): Structural and quantitative studies on the normal C3H and lurcher mutant mouse. *Phil. Trans. R. Soc. (Lond.)*, 287:167–201.
6. Caddy, K. W. T., and Sidman, R. L. (1981): Purkinje cells and granule cells in the cerebellum of the stumbler mutant mouse. *Dev. Brain Res.*, 1:221–236.
7. Caddy, K. W. T., Sidman, R. L., and Eicher, E. M. (1981): Stumbler, a new mutant mouse with cerebellar disease. *Brain Res.*, 208:251–255.
8. Carroll, W. M., Jay, B. S., McDonald, W. I., and Halliday, A. M. (1980): Two distinct patterns of visual evoked response asymmetry in human albinism. *Nature*, 286:604–606.
9. Caviness, V. S., Jr., and Rakic, P. (1978): Mechanisms of cortical development: A view from mutations in mice. *Annu. Rev. Neurosci.*, 1:297–326.
10. Caviness, V. S., Jr., So, D. K., and Sidman, R. L. (1972): The hybrid reeler mouse. *J. Hered.*, 63:241–246.
11. Coleman, D. (1978): Obese and diabetes: Two mutant genes causing diabetes–obesity syndromes in mice. *Diabetologia*, 14:141–148.
12. Coleman, D. L., and Hummel, K. P. (1969): Effects of parabiosis of normal with genetically diabetic mice. *Am. J. Physiol.*, 217:1298–1304.
13. Cooper, M. L., and Pettigrew, J. D. (1979): The retinothalamic pathways in siamese cats. *J. Comp. Neurol.*, 187:313–348.
14. Cowan, W. M. (1970): Anterograde and retrograde transneuronal degeneration in the central and peripheral nervous systems. In: *Contemporary Research Methods in Neuroanatomy,* edited by W. J. H. Nauta and S. O. E. Ebbesson, pp. 217–251. Springer-Verlag, New York.
15. Cowan, W. M. (1973): Neuronal death as a regulative mechanism in the control of cell number in the nervous system. In: *Development and Aging in the Nervous System,* edited by M. Rockstein, pp. 19–41. Academic Press, New York.
16. Cowen, J., and Sidman, R. (1981): Tortured. Personal communication. *Mouse News Lett.*, 65:16.
17. Creel, D., O'Donnell, F. G., Jr., and Witkop, C. J., Jr. (1978): Visual system anomalies in human ocular albinos. *Science*, 189:931–933.
18. Creel, D., Witkop, C. J., Jr., and King, R. A. (1974): Asymmetric visually evoked potentials in human albinos: Evidence for visual system anomalies. *Invest. Ophthalmol.*, 13:430–440.
19. Crepel, F., Delhayeb, N., Guastavi, J. M., and Sampaio, I. (1980): Multiple innervation of cerebellar Purkinje cells by climbing fibers in staggerer mutant mouse. *Nature*, 283:483–484.
20. Crepel, F., and Mariani, J. (1975): Anatomical, physiological and biochemical studies on the cerebellum from mutant mice. I. Electrophysiological analysis of cerebellar cortical neurons in the staggerer mouse. *Brain Res.*, 98:135–147.
21. Deol, M. S. (1964): The abnormalities of the inner ear in *Kreisler* mice. *J. Embryol. Exp. Morphol.*, 12:475–490.
22. Drager, U. C., and Olsen, J. F. (1980): Origins of crossed and uncrossed retinal projections in pigmented and albino mice. *J. Comp. Neurol.*, 191:383–412.
23. Feder, N. (1976): Solitary cells and enzyme exchange in tetra-parental mice. *Nature*, 263:67–69.
24. Goldowitz, D., and Mullen, R. J. (1980): Weaver mutant granule cell defect expressed in chimeric mice. *Neurosci. Abstr.*, 6:743.
25. Goldowitz, D., and Mullen, R. J. (1982): Nuclear morphology of ichthyosis mutant mice as a cell marker in chimeric brain. *Dev. Biol.*, 89:261–267.
26. Green, M. C., and Sidman, R. L. (1962): Tottering, a neuromuscular mutation in the mouse and its linkage with oligosyndactylism. *J. Hered.*, 53:233–237.
27. Green, M. C., Snell, G. D., and Lane, P. W. (1963): Linkage group XVIII of the mouse. *J. Hered.*, 54:245–247.
28. Guillery, R. W., Okora, A. N., and Witkop, C. J., Jr. (1975): Abnormal visual pathways in the brain of the human albino. *Brain Res.*, 96:373–377.
29. Guillery, R. W., Scott, G. L., Cattanach, B. M., and Deol, M. S. (1973): Genetic mechanisms determining the central visual pathways of mice. *Science*, 179:1014–1016.

30. Hall, J. C., Greenspan, R. J., and Harris, W. A. (1982): *Genetic Neurobiology.* The MIT Press, Cambridge.
31. Hatten, M. E., and Messer, A. (1978): Postnatal cerebellar cells from staggerer mutant mice express embryonic cell surface characteristic. *Nature*, 276:504–506.
32. Hearing, V. J. (1973): Tyrosinase activity in subcellular fractions of black and albino mice. *Nature [New Biol.]*, 245:81–83.
33. Heininger, H.-J., and Dorey, J. J. (1980): *Handbook of Genetically Standardized JAX Mice, Third Edition.* The Jackson Laboratory, Bar Harbor, Maine.
34. Herrup, K., and Mullen, R. J. (1979): Regional variation and absence of large neurons in the cerebellum of the staggerer mouse. *Brain Res.*, 172:1–12.
35. Herrup, K., and Mullen, R. J. (1979): Staggerer chimeras: Intrinsic nature of Purkinje cell defects and implications for normal cerebellar development. *Brain Res.*, 178:443–457.
36. Herrup, K., and Mullen, R. J. (1981): Role of the staggerer gene in determining Purkinje cell number in the cerebellar cortex of mouse chimeras. *Dev. Brain res.*, 1:475–485.
37. Herrup, K., and Wilczynski, S. L. (1982): Cerebellar cell degeneration in the leaner mutant mouse.
38. Hirano, A., and Dembitzer, H. M. (1974): Observations on the development of the weaver mouse cerebellum. *J. Neuropathol. Exp. Neurol.*, 33:354–364.
39. Hirano, A., and Dembitzer, H. M. (1975): The fine structure of staggerer cerebellum. *J. Neuropathol. Exp. Neurol.*, 34:1–11.
40. Hirano, A., Dembitzer, H. M., Ghatak, N. R., Fan, K. J., and Zimmerman, H. M. (1973): On the relationship between human and experimental granule cell type cerebellar degeneration. *J. Neuropathol. Exp. Neurol.*, 32:493–502.
41. Hirano, A., Llena, J. F., French, J. H., and Ghatak, N. R. (1977): Fine structure of the cerebellar cortex in Menkes kinky-hair disease. *Arch. Neurol.*, 34:52–56.
42. Hollander, W. F., and Waggie, K. S. (1977): Meander tail, a recessive mutant located in chromosome 4 of the mouse. *J. Hered.*, 68:403–406.
43. Horn, N. (1981): Abnormal copper metabolism in cultured cells of mottled mice. Personal communication. *Mouse News Lett.*, 65:14–15.
44. Hummel, K. P., Dickie, M. M., and Coleman, D. L. (1966): Diabetes, a new mutation in the mouse. *Science*, 153:1127–1128.
45. Hunt, D. M. (1974): Primary defect in copper transport underlies mottled mutants in the mouse. *Nature*, 249:852–854.
46. Hunt, D. M. (1976): A study of copper treatment and tissue copper levels in the murine congenital copper deficiency, mottled. *Life Sci.*, 19:1913–1920.
47. Hunt, D. M., and Johnson, D. R. (1972): An inherited deficiency in noradrenaline biosynthesis in the brindled mouse. *J. Neurochem.*, 19:2811–2819.
48. Jervis, G. A. (1950): Early familial cerebellar degeneration (report of 3 cases in one family). *J. Nerv. Ment. Dis.*, 111:398–407.
49. Johnson, L. M., and Sidman, R. L. (1979): A reproductive endocrine profile in the diabetes *(db)* mutant mouse. *Biol. Reprod.*, 20:552–559.
50. Kaas, J. H., and Guillery, R. W. (1973): The transfer of abnormal visual field representations from the dorsal lateral geniculate nucleus to the visual cortex in Siamese cats. *Brain Res.*, 59:61–95.
51. Kaplan, B. J., Seyfried, T. N., and Glaser, G. H. (1979): Spontaneous polyspike discharges in the epileptic mutant mouse (tottering). *Exp. Neurol.*, 66:577–586.
52. Landis, D. M. D., and Sidman, R. L. (1978): Electron microscopic analysis of postnatal histogenesis in the cerebellar cortex of staggerer mutant mice. *J. Comp. Neurol.*, 179:831–863.
53. Landis, S. C. (1973): Ultrastructural changes in the mitochondria of cerebellar Purkinje cells of nervous mutant mice. *J. Cell. Biol.*, 57:782–797.
54. Landis, S. C., and Mullen, R. J. (1978): The development and degeneration of Purkinje cells in pcd mutant mice. *J. Comp. Neurol.*, 177:125–143.
55. Lane, P. W., and Deol, M. S. (1974): Mocha, a new coat color and behavior mutation on chromosome 10 of mouse. *J. Hered.*, 65:362–364.
56. LaVail, J. H., Nixon, R. A., and Sidman, R. L. (1978): Genetic control of retinal ganglion cell projections. *J. Comp. Neurol.*, 182:399–421.
57. Levitt, P., and Noebels, J. L. (1981): Mutant mouse tottering. Selective increase in locus coeruleus axons in a defined single locus mutation. *Proc. Natl. Acad. Sci. U.S.A.*, 78:4630–4634.
58. Lockman, L. A., Kennedy, W. R., and White, J. G. (1967): The Chediak–Higashi syndrome: Electrophysiologic and electron microscopic observations on the peripheral neuropathy. *J. Pediatr.*, 70:942–951.

59. Lund, R. D. (1978): *Development and Plasticity of the Brain*. Oxford University Press, New York.
60. Mallet, J. (1980): Biochemistry and immunology of neurological mutants in the mouse. *Curr. Top. Dev. Biol.*, 15:41–65.
61. Mann, J. R., Camakaris, J., and Danks, D. M. (1979): Copper metabolism in mottled mouse mutants. Distribution of ^{64}Cu in brindled *(Mo^{br})* mice. *Biochem. J.*, 180:613–619.
62. Mann, J. R., Camakaris, J., Danks, D. M., and Walliczek, E. G. (1979): Copper metabolism in mottled mouse mutants. Copper therapy of brindled *(Mo^{br})* mice. *Biochem. J.*, 180:605–612.
63. Mariani, J., Crepel, F., Mikoshiba, K., Changeux, J.-P., and Sotelo, C. (1977): Anatomical, physiological and biochemical studies of the cerebellum from reeler mutant mouse. *Phil. Trans. R. Soc. (Lond.)*, 281:1–28.
64. Meier, H., and MacPike, A. D. (1971): Three syndromes produced by two mutant genes in the mouse. Clinical, pathological and ultrastructural bases of tottering, leaner and heterozygous mice. *J. Hered.*, 62:297–302.
65. Messer, A. (1978): Abnormal staggerer cerebellar cell interactions and survival *in vitro*. *Neurosci. Lett.*, 9:185–188.
66. Messer, A., and Smith, D. M. (1977): *In vitro* behavior of granule cells from staggerer and weaver mutants of mice. *Brain Res.*, 130:13–23.
67. Mullen, R. J. (1977): Genetic dissection of the CNS with mutant–normal mouse and rat chimeras. *Soc. Neurosci. Symp.*, 11:47–65.
68. Mullen, R. J., Eicher, E. M., and Sidman, R. L. (1976): Purkinje cell degeneration, a new neurological mutant in the mouse. *Proc. Natl. Acad. Sci. U.S.A.*, 73:208–212.
69. Mullen, R. J., and Herrup, K. (1979): Chimeric analysis of mouse cerebellar mutants. In: *Neurogenetics: Genetic Approaches to the Nervous System*, edited by X. O. Breakfield, pp. 173–196. Elsevier, New York.
70. Mullen, R. J., and LaVail, M. M. (1975): Two new types of retinal degeneration in cerebellar mutant mice. *Nature*, 258:528–530.
71. Nagara, H., Yajima, K., and Suzuki, K. (1981): The effect of copper supplementation on the brindled mouse. A clinico–pathological study. *J. Neuropathol. Exp. Neurol.*, 40:428–446.
72. Nagatsu, I., Kondo, Y., Inagaki, S., Oda, S., and Nagatsu, T. (1980): Dopamine-beta-hydroxylase and tyrosine hydroxylase activities in brain regions of rolling mouse Nagoya. *Biomed. Res.*, 1:88–90.
73. Nakane, K. (1976): Postnatal development of brain in mice with congenital ataxia, rolling (rol) and tottering (tg). *Teratology*, 14:248–249.
74. Noebels, J. L. (1979): Analysis of inherited epilepsy using single locus mutations in mice. *Fed. Proc.*, 38:2405–2410.
75. Noebels, J. L., and Sidman, R. L. (1979): Inherited epilepsy: Spike–wave and focal motor seizures in the mutant mouse tottering. *Science*, 204:1334–1336.
76. Norman, R. M. (1940): Primary degeneration of the granular layer of the cerebellum: An unusual form of familial cerebellar atrophy occurring in early life. *Brain*, 63:365–379.
77. Oda, S. (1981): A new allele of the tottering locus, rolling mouse Nagoya, on chromosome no. 8 in the mouse. *Jpn. J. Genet.*, 56:295–299.
78. O'Gorman, S. V., and Sidman, R. L. (1980): Cell loss in diencephalic nuclei of Purkinje cell degeneration (pcd) mutant mice. *Soc. Neurosci. Abstr.*, 6:81.
79. Paigen, K. (1979): Acid-hydrolases as models of genetic control. *Annu. Rev. Genet.*, 13:417–466.
80. Paterson, M. C., and Smith, P. J. (1979): Ataxia telangectasia. Inherited human disorder involving hypersensitivity to ionizing radiation and related DNA-damaging chemicals. *Annu. Rev. Genet.*, 13:291–318.
81. Penfield, W., and Jasper, H. H. (1954): *Epilepsy and the Functional Anatomy of the Human Brain*. Little, Brown, Boston.
82. Prestige, M. C. (1970): Differentation, degeneration, and the role of the periphery: Quantitative considerations. In: *The Neurosciences, Second Study Program*, edited by F. O. Schmitt, pp. 73–82. Rockefeller University Press, New York.
83. Prins, H. W., and Van den Hamer, C. J. A. (1980): Abnormal copper–thionein synthesis and impaired copper utilization in mutated brindled mice: Model for Menkes' disease. *J. Nutr.*, 110:151–157.
84. Purpura, D. P., Hirano, A., and French, J. H. (1977): Polydendritic Purkinje cells in X-chromosome linked copper malabsorption: A Golgi study. *Brain Res.*, 117:125–129.

85. Rakic, P. (1976): Synaptic specificity in the cerebellar cortex: Study of anomalous circuits induced by single gene mutations in mice. *Cold Spring Harbor Symp. Quant. Biol.*, 40:333–346.
86. Rakic, P. (1979): Genetic and epigenetic determinants of local neuronal circuits in the mammalian central nervous system. In: *The Neurosciences. Fourth Study Program*, edited by F. O. Schmitt and F. G. Worden, pp. 109–127. The MIT Press, Cambridge.
87. Rakic, P., and Sidman, R. L. (1973): Sequence of developmental abnormalities leading to granule cell deficit in cerebellar cortex of weaver mutant mice. *J. Comp. Neurol.*, 152:103–132.
88. Rakic, P., and Sidman, R. L. (1973): Organization of cerebellar cortex secondary to deficit of granule cells in weaver mutant mice. *J. Comp. Neurol.*, 152:133–162.
89. Rezai, A., and Yoon, C. H. (1972): Abnormal rate of granule cell migration in the cerebellum of "weaver" mutant mice. *Dev. Biol.*, 29:17–26.
90. Roderick, T., and Davisson, M. (1981): Linkage map of the mouse. Personal communication. *Mouse News Lett.*, 65:5.
91. Roffler-Tarlov, S., Beart, P. M., O'Gorman, S., and Sidman, R. L. (1979): Neurochemical and morphological consequences of axon terminal degeneration in cerebellar deep nuclei of mice with inherited Purkinje cell degeneration. *Brain Res.*, 168:75–95.
92. Roffler-Tarlov, S., and Herrup, K. (1981): Quantitative examination of the deep cerebellar nuclei in the staggerer mutant mouse.
93. Roffler-Tarlov, S., and Sidman, R. L. (1978): Concentrations of glutamic acid in cerebellar cortex and deep nuclei of normal mice and weaver, staggerer and nervous mutants. *Brain Res.*, 142:269–283.
94. Sanderson, K. J., Guillery, R. W., and Shackelford, R. M. (1974): Congenitally abnormal visual pathways in mink *(Mustela vision)* with reduced retinal pigment. *J. Comp. Neurol.*, 154:225–248.
95. Schmitt, F. O., Bloom, F. E., and Bird, S. (1982): *Molecular Genetics and Neuroscience: A New Hybrid*. Raven Press, New York.
96. Shatz, C. J. (1977): A comparison of visual pathway in Boston and Midwestern Siamese cats. *J. Comp. Neurol.*, 171:205–228.
97. Shatz, C. J., and LeVay, S. (1979): Siamese cat: Altered conditions of visual cortex. *Science*, 204:328–330.
98. Sheridan, C. L. (1965): Interocular transfer of brightness and pattern discriminations in normal and corpus callosum-transected rats. *J. Comp. Physiol. Psychol.*, 59:292–294.
99. Sidman, R. L. (1967): Cerebellar outflow degeneration. Personal communication. *Mouse News Lett.*, 36:33.
100. Sidman, R. L. (1968): Development of interneuronal connections in brains of mutant mice. In: *Society of General Physiologists Monograph, Physiological and Biochemical Aspects of Nervous Integration*, edited by F. D. Carlson, pp. 163–193. Prentice-Hall, Engelwood Cliffs, New Jersey.
101. Sidman, R. L. (1972): Cell interactions in developing mammalian central nervous system. In: *Third Lepetit Colloquium on Cell Interaction, London*, edited by L. G. Silvestri, pp. 1–13. North-Holland, Amsterdam.
102. Sidman, R. L. (1974): Contact interaction among developing mammalian brain cells. In: *The Cell Surface in Development*, edited by A. A. Moscona, pp. 221–253. John Wiley & Sons, New York.
103. Sidman, R. L. (1980): Cerebellar outflow degeneration. Personal communication. *Mouse News Lett.*, 63:13.
104. Sidman, R. L. (1982): Mutations affecting the central nervous system in the mouse. In: *Molecular Genetics and Neuroscience: A New Hybrid*, edited by F. O. Schmidt, F. E. Bloom, and S. Bird. Raven Press, New York *(in press)*.
105. Sidman, R. L., Cowen, J. S., and Eicher, E. M. (1979): Inherited muscle and nerve diseases in mice: A tabulation with commentary. *Ann. N.Y. Acad. Sci.*, 317:497–505.
106. Sidman, R. L., and Green, M. C. (1970): "Nervous," a new mutant mouse with cerebellar disease. In: *Symposium, Centre National de la Recherche Scientifique, Les Mutants Pathologiques Chez l'Animal, Leur Interet pour la Recherche Biomedicale*, edited by M. Sabourday, pp. 69–79. CNRS, Paris.
107. Sidman, R. L., Green, M. C., and Appel, S. H. (1965): *Catalog of the Neurological Mutants of the Mouse*. Harvard University Press, Cambridge.
108. Sidman, R. L., Lane, P. W., and Dickie, M. (1962): Staggerer, a new mutation in the mouse affecting the cerebellum. *Science*, 137:610–612.
109. Sidman, R. L., and O'Gorman, S. V. (1981): Cellular interactions in Schwann cell development. In: *Neurofibromatosis: Genetics, Cell Biology, and Biochemistry*, edited by J. J. Mulvihill and V. M. Riccardi, pp. 213–234. Raven Press, New York.

110. Silver, J., and Sapiro, J. (1981): Axonal guidance during development of the optic nerve: The role of pigmented epithelia and other extrinsic factors. *J. Comp. Neurol.*, 202:521–538.
111. Skolnick, P., Syapin, P. J., Paugh, B. A., and Paul, S. M. (1979): Reduction in benzodiazepine receptors associated with Purkinje cell degeneration in "nervous" mutant mice. *Nature*, 277:397–399.
112. Sotelo, C. (1975): Anatomical, physiological and biochemical studies of cerebellum from mutant mice. II. Morphological study of cerebellar cortical neurons and circuits in the weaver mouse. *Brain Res.*, 94:19–44.
113. Sotelo, C. (1980): Anatomical and electrophysiological studies with the mouse mutant hyperspiny Purkinje cell (hpc). Personal communication. *Mouse News Lett.*, 62:71.
114. Sotelo, C., and Changeux, J.-P. (1974): Transsynaptic degeneration "en cascade" in the cerebellar cortex of staggerer mutant mice. *Brain Res.*, 67:519–526.
115. Sotelo, C., and Changeux, J.-P. (1974): Bergmann fibers and granular cell migration in the cerebellum of homozygous weaver mutant mouse. *Brain Res.*, 77:484–491.
116. Sotelo, C., and Triller, A. (1979): Fate of presynaptic afferents to Purkinje cells in the adult nervous mouse: A model to study presynaptic stabilization. *Brain Res.*, 175:11–36.
117. Staats, J. (1975): Diabetes in the mouse due to two mutant genes—a bibliography. *Diabetologia*, 11:325–327.
118. Sweet, H. O. (1981): Wasted. Personal communication. *Mouse News Lett.*, 65:27.
119. Trenkner, E. (1979): Postnatal cerebellar cells of staggerer mutant mice express immature components on their surface. *Nature*, 277:566–567.
120. Trenkner, E., Hatten, M. E., and Sidman, R. L. (1978): Ether-soluble serum components affect *in vitro* behavior of immature cerebellar cells in weaver mutant mice. *Neuroscience*, 3:1093–1100.
121. Trenkner, E., and Sidman, R. L. (1977): Histogenesis of mouse cerebellum in microwell culture. Cell reaggregation and migration, fiber and synapse formation. *J. Cell Biol.*, 75:915–940.
122. Tsuji, S., and Meier, H. (1971): Evidence for allelism of leaner and tottering in mouse. *Genet. Res.*, 17:83.
123. Ule, G. (1952): Kleinhirnrindenatrophie vom Körnertyp. *Dtsch. Z. Nervenheilkd.*, 168:195–226.
124. Vogt, H., and Astwazaturow, M. (1911): Üeber angeborene Kleinhirnerkrankungen mit Beitragen zur Entwicklungsgeschichte des Kleinhirns. *Arch. Psychiatr. Nernenkr.*, 49:75–203.
125. Weimar, W. R., Lane, P. W., and Sidman, R. L. (1982): Vibrator (vb): A spinocerebellar system degeneration with autosomal recessive inheritance in mice.
126. Weimar, W. R., and Sidman, R. L. (1978): Neuropathology of "vibrator"—a neurological mutation of the mouse. *Soc. Neurosci. Abstr.*, 4:401.
127. Wetts, R., and Herrup, K. (1980): Lurcher↔wild-type chimeric mice: Cerebellar Purkinje cells are primary site of gene action. *Neurosci. Abstr.*, 6:142.
128. Williams, R. S., Marshall, P. C., Lott, I. T., and Caviness, V. S., Jr. (1978): The cellular pathology of Menkes steely hair syndrome. *Neurology (Minneap.)*, 28:575–583.
129. Willinger, M., Margolis, D. M., and Sidman, R. L. (1981): Neuronal differentiation in cultures of weaver (wv) mutant mouse cerebellum. *J. Supramol. Struct. Cell Biochem.*, 17:79–86.
130. Willinger, M., Margolis, D. M., and Sidman, R. L. (1981): Granule cell behavior in dissociated cultures of Weaver *(wv)* mutant cerebellum. *Soc. Neurosci. Abstr.*, 7:346.
131. Witkop, C. J., Jr., Quevedo, W. C., Jr., and Fitzpatrick, T. B. (1978): Albinism. In: *The Metabolic Basis of Inherited Disease, Fourth Edition*, edited by J. B. Stanbury, J. B. Wyngaarden, and D. S. Fredrickson, pp. 283–316. McGraw-Hill, New York.
132. Yajima, K., and Suzuki, K. (1979): Neuronal degeneration in the brain of the Brindled mouse—a light microscope study. *J. Neuropathol. Exp. Neurol.*, 38:35–46.
133. Yajima, K., and Suzuki, K. (1979): Neuronal degeneration in the brain of the brindled mouse. An ultrastructural study of cerebral cortical neurons. *Acta Neuropathol. (Berl.)*, 45:17–25.
134. Yen, T. T. T., Steinmetz, J., and Simpson, P. J. (1974): The response of obese *(ob/ob)* and diabetic *(db/db)* mice to treatments that influence body temperature. *Comp. Biochem. Physiol.*, 49:377–385.
135. Yoon, C. H. (1969): Disturbances in developmental pathways leading to a neurological disorder of genetic origin, "leaner," in mice. *Dev. Biol.*, 20:158–181.
136. Yoon, C. H. (1972): Developmental mechanism for changes in cerebellum of "staggerer" mouse, a neurological mutant of genetic origin. *Neurology (Minneap.)*, 33:743–754.
137. Yoon, C. H. (1976): Pleiotropic effect of the staggerer gene. *Brain Res.*, 109:206–215.

Genetics of Neurological and Psychiatric Disorders, edited by Seymour S. Kety, Lewis P. Rowland, Richard L. Sidman, and Steven W. Matthysse. Raven Press, New York © 1983.

Genetic Control of Developmental Antigens

Dorothea Bennett

Sloan-Kettering Institute for Cancer Research, New York, New York 10021

A major question confronting modern biology is an understanding of the process of mammalian embryonic differentiation and the mechanisms that control it. We can readily define differentiation as the process by which a fertilized egg transforms with great rapidity and accuracy into an entity with hundreds or thousands of different cell types, all with precise morphological and functional relationships to one another. Yet we know almost nothing, at any level, of how the process works.

A few obvious facts can be presented. First of all, one fundamental connotation of the term differentiation is "different." The crucial implications for cell differentiation were defined elegantly in 1963 by Jacob and Monod (8) who said "Two cells are differentiated from one another when they harbor the same genome but synthesize different sets of proteins." Another connotation implicit in the term differentiation is that of "maturation." In this sense, it means an increase in the complexity and organization of cells and tissues during development which is accompanied by differential gene expression and which results in communities of cells whose individual potentialities are restricted to a rigid constellation of functions.

It is worth emphasizing some important concepts here. First, the organization constructed by interacting cells produces differential gene expression which is stable in the sense that, as long as normal organization is maintained, cells do not revert to an earlier stage of differentiation. It is an interesting corollary that malignant differentiation may represent an escape from the orderly controls provided by normal morphological architecture. It has been recognized for some years now (3) that the elicitation and maintenance of differential gene function in maturing cells depends heavily on environmental factors such as other cells and extracellular matrix components. This in turn implies that cells must have receptors capable of sensing information about their environment and transducing it internally so as to activate or repress specific batteries of genes.

It can be suggested, therefore, that for purposes of understanding development and differentiation mechanistically, one should concentrate on studying the cell surface, since that is where information-gathering receptors must be and, likewise, where cells must display some sort of instructional signals to identify themselves to others (6). Thus, the simple diagram in Fig. 1 illustrates the essential question to be addressed.

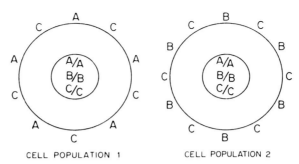

FIG. 1. Different populations of cells in the same individual do not display the same sets of cell surface components. (From Bennett et al., ref. 3, with permission.)

A closer look at the problem of understanding mechanisms of differentiation about which we know so little reveals that one of the most important reasons for our lack of insight no doubt results from the fact that development is comprised of a series of complexly interlocking small changes which occur very rapidly. Thus, it has so far been almost impossible to identify the key events that are presumbly responsible for initiating whole cascades of events. Theoretically, one way to identify such key events is to study the effects of mutations that interrupt fundamental processes of embryonic development and differentiation and therefore produce embryonic abnormalities and death. The use of mutant genes as experimental tools should make it possible to identify unique single critical events during embryogenesis, since a single gene presumably has only one primary effect.

The genetic system focused on in this chapter is a series of mutations in the mouse *T/t* complex. About eight different recessive lethal mutations have been identified in this region of chromosome 17 because of their phenotypic interaction with a dominant mutation, *T*. Very briefly, heterozygotes (+/*t*) for these recessive mutations are quite normal morphologically, but matings to dominant heterozygotes (+/*T*), which are abnormal only in being short-tailed, produce a new class of progeny, *T/t* animals, which are entirely tailless. Although abnormal tails in mice may seem trivial although peculiar, these tailless mice have been instrumental in defining a set of recessive mutations with profound effects on very early development that would otherwise certainly have escaped detection. As shown in Table 1 tailless mice carrying the same recessive mutations "breed true"; that is, they produce at birth only tailless progeny like themselves. Genetically speaking, this constitutes a balanced lethal system in which both homozygous classes die. The balanced lethal system has also been informative in defining different classes of *t* mutations, since, where they are different, genetic complementation occurs, and a viable class of progeny appears (Table 1) (2).

Over the past 50 years, a combination of complementation analysis and embryological study has defined eight genetically distinct *t* mutations, each of which has specific and different effects on early embryonic development. These mutations have been especially useful because they all affect development at very early stages,

TABLE 1. *Representation of balanced lethal breeding scheme* (top) *and genetic complementation between two different lethal t-mutations* (bottom)

Balanced lethal system	T/t⁰	×	T/t⁰
		↓	
	T/T	T/t⁰	t⁰/t⁰
	Lethal	Tailless	Lethal
Complementation between different t-mutations	T/t⁰	×	T/t¹
		↓	
	T/T	T/t⁰, T/t¹	t¹/t⁰
	Lethal	Tailless	Normal tailed

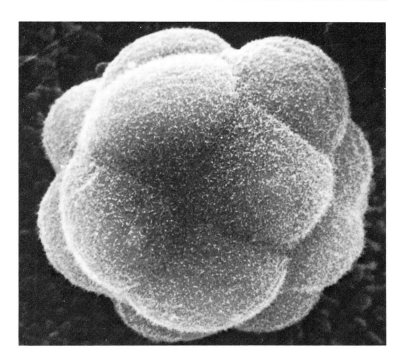

FIG. 2. A: Normal morula. Embryo flushed from oviduct at 2½ days post-fertilization and cultured for 7 hr. ×1,600.

when embryos are relatively simple both in terms of cell number and organizational complexity. Equally important is the fact that crucial events of determination and commitment occur during these early stages. For reference, Fig. 2 shows normal embryos at 3 and 8 days, the time point marking the earliest and latest stages when the initial effects of *t*-mutation homozygosity become manifest.

The first necessary step in understanding the pathology produced by these various mutations was, of course, the morphological study of afflicted embryos. Their histological analysis showed that each class of lethal homozygote had a specific

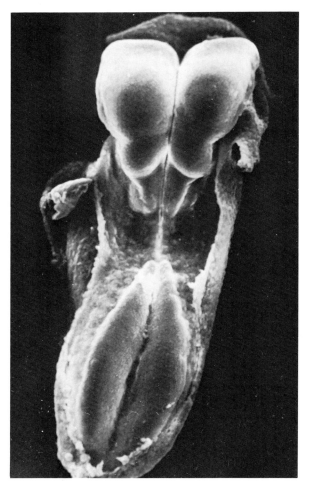

FIG. 2. B: Normal embryo dissected from the uterus and decidual capsule at early 8 days post-fertilization. View is of dorsal surface; extraembryonic membranes have been removed. ×90.

type of abnormality that, at least initially, was limited to only one cell or tissue type. In general, the effects associated with each mutation also had a rather precise time of onset and appeared to affect transition points at which specific events of normal differentiation should have occurred (1). More detailed studies with the electron microscope revealed some instances of striking evidence for abnormal cell–cell interaction. In t^{w18} homozygotes, for example, nascent mesoderm cells emerging from the primitive streak neither produce the extensive filopodia by which normal mesoderm cells become interconnected nor generate focal junctions where they contact one another, although normal cells produce junctions at virtually every point of contact (9). This suggested that either recognition or response to recognition is specifically defective in t^{w18}/t^{w18} mesoderm cells, although these processes are normal in all other cell types.

Another dramatic example of ineffective cell–cell interaction occurs in homozygotes for the mutation t^{12}. As shown in Fig. 3, the blastomeres of 3-day-old mutant embryos at the morula stage fail to adhere properly to one another (compare with Fig. 2a), and these embryos die without making the transition to the blastocyst stage. The effects of this mutation are especially interesting, since blastocyst formation is the very first step in differentiation undertaken by mammalian embryos and entails the differentiation, from the totipotential cells of the morula, of two cell types: the trophectodermal epithelium and the still totipotential cells of the inner cell mass. The hope is, of course, that the reason for the mutation-induced failure of this relatively simple step in such a simple embryo will be amenable to analysis in molecular terms.

FIG. 3. Abnormal morula from litter segregating t^{12}. Embryo flushed from oviduct at 2½ days post-fertilization and cultured for 7 hr. ×1,600.

Efforts to generate probes for the biochemical study of the effects of T/t-complex mutations began more than 10 years ago, when we started to use immunological techniques to try to identify as antigens the abnormal cell surface components we suspected to reside on mutant cells (5). For this purpose, we took advantage of a peculiar effect of recessive lethal t mutations which has not yet been mentioned. In addition to their deleterious effects on embryos when homozygous, these mutations also affect the function of spermatozoa; heterozygous $+/t$ males transmit their abnormal gene to a very high (over 90%) proportion of progeny (4). Although the physical basis for this surprising abrogation of Mendel's rules has not yet been defined, and no abnormalities of spermatozoal morphology or numbers have been detected, we reasoned that the spermatozoa from such males might display the same genetically determined abnormal cell surface components that we suspected to exist on embryos. Accordingly, we began a set of allogenic immunizations to test this point. It will suffice here to say that the experiments succeeded in the serological definition on sperm of specific different abnormal antigens associated with each of the lethal mutations we had studied (10). These antigens met the criteria for true differentiation antigens (6) since they were found to be present only on spermatozoa and no other cells of adult animals. Thus, the t mutations appear to be involved in two important processes of differentiation, embryogenesis and spermatogenesis.

Recent work has addressed the important question of the nature of the t-associated antigens. Cell surface antigens are typically either glycoproteins or glycolipids, so experiments were undertaken to treat testicular cells with a battery of specific glycosidases and subsequently assay them for antigenicity. These experiments showed not only that the t-antigenic specificity resides in carbohydrate but that different antigens have different carbohydrate determinants. For example, the antigenic determinant on $+/t^{12}$ testicular cells can be removed by treatment with β-galactosidase, whereas the same treatment has no effect on testicular cells from other t-genotypes. This suggests that an important part of the t^{12} antigen is in terminal galactose molecules (7).

Knowing something of the biochemical nature of the t^{12} antigen has provided ways of attempting to purify the antigen in quantities sufficient for more detailed biochemical study. Work in progress by C. C. Cheng and K. Artzt entails the use of affinity columns of Ricinus II, a lectin that binds specifically to terminal galactose residues. Affinity purification of lysates of $+/t^{12}$ testicular cells has enabled us to obtain molecular populations greatly enriched for molecules with terminal galactose moieties. When the enriched fraction was used to immunize rabbits, they produced a potent antiserum with cytotoxic specificity for $+/t^{12}$ testicular cells. This antiserum is currently being used in immunoprecipitation studies which should permit the final and positive identification of the t^{12} antigen. We know so far only that the molecule is a glycoprotein, but future structural studies of both the protein and carbohydrate moieties will identify the site of the genetic lesion we are dealing with. Two possibilities are obvious. The t^{12} mutation may code for an abnormal protein which, because of conformational or other irregularities, is subject to incorrect glycosylation by normal glycosyltransferase enzymes. Alternatively, of course,

t^{12} may produce an abnormal galactosyltransferase that inappropriately galactosylates a normal protein backbone. The definition of which of these alternatives is correct is important, since it bears on our understanding of how genes with related functions may operate to control differentiation.

ACKNOWLEDGMENTS

This chapter is based on work supported by grants from NIHCD, NCI, and NSF, and a contract from DOE. The author is grateful to Drs. K. Artzt, C. Cheng, and M. Spiegelman for access to unpublished results of work in progress.

REFERENCES

1. Bennett, D. (1964): Abnormalities associated with a chromosome region in the mouse. II. The embryological effects of lethal alleles at the t-region. *Science*, 144:263–267.
2. Bennett, D. (1978): The T-complex in the mouse: An assessment after 50 years of study. *Harvey Lect.*, 74:1–21.
3. Bennett, D., Boyse, E. A., and Old, L. J. (1972): Cell surface immunogenetics in the study of morphogenesis. In: *Proceedings III Lepetit Colloquium, Cell Interactions*, edited by L. G. Silvestri, pp. 247–263. North Holland, Amsterdam.
4. Bennett, D., and Dunn, L. C. (1971): Transmission ratio distorting genes on chromosome IX and their interactions. In: *Proceedings of a Symposium on Immunogenetics of the H-2 System*, edited by A. Lengerova and M. Vojtiskova, pp. 90–103. Karger, Basel.
5. Bennett, D., Goldberg, E., Dunn, L. C., and Boyse, E. A. (1972): Serological detection of a cell surface antigen specified by the T (Brachyury) mutant gene in the house mouse. *Proc. Natl. Acad. Sci. U.S.A.*, 69:2076–2080.
6. Boyse, E. A., and Old, L. J. (1969): Some aspects of normal and abnormal cell surface genetics. *Annu. Rev. Genet.*, 3:269–290.
7. Cheng, C. C., and Bennett, D. (1980): Nature of the antigenic determinants of T-locus antigens. *Cell*, 19:537–544.
8. Jacob, F., and Monod, J. (1963): Genetic repression, allosteric inhibition, and cellular differentiation. In: *Cytodifferentiation and Macromolecular Synthesis*, edited by M. Locke, p. 30. Academic Press, New York.
9. Spiegelman, M., and Bennett, D. (1974): Fine structural study of cell migration in the early mesoderm of normal and mutant mouse embryos (T-locus: t^9/t^9). *J. Embryol. Exp. Morphol.*, 32:723–738.
10. Yanagisawa, K., Bennett, D., Boyse, E. A., Dunn, L. C., and DiMeo, A. (1974): Serological identification of sperm antigens specified by lethal t-alleles in the mouse. *Immunogenetics*, 1:57–67.

Genetics of Neurological and Psychiatric Disorders, edited by Seymour S. Kety, Lewis P. Rowland, Richard L. Sidman, and Steven W. Matthysse. Raven Press, New York © 1983.

Genetic Control of the Number of Dopamine Neurons in the Brain: Relationship to Behavior and Responses to Psychoactive Drugs

Donald J. Reis, J. Stephen Fink, and Harriet Baker

Laboratory of Neurobiology, Department of Neurology, Cornell University Medical College, New York, New York 10021

Over the past several years our laboratory has been investigating the mechanisms which might account for variations in neurotransmitter-related macromolecules in the mammalian brain. In our studies we have focused on the mechanisms which underlie the strain-dependent differences in the activity of the catecholamine biosynthetic enzyme tyrosine hydroxylase (TH), first reported by Ciaranello and associates (5) in inbred mice. The problem is of particular interest since the neurochemical differences represent a genetic variation of an enzyme of importance for normal behaviors, such as motility in rodent strains which are otherwise neurologically and behaviorally "normal." The questions posed by the biochemical differences between the strains are numerous. Why does enzyme activity differ between strains? Does the enzyme differ in its distribution? In molecular forms? In catalytic activity or amount? Are the differences generalized to all brain dopamine (DA) systems? Are variations between the strains in the enzyme which synthesizes DA reflected in differences in behaviors or drug responses whose expression is mediated by this transmitter? What are the developmental bases of the strain differences? Are they prenatally determined or do they appear during early postnatal development?

This paper will review our studies. We shall demonstrate that: (a) the differences of TH activity in the brain DA systems are due to variations in the number of DA neurons; and (b) the variations of DA neurons relating to differences in the morphology and biochemistry of target areas, correlate with the variability in those drug-induced and spontaneous behaviors mediated in large measure through the nigrostriatal DA system which appear to develop postnatally. These studies raise interesting questions with respect to genetically associated neuronal mechanisms which may underlie disorders of thought, mood, and behavior in man.

REGIONAL DIFFERENCES IN TYROSINE HYDROXYLASE ACTIVITY

Midbrain Dopaminergic Systems

In our first investigations, (2,36,37) we sought to determine, by regional dissections of the brains of mice of the BALB/cJ and CBA/J strains, whether the greater tyrosine hydroxylase (TH) activity within the whole brain of BALB/cJ animals was restricted to either of the two principal classes of central catecholamine neurons (24,46), i.e., those synthesizing dopamine (DA) and those synthesizing norepinephrine (NE), and, if so, whether the differences were preferentially localized to cell bodies or to axon terminals of the neurons. Initially we limited our studies to comparisons of TH in the two principal midbrain DA systems, the nigrostriatal (A9) and mesolimbic and mesocortical (A10) systems (24,26), and, as a control, the pontine NE nucleus, the locus ceruleus (A6).

The results (Table 1) indicated that TH activity of BALB/cJ mice was greater in regions containing the cell bodies of DA neurons [including A10–substantia nigra regions assayed *in toto* or the A9 or A10 areas assayed in microdissections (2)] and regions containing only DA terminals (e.g., striatum, nucleus accumbens, olfactory tubercle). In contrast, the activity of TH in the locus ceruleus did not differ between strains, nor did the activity of DA-β-hydroxylase, a marker of NE neurons (36,37).

Other DA Systems

Studies were then undertaken to determine whether the strain differences in TH activity were restricted to only DA systems of the midbrain or were shared by other

TABLE 1. *Regional TH activity in brain of two mouse strains*[a]

	Strains					
Area	BALB/cJ	n	CBA/J	n	CBA/BALB	p
Substantia nigra—A10[b,f]	3.98 ± 0.007	32	3.19 ± 0.04	32	0.80	<0.001
A9[c,e]	6.28 ± 0.89	6	3.60 ± 0.47	6	0.57	<0.05
A10[c,f]	17.76 ± 1.89	5	9.96 ± 1.01	6	0.56	<0.01
Corpus striatum[c,f]	12.49 ± 0.39	30	10.31 ± 0.21	32	0.82	<0.001
Nucleus accumben[c,f]	4.43 ± 0.29	6	3.36 ± 0.18	5	0.83	<0.01
Olfactory tubercle[c,f]	3.48 ± 0.13	6	2.56 ± 0.37	5	0.74	<0.05
Hypothalamus (whole)[b]	0.59 ± 0.03	5	0.365 ± 0.02	5	0.62	<0.001
Zona incerta (ZI) and Posterior periventricular region (A13, A14)[b]	0.47 ± 0.021	8	0.307 ± 0.014	8	0.65	<0.001
Arcuate nucleus (A12)[b]	0.274 ± 0.010	8	0.141 ± 0.006	8	0.51	<0.001
Preoptic region (A14)[b]	0.189 ± 0.015	8	0.104 ± 0.008	8	0.55	<0.002
Olfactory bulb[c]	15.41 ± 0.64	8	10.47 ± 0.34	8	0.68	<0.001
Retina[c]	.766 ± 0.04	8	.313 ± 0.01	8	0.41	<0.001
Locus ceruleus[d,f]	12.4 ± 3.3	5	10.9 ± 2.0	5	0.88	ns

[a]TH activity was measured as in Baker et al. (2) and Coyle (7).
[b]TH activity expressed as nmole dopa/region/hr.
[c]TH activity expressed as nmole dopa/mg protein/hr.
[d]TH activity expressed as pmole dopa/locus ceruleus/hr.
[e]From Baker et al. (2).
[f]From Reis et al. (36).

principal DA systems of the CNS (24,31), including those of the hypothalamus, olfactory bulb, and retina.

Regional assays of TH activity indicated (Table 1) that TH activity differed between BALB/cJ and CBA/J mice in all of the major groups of DA neurons (24,31), including those representative of the nigrostriatal, mesolimbic, and various hypothalamic systems, as well as those of the retina and olfactory bulb. Thus, strain differences in TH activity appear to be generally distributed among DA systems within the CNS.

MOLECULAR MECHANISMS ACCOUNTING FOR THE VARIATIONS IN TH ACTIVITY

Two principal molecular mechanisms could account for the difference in TH activity in the DA systems of the two mouse strains: (a) variations in the amount of enzyme protein, or (b) differences in the catalytic activity of the enzyme (36). The latter would suggest strain differences in the molecular structure of the enzyme.

Immunotitration (18) with a specific antibody to TH demonstrated that the differences in TH activity in the nigrostriatal systems were attributable to differences in the amount of TH enzyme protein (37) (Fig. 1). This fact and the observation that no differences could be detected by standard immunochemical procedures between TH in the two strains *(unpublished data)* suggest that the TH protein is

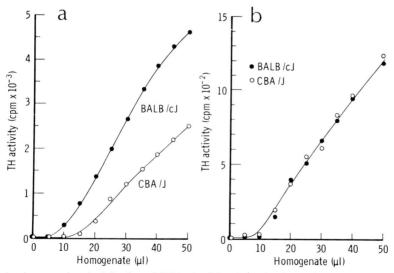

FIG. 1. Immunochemical titration of TH in the **(a)** substantia nigra—A10 area and **(b)** locus ceruleus of BALB/cJ *(closed circles)* and CBA/J *(open circles)* mice. In **(a)** the equivalence point for the substantia nigra—A10 area from the CBA/J mice (13.4 μl) is to the right of that of the BALB/cJ strain (7.0 μl). This demonstrates that the midbrain DA neurons of the CBA/J strain contain 47% less enzyme protein than the identical region of the BALB/cJ strain. The equivalence points for TH in the locus ceruleus **(b)** from both strains are not statistically different, indicating that the structure contains the same amount of enzyme protein. (From Ross et al., ref. 37.)

the same between strains, the activity differences reflecting differences only in the number of enzyme molecules.

STRAIN DIFFERENCES IN NUMBER OF DA NEURONS

Midbrain DA Systems

Variations in the amount of TH could reflect increased number of TH molecules per neuron or differences in the number of cell bodies and/or their processes in which the enzyme is contained (36). To distinguish between these possibilities, we stained the brains of mice of the two strains with antibodies to TH and, by standard morphometric techniques, counted the numbers of neurons containing the enzyme. Serial sections were taken through the entire extent of a particular nucleus to assure that strain differences and cell number were not a consequence of differences in configuration of the nucleus. Cell counts were done by standard procedures (20).

In the midbrain, differences in the density of TH-containing neurons of BALB/cJ and CBA/J mice were obvious by inspection (2) (Fig. 2). Quantitatively, mice of the CBA/J strain, overall, had approximately 20% fewer neurons that contained the enzyme in the midbrain than BALB/cJ mice (Table 2). Calculation of TH activity per neuron, however, demonstrated no differences between strains (Table 2). Thus, on the average, DA neurons of each strain contained equal enzymic activity.

Interestingly, the differences in the number of TH neurons between strains were not uniformly distributed throughout the midbrain but varied along its rostrocaudal plane (2). (Fig. 3). Differences were present only in the middle one-third of the midbrain with, at maximum, CBA/J mice having approximately 60% of the number of neurons found in the same region in BALB/cJ mice. Both A9 and A10 regions contributed to the observed strain differences in the number of midbrain neurons. The regional differences in the number of TH-containing neurons in this portion of the midbrain were paralleled by differences of a comparable magnitude in the activity of TH assayed in a small punch through either the A9 or A10 midbrain region (Table 1).

Hypothalamic DA Systems

The strain-dependent differences of TH-containing neurons were also seen in all areas of the hypothalamus encompassing all of the major hypothalamic groups (Table 2). With one exception, the difference in cell number entirely accounted for enzyme activity. The exception was the A12—arcuate system in which the difference in enzyme activity was greater than that for neuron number. Since TH activity in this area may be hormonally regulated (30), however, the disproportionate difference in TH activity may in fact reflect differences in endocrine control between the two strains, in turn possibly mediated by DA.

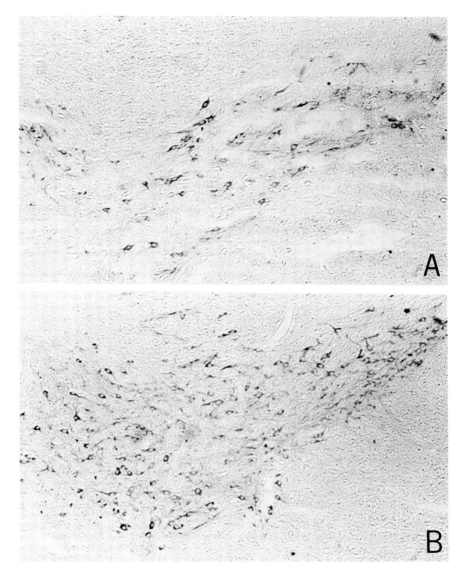

FIG. 2. Immunohistochemical localization of TH-containing neurons in the same region of the ventral tegmentum of CBA/J strain **(A)** and BALB/cJ strain **(B)**, demonstrating the lower density of neurons in the former. Sections are in the coronal plane at the most anterior pole of the interpeduncular nucleus; left is medial ×70. (From Baker et al., ref. 2.)

The differences in the number of TH-containing neurons of the DA systems of the two strains were not reflected in a NE system: CBA/J and BALB/cJ mice had

TABLE 2. *Number of neurons stained for TH and their estimated enzyme activity in regions of two mouse strains*

	BALB/cJ	n	CBA/J	n	CBA/ BALB	p
Substantia nigra (A10)						
Number of neurons	7,849 ± 487	6	6,223 ± 151	5	0.79	< 0.01
TH activity	3.98 ± 0.07	32	3.19 ± 0.04	32	0.80	< 0.001
TH activity/neuron (nmole dopa/hr/neuron × 10^{-4})	5.07		5.13		0.99	ns
Locus ceruleus						
Number of neurons	780 ± 6	4	774 ± 26	4	0.99	ns
TH activity	12.4 ± 3.3	5	10.9 ± 2.0	5	0.88	ns
TH activity/neuron ($\times 10^2$)	1.58		1.41		0.89	ns
Hypothalamus						
Number of neurons	6,607 ± 400	6	4,451 ± 313	5	0.67	< 0.002
TH activity	0.59 ± 0.03	5	0.365 ± 0.02	5	0.62	< 0.001
TH activity/neuron ($\times 10^4$)	0.892		0.820		0.92	ns
ZI and posterior peri-ventricular nucleus						
Number of neurons	3,257 ± 237	5	2,008 ± 141	6	0.62	< 0.002
TH activity	0.47 ± 0.021	8	0.307 ± 0.014	8	0.65	< 0.001
TH activity/neuron ($\times 10^4$)	1.443		1.528		1.06	ns
Arcuate nucleus (A12)						
Number of neurons	3,621 ± 265	5	2,695 ± 151	6	0.74	< 0.002
TH activity	0.274 ± 0.010	8	0.141 ± 0.006	8	0.51	< 0.001
TH activity/neuron ($\times 10^4$)	0.757		0.523		0.69	ns
Preoptic region (A14)						
Number of neurons	3,949 ± 338	6	1,929 ± 175	5	0.49	< 0.002
TH activity	0.189 ± 0.015	8	0.104 ± 0.008	8	0.55	<0.001
TH activity/neuron ($\times 10^4$)	0.479		0.539		1.12	ns

the same number of TH-containing neurons within the nucleus locus ceruleus (37) and the same TH activity per neuron (Table 2).

Thus the close correspondence between the differences in enzyme activity and the number of TH-containing cells indicate that (a) the strain-dependent differences in the amount and activity of TH in mouse brain activity are entirely attributable to differences in the number of DA neurons, and (b) the differences affect all DA systems of the brain.

RELATIONSHIP OF STRAIN DIFFERENCES IN NUMBER OF NEURONS TO TARGET ORGAN MORPHOLOGY

Size of Target Organ and Number of Constituent Neurons

The number of neurons within specific nuclei of the brain and spinal cord is often directly related to the size of the target organ (e.g., 17). This fact raises the interesting question of whether target areas innervated by different numbers of DA neurons differ in their morphology or neurochemistry.

We therefore compared the size and neuronal density of the striatum in BALB/cJ and CBA/J mice (2), the principal target of DA neurons of the substantia nigra

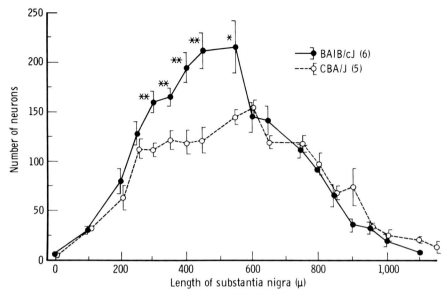

FIG. 3. The strain-dependent differences in the number of TH-containing neurons in the ventral tegmentum (including the substantia nigra—A10 and A8 regions) are shown with the most rostral section containing immunocytochemically stained neurons indicated at 0 μ. The strain differences were restricted to the midregions of the nucleus from 200–600 μ. *:$p<.05$; **:$p<.01$. (From Baker et al., ref. 2).

(A9) (24,26). The volume of the striatum was established by planimetric analysis of serial sections taken throughout its whole extent. Neuronal (packing) density was determined by standard morphometry. The total number of neurons per striatum was established as the product of the volume (mm^3) and neuronal density (number of neurons/mm^3).

As indicated in Table 3, there is a significant difference in the size of the caudate nucleus between the two strains of mice: the caudate is 16% smaller in CBA/J than in BALB/cJ mice. Of interest is the fact that the difference in striatal size is restricted to only the caudal one-half of the body of the striatum (2) (Fig. 4), a region topographically innervated by those regions of the midbrain in which differences in DA neurons are maximal (31).

The number of neurons per unit of the striatum in both strains did not differ (Table 3). Because of a larger volume and constant cell density, the total number of neurons in the striatum of the two strains varied by 16%. Thus, the larger number of DA neurons in the substantia nigra of mice of the BALB/cJ strain innervates a larger caudate nucleus.

Neurotransmitters in the Target Organs

The majority of neurons in the striatum are intrinsic and are probably heterogeneous with respect to their neurotransmitters (27,28,34). A substantial number

TABLE 3. *Volume, neuronal number, and CAT activity in corpus striatum of two mouse strains*

	BALB/cJ	n	CBA/J	n	CBA/BALB	p
Volume (mm^3)	7.86 ± 0.26	6	6.61 ± 0.28	6	0.84	<0.05
Neuronal density (neurons/mm^3)	90,593 ± 5,901	6	89,722 ± 5,164	6	0.99	ns
Total number neurons	710,409 ± 40,789	5	598,392 ± 21,264	5	0.84	<0.05
CAT/gCN[a]	17.6 ± 0.92	26	18.03 ± 0.96	22	1.02	ns
CAT/CN	138.3 ± 7.23		119.2 ± 6.34		0.86	ns
TH activity/mg prot./CN	12.49 ± 0.39	30	10.31 ± 0.21	30	0.82	<0.001
TH activity/CN	98.17		68.14		0.69	

[a]CN = striatum.
From Baker et al. (2).

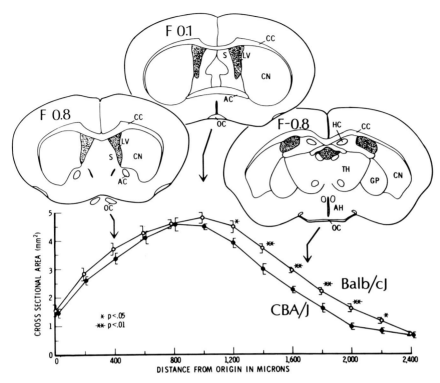

FIG. 4. Cross-sectional areas of the striatum of BALB/cJ and CBA/J mice, demonstrating that the volume difference was restricted to the posterior head and the body of the striatum. Measurements commenced where the striatum could first be visualized in the cresyl violet-stained sections. AC: anterior commissure; AH: anterior hypothalamus; CC: corpus callosum; CN: striatum; GP: globus pallidus; HC: hippocampus; LV: lateral ventricle; OC: optic chiasm; S: septum; TH: thalamus. Results are shown as a mean ± SEM. *:$p<0.05$; **:$p<0.01$. (From Baker et al., ref. 2.)

of these small intrinsic (Golgi Type II) neurons, however, are cholinergic (27). To determine if the strain-linked differences in caudate size are associated with differences in the number of cholinergic neurons, we assayed the caudate nucleus in both mouse strains for the activity of choline acetyltransferase (CAT), the enzyme catalyzing the biosynthesis of acetylcholine and, hence, a marker for this class of neuron. The results are indicated in Table 3.

Although the activity of CAT per mg of caudate is similar for both strains, the calculated activity is greater in the whole striatum of BALB/cJ mice. The most parsimonious interpretation of this result is that the number of cholinergic neurons per unit area of striatum is similar in both strains. However, since the volume of the striatum of BALB/cJ mice is larger, it therefore contains more cholinergic neurons than does the striatum of CBA/J mice.

DA Receptor Density

We also investigated whether or not differences in target organ size and composition are matched to differences in purported DA receptor density. We measured binding of the DA agonist and neuroleptic ^{3}H-spiroperidol (^{3}H-spiro) in homogenates of the striatum from brains of BALB/cJ and CBA/J mice (16). Binding was measured by the method of Creese et al. (8) using 0.05 to 2.0 nM of ^{3}H-spiro. The maximum number of ^{3}H-spiro binding sites, as determined by Scatchard analysis, was 24% greater in BALB/cJ than in CBA/J mice (B_{max} in BALB/cJ = 279 ± 4; CBA/J = 224 ± 11 fmole/mg protein; $n = 15$; $p<0.01$). There was no difference in the K_D between the two strains (K_D: BALB/cJ = 0.22 ± 0.05; CBA/J = 0.22 ± 0.03 nm; $p>0.05$). These results suggest that there is a strain-dependent difference in the number of ^{3}H-spiro binding sites within the striatum. Thus, one class of putative DA receptors may vary in two strains—the difference being a function of the number of binding sites, not of affinity of the ligand to the receptor.

STRAIN DIFFERENCES IN DRUG-INDUCED AND SPONTANEOUS BEHAVIORS MEDIATED BY DA NEURONS

The differences between mice of the BALB/cJ and CBA/J strains with respect to the number of DA neurons raise the important question: Are the morphological differences paralleled by variations in the behaviors whose expression depends on the integrity of the DA neurons? To investigate this question, we compared strains with respect to drug-induced and spontaneous behaviors believed to depend on the integrity of the DA systems (6,10,12,14,19,23).

Response to a DA Agonist: *d*-Amphetamine

We first examined the behavioral responses of mice to the DA agonist *d*-amphetamine. *d*-Amphetamine acts primarily by releasing DA from presynaptic terminals (23). In the lower dose range (<5 mg/kg), it increases locomotor activity primarily by release of DA in the terminal fields of mesolimbic neurons (19). Higher doses produce stereotyped patterns of motor behavior which, in the rodent, consist of sniffing, grooming, and gnawing. These behaviors are largely mediated by the nigrostriatal system (19).

Mice of the BALB/cJ strain were more sensitive to the actions of *d*-amphetamine than CBA/J mice (Table 4). Thus, in the lower dose range (<5 mg/kg), they exhibited a lower threshold to the effects of amphetamine on locomotor activity and exhibited greater amounts of spontaneous locomotion in each dose up to that at which stereotyped behavior appeared. At an intermediate dose (3.5 mg/kg), the drug-elicited activity of the CBA/J mice was only one-third that of the BALB/cJ animals (Table 4; Fig. 5). BALB/cJ mice had a lower threshold and exhibited a more intense stereotypy over the entire range (5 to 20 mg/kg) than did CBA/J mice (Table 4). Despite differences in the magnitude of both types of responses, their time course was similar (see, for example, Fig. 5).

TABLE 4. *DA-mediated, drug-induced, and spontaneous behaviors in BALB/cJ and CBA/J mice*[a]

		Strain		
Drug-induced behaviors	n	BALB/cJ	CBA/J	p
d-Amphetamine[b]				
Locomotion (square crossed, 3.5 mg/kg)	10	326 ± 52	110 ± 40	< 0.001
Stereotypy score (7.5 mg/kg)	10	16.0	7.5	< 0.05
Spiroperidol[c]				
Catalepsy scores (sec/300-min test, 1 mg/kg)	10	849 ± 239	131 ± 85	< 0.05
Haloperidol[d]				
Catalepsy scores (sec/300-min test, 4 mg/kg)	10	755 ± 292	39 ± 26	<0.05
Spontaneous behaviors				
Exploration[d]				
Approaches to object	10	8.6 ± 1.8	2.6 ± 0.8	< 0.001
Duration of investigation (sec)	10	10.3 ± 1.9	5.2 ± 1.9	< 0.01
Activity[b]				
Locomotion (squares crossed)	20	55 ± 5	32 ± 4	< 0.01
Rears	20	15 ± 2	6 ± 5	< 0.001

[a]Data are presented as mean ± SEM and group comparison made by Student's t-test, except for the stereotypy scores, which are expressed as medians and compared by the Mann-Whitney U test.
[b]After a 30-min habituation period, locomotion after amphetamine and spontaneous activity were measured in plexiglass cages by summing the number of squares traversed in six 45-sec observation periods at 10-min intervals. Behavior was rated during these observation periods using the following criteria: 1 = inactive, 2 = locomotion, 3 = discontinuous stereotypy, 4 = continuous stereotypy. The stereotypy scores are the "3" and "4" ratings summed for the six observation periods.
[c]Neuroleptic drugs were administered i.p. at t = 0 following a 30-min adaptation period. The degree of catalepsy (defined as the number of sec the front paws of an animal remained on a 5 × 5 cm wooden block) was determined at t = 15 and t = 30 min and thereafter up to t = 300 min.
[d]Investigatory exploration of a novel object was measured in a 3-min trial by the method of Berlyne and Slater (3) as modified by Fink and Smith (12).
From Reis et al. (36).

The greater sensitivity of the BALB/cJ mice to the behavioral effects of d-amphetamine was not due to differences in brain levels of the drug, which were identical in both strains (11). Rather, the study suggests that a greater presynaptic innervation of the target area by the DA-containing terminals in BALB/cJ mice is associated with an enhanced response to an agonist that releases transmitter from the presynaptic unit.

Response to DA Antagonists

Some drugs of the neuroleptic class exert their cataleptic activity primarily by interaction with DA receptors largely within the nigrostriatal system (32). We sought to determine whether or not differences in the number of DA neurons and/or the size and neuronal numbers of the target areas are reflected by variations in the sensitivity to the behavioral action of these drugs.

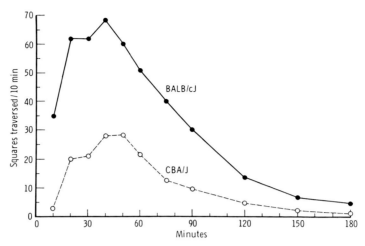

FIG. 5. Time-action curve of the effect of a low dose of d-amphetamine sulfate (3.5 mg/kg, i.p.) on locomotion in BALB/cJ mice (*solid circles*, n=10). The number of squares traversed during a 45-sec observation period was recorded at 10-min intervals. Data are expressed as medians. (From Baker et al., ref. 1.)

Neuroleptic drugs of three chemical classes, butyrophenones (spiroperidol and haloperidol), phenothiazines (trifluperazine), and diphenylbutylpiperidines (pimozide), were tested. We found that all of these such drugs produced 3 to 20 times greater catalepsy in BALB/cJ than CBA/J mice (16). Of particular interest was the response to spiroperidol. This drug in a dose of 1 mg/kg produced a catalepsy score of 849 ± 239 in BALB/cJ mice in contrast to 131 ± 85 in CBA/J mice, a 6.5-fold difference. The time-action curve for the drugs was similar between strains. Thus the BALB/cJ strain, with a greater number of ^{3}H-spiro binding sites, had a much greater response to this drug than did the CBA/J counterparts.

Spontaneous Activities

We next investigated if variations in the number of DA neurons in the midbrain were associated with differences between the two mouse strains in expression of spontaneous motor activity and exploration, behavior whose expression depends on the integrity of forebrain DA pathways (10–14). As shown in Table 4, the differences in drug-induced behaviors were paralleled by differences in spontaneous activities mediated by DA systems. Thus, mice of the BALB/cJ strain exhibited more intense exploratory behavior as well as spontaneous motor activity in the test cage, when compared with their CBA/J counterparts (Fig. 6). These results suggest that the animals with the greater number of DA neurons exhibit a greater magnitude of spontaneous activities mediated by DA neurons.

ONTOGENY OF STRAIN DIFFERENCES OF MIDBRAIN TH ACTIVITY

A major question posed by these studies relates to the cellular mechanism accounting for the differences in the number of DA neurons between the BALB/cJ

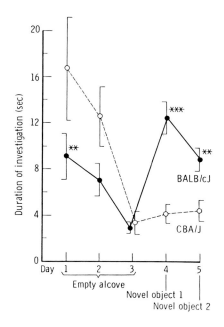

FIG. 6. Duration of investigation of a novel object in BALB/cJ *(solid circles)* and CBA/J *(open circles)* mice during a 3-min trial on 5 consecutive days. Design of experiment is described in legend to Table 4. Data are expressed as mean ± SEM. **:$p<0.05$; ***:$p<0.001$. (From Fink and Reis, ref. 11.)

and CBA/J strains. Do the strains differ in the number of DA neurons formed during neurogenesis in the fetus, or do they somehow reflect differences in the survival or expression of the enzyme during postnatal growth and development? Since the neurogenesis of DA neurons in the rodent brain is completed early in the third trimester of prenatal development (21,32,39,40), it would be expected that if this were the mechanism, differences in TH activity should be detectable at birth.

We therefore measured TH activity in the midbrain regions containing cell bodies of the A9 and A10 neurons and also in the principal target area of the A9 neurons, the corpus striatum, at various times during postnatal maturation of mice of both strains. As illustrated in Fig. 7, TH activity in both strains rapidly increases postnatally. At the earliest time measured, day 3, enzyme activity did not differ. By day 9, however, TH activity was significantly greater in the BALB/cJ strains and remained elevated throughout development into adulthood.

A comparable postnatal differentiation of TH activity also occurred in the striatum. At the earliest time of measurement, day 3, TH activity did not differ between the two strains. During subsequent development, however, differentiation occurred, although 2 days later than in the region of the cell bodies (1). The changes in TH activity in the nigrostriatal system were not reflected by any strain differences in the postnatal differentiation of CAT within portions of the nigrostriatal system (1).

It should be cautioned that, at this time, we only have evidence that in the two strains TH activity is similar at birth, differing postnatally. Whether the changes in development in enzyme activity reflect differences in the number of TH neurons will have to be determined by careful counting of labeled neurons. However, given

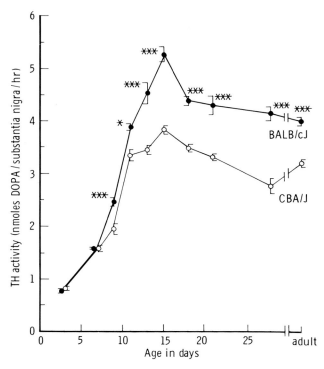

FIG. 7. Developmental changes of the TH activity in the substantia nigra—A10 region in two strains of mice. TH activity is expressed as mean ± SEM. *:$p<.01$.; ***:$p<.001$; $n = 5-7$ animals. Note that strain differences do not appear until day 9 of postnatal life.

the extremely close concordance between enzyme activity and cell number in virtually all regions of the DA system in the adult mouse, the most likely interpretation of the results, pending proof, is that enzyme differences do indeed reflect cell number.

It is of considerable interest that the differences in TH activity between strains occurred in the nigrostriatal system in advance of differences in spontaneous locomotion and the responses to amphetamine in the two mouse strains. As illustrated in Fig. 8, the earliest onset of significant forward locomotion (i.e., motility) is around the 20th day of postnatal development, at which time the strain differences characteristic of the adult are well established. Similarly, the strain differences in locomotor responses to amphetamine do not appear until well after biochemical differentiation of midbrain and striate TH activity (Fig. 8). The results therefore suggest that strain-dependent differences in locomotion and drug-induced behavior follow, rather than lead, the biochemical and cellular events and, in all likelihood, are not the cause of the differences in the number of DA neurons, but possibly the result.

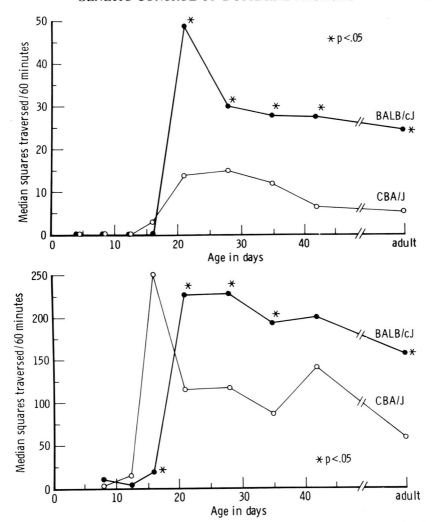

FIG. 8. Comparison of development of strain differences in spontaneous locomotion **(top)** and amphetamine-induced (3.5 mg/kg) motor activity **(bottom)** in BALB/cJ and CBA/J mice. Activity at various ages was measured in an open field. Strain differences in both behaviors characteristic of the adult were first observed between 16 and 21 days postnatally.

SUMMARY AND CONCLUSIONS

Differences in Brain TH Are Restricted to DA Systems in Two Inbred Strains of Mice

The present study has demonstrated that the strain differences in the activity of the catecholamine biosynthetic enzyme TH in whole brain of mice of the BALB/cJ and CBA/J strains (5) are restricted to regions of the brain containing the cell bodies and terminals of DA neurons. The difference of TH activity involves all DA systems of the CNS, including the major midbrain (the nigrostriatal and mesolimbic) and hypothalamic systems and those of the olfactory bulb and retina. In contrast, no differences in the activity and/or amount of TH were found in NE neurons of the locus ceruleus or in its projections. These variations in enzyme activity are due to differences in amount of enzyme protein and not to differences in molecular configuration.

The results indicate that the genetic signal that produces strain differences in enzyme activity does not govern the expression or configuration of molecular TH in all neurons. Rather, it is limited to only one chemical class of catecholamine neurons, those synthesizing DA.

Differences in TH Activity Are Due to Variations in Cell Number

By immunocytochemically staining brains of both strains for TH and carefully counting neurons containing the enzyme in the midbrain and hypothalamus, we have discovered that differences in the amount of TH can be entirely attributed to variations in the number of DA neurons. Indeed, with only the exception of the arcuate DA system (A12), the average activity of enzyme per neuron in the midbrain and hypothalamic systems is exactly the same in animals of either strain. Presumably the strain differences in the number of DA perikarya are also responsible for a proportional variation in the density of terminal arborizations. This, in turn, is reflected by differences in TH activity in terminal fields, such as the striatum and olfactory tubercle.

The generalized differences in numbers of DA neurons, it should be emphasized, are not part of a generalized difference in neuronal numbers between strains. Thus, large portions of the striatum are of similar size and neuronal density, and the nucleus locus ceruleus in both BALB/cJ and CBA/J mice contains equal numbers of neurons. Rather, our observations indicate a genetic control over the number of neurons of a particular neurochemical class in the CNS, those of the DA system, which influences their expression irrespective of the region in which they are contained.

Variations in DA Neuron Number May Be Associated With Proportional Variations in Size, Neuronal Number, and Neurochemistry of Target Areas

The findings that the variations in DA neuron number correlate with a larger target organ, the striatum, indicate that variations in cell number of a particular class of neurons may not exist in isolation. Rather, it may be a manifestation of an interaction between neurons and their targets with respect to size and neuronal number. Such interactions may be subtle and may result in secondary variations in amounts of neurotransmitters and/or their biosynthetic enzymes. Conceivably such interactions may impose, by variations in the innervation density, a control over the expression of receptors within the innervated cell, operating, perhaps, to regulate receptor density in pre- and postsynaptic units.

Strain Differences in TH Activity and Possibly DA Cell Number Appear Postnatally

The cellular mechanisms leading to differences in the number of DA neurons between the CBA/J and BALB/cJ strains are at present unknown. The fact that strain differences in enzyme activity in the midbrain do not appear until after 1 week of postnatal life suggests that variation in the production of DA neurons is probably not the case, since neurogenesis of DA neurons in the rodent is completed early in the third trimester of pregnancy (21,33,39,40). More likely postnatal events result in the differences in DA cell number in adults of the two strains.

The most probable mechanism for postnatal differentiation relates to differences in cell survival, a most common and almost universal form of developmental modeling of neuronal systems in the CNS (17). This interpretation assumes that an overabundance of DA cells is produced and survives into early postnatal life. In the course of neuronal remodeling, however, some of these cells are destined to die, more in the CBA/J than in the BALB/cJ mice. Conceivably, differences in neuronal survival could be linked to an interaction of DA neurons with target areas. That this might be the case is suggested by the fact that more DA neurons in the substantia nigra of BALB/cJ animals innervate a corpus striatum with more neurons. Whether targets of DA neurons other than those of the nigrostriatal system also differ in size and/or neuronal number remains to be established, however. A process that affects all DA neurons of the brain must also involve a wide multiplicity of target areas, and raises the question of whether some common bond links targets of all DA neurons.

An alternative explanation is that all DA neurons survive, but some fail to maintain the phenotypic expression of TH. Although not yet observed in brain, a transient expression of TH occurs in peripheral cells derived from the neural crest that invade the gut and pancreas (41–43). However, that some neurons in the CNS may lose their phenotype with respect to a purported neuromodulator is suggested by the so-called Brattleboro rat (48). This species lacks the neuropeptide vasopressin in the magnocellular neurons of its hypothalamus, although these cells seem morphologically intact.

It is obvious that extensive developmental studies, including careful cell counts, must be undertaken to elaborate on those mechanisms accounting for postnatal appearance of the differences in the number of DA cells in the brain.

Differences in Number of DA Neurons Are Directly Paralleled by Differences in Expression of Behaviors Dependent on DA Systems

Since the BALB/cJ and CBA/J strains have never been considered neurologically or behaviorally abnormal mutants, the question is posed: What differences in the organism do variations in the number of DA neurons produce? Our studies indicate that BALB/cJ mice, harboring more DA neurons and richer terminal fields, exhibit a greater sensitivity to the behavioral actions of *d*-amphetamine and more intense motor and exploratory behaviors. Since these behaviors depend on the integrity of DA systems for their expression (6,10,12,14,15,19,23,32), the findings would be predicted by the differences in DA neurons in these strains.

However, given the complexity and number of control mechanisms involved in the expression of a simple behavior, such as motility, a far more critical evaluation of the hypothesis is required. Do other functions controlled by DA cells differ in comparable ways? For example, are the strain-dependent differences in the numbers of DA neurons in the arcuate nucleus reflected by peripheral variations in levels of prolactin, the pituitary hormone whose release is regulated by these DA neurons (31)?

Obviously, a definitive assessment of the relationship between cell number and behavior requires a much more extensive analysis by genetic techniques, examining for cosegregation of the traits.

General Implications

These studies lead to two general implications of relevance in biology and medicine.

Genetic Control of Cell Number: A Generalized Phenomenon?

Our studies have demonstrated that the mechanism accounting for strain-dependent differences in the activity of brain TH in BALB/cJ and CBA/J mice is due to variations in the number of DA neurons, and that all DA neuronal systems are comparably affected. The observation suggests a new factor governing the amount of neurotransmitter synthesizing enzymes in brain: cell number. Presumably, although not yet proved, it might account for differences in the amount of the neurotransmitter itself and/or its metabolites.

Our observations raise the obvious question of whether similar mechanisms account for strain-dependent variations in other neurotransmitters or of their biosynthetic enzymes (9,25,29,38,44,45,47). Are the genetically determined differences in central NE, cholinergic, GABAergic, serotonergic, and possibly peptidergic systems, which have been described by others, due to differences in cell number? The answer to this question will require comparable application of immunochemical and immunocytochemical techniques to the problem.

Implications for Neurological and Psychiatric Diseases in Humans

Is it possible that genetic determinants in the number of cells play a role with respect to the expression of diseases, including those associated with aging? For example, it is now established that there is a progressive decrease in TH in the nigrostriatal system in aging humans (4,27). This appears to be associated with reduction of the number of DA neurons in the substantia nigra. Individuals endowed with more DA neurons would be expected not to express symptoms of such degeneration, i.e., Parkinsonism, until a more advanced age than those born with a fewer number of cells, assuming that the rate of decline of cells is comparable. Likewise in other neurological diseases, a large reservoir of cells may be protective against the expression of symptoms when brain cells are progressively lost. Could variations in cell number be one genetic element of diseases such as schizophrenia, manic depressive illnesses, and Parkinson's disease, in which catecholamine neurons or their receptors are believed to be involved (4,22)?

Finally, the question remains concerning the relationship between variations in cell numbers of a particular neurochemical class and the expression of those traits of mind and behavior that endow individuals or species with their personality or character. Is it possible that variations in a chemical class of neurons are the cellular basis for genetic differences in traits of aggressivity, tameness, obedience, or intelligence, which distinguish different strains of mice, rats, and dogs? Even more intriguing is the question: Could this mechanism in the mixed gene pool of humans underlie those variations in temperament that endow human beings with their variety and uniqueness?

These questions will not be easy to answer. Perhaps efforts to correlate measurements of personality and mental capacities with products of neurotransmitter systems in urine and blood, and with brain morphology assessed by, for example, computerized tomography (CAT) scans, could be starting points.

ACKNOWLEDGMENTS

We thank Dr. Tong H. Joh for continuing support in biochemistry. Alissa Swerdloff provided excellent technical assistance. Supported by research grants from NIMH—(MH 33190) and NHLBI (HL 18974).

REFERENCES

1. Baker, H., Fink, J. S., Joh, T. H., Swerdloff, A., and Reis, D. J. (1979): Ontogeny of strain differences of nigrostriatal tyrosine hydroxylase activity and spontaneous or drug-induced behaviors. *Neurosci. Abstr.*, 5:641.
2. Baker, H., Joh, T. H., and Reis, D. J. (1980): Genetic control of the number of midbrain dopaminergic neurons in inbred strains of mice: Relationship to size and neuronal density of the striatum. *Proc. Natl. Acad. Sci. USA*, 77:4,369–4,373.
3. Berlyne, D. E., and Slater, J. (1955): The arousal and satiation of perceptual curiosity in the rat. *J. Comp. Physiol. Psychol.*, 48:238–246.
4. Carlsson, A. (1979): The impact of catecholamine research on medical science and practice. In: *Catecholamines: Basic and Clinical Frontiers*, edited by E. Usdin, I. J. Kopin, and J. Barchas, pp. 4–19. Pergamon Press, New York.
5. Ciaranello, R. D., Barchas, R., Kessler, S., and Barchas, J. D. (1972): Catecholamines: Strain differences in biosynthetic enzyme activity in mice. *Life Sci.*, 2:565–572.
6. Costall, B., Marsden, C. D., Naylor, R. J., and Pycock, C. J. (1977): Stereotyped behavior patterns and hyperactivity induced by amphetamine and apomorphine after discrete 6-hydroxydopamine lesions of extrapyramidal and mesolimbic nuclei. *Brain Res.*, 123:89–111.
7. Coyle, J. T. (1972): Tyrosine hydroxylase in rat brain—Cofactor requirements, regional and subcellular distribution. *Biochem. Pharmacol.*, 21:1,935–1,944.
8. Creese, I., Prosser, T., and Snyder, S. H. (1978): Dopamine receptor binding: Specificity, localization, and regulation by ions and guanylnucleotides. *Life Sci.*, 23:495–500.

9. Eleftheriou, B. E. (1974): A gene influencing hypothalamic norepinephrine levels in mice. *Brain Res.*, 70:538–540.
10. Fink, J. S. (1977): Studies on the role of forebrain catecholamine neurons in exploratory behavior in rat. Ph.D. dissertation. Cornell University Medical College.
11. Fink, J. S., and Reis, D. J. (1982): Genetic variations in midbrain dopamine cell number: Correlation with differences in responses for dopaminergic agonists and in naturalistic behavior mediated by central dopaminergic systems. *Brain Res. (in press).*
12. Fink, J. S., and Smith, G. P. (1979): Decreased locomotor and investigatory exploration after denervation of catecholamine terminal fields in the forebrain of rats. *J. Comp. Physiol. Psychol.*, 93:34.
13. Fink, J. S., and Smith, G. P. (1980a): Mesolimbic and mesocortical dopaminergic neurons are necessary for normal exploratory behavior in rats. *Neurosci. Lett.*, 17:61–66.
14. Fink, J. S., and Smith, G. P. (1980b): Relationships between selective denervation of dopamine terminal fields in the anterior forebrain and behavioral responses to amphetamine and apomorphine. *Brain Res.*, 201:107–127.
15. Fink, J. S., and Smith, G. P. (1982): Mesolimbicocortical dopamine terminal fields are necessary for normal locomotor and investigatory exploration in rats. *Brain Res. (in press).*
16. Fink, J. S., Swerdloff, A., Joh, T. H., and Reis, D. J. (1979): Genetic differences in ^{3}H-spiroperidol binding in caudate nucleus and cataleptic response to neuroleptic drugs in inbred mouse strains with different numbers of midbrain dopamine neurons. *Neurosci. Abstr.*, 5:647.
17. Hamburger, V. (1977): The developmental history of the motor neuron. In: *Neurosci. Res. Program Bull.*, 15: Suppl.1–37.
18. Joh, T. H., Geghman, C., and Reis, D. J. (1973): Immunochemical demonstration of increased accumulation of tyrosine hydroxylase protein in sympathetic ganglia and adrenal medulla elicited by reserpine. *Proc. Natl. Acad. Sci. USA*, 70:2,767–2,771.
19. Kelly, D. H., Seviour, P. W., and Iverson, S. D. (1975): Amphetamine and apomorphine responses in the rat following 6-OHDA lesions of the nucleus accumbens septi and corpus striatum. *Brain Res.*, 94:507–522.
20. Konigsmark, B. W. (1970): Methods for counting of neurons. In: *Contemporary Research Methods in Neurochemistry*, edited by W. J. H. Nauta and S. O. E. Ebbesson, p. 315. Springer, New York.
21. Lauder, J., and Bloom, F. (1974): Ontogeny of monoamine neurons in the locus coeruleus, raphe nuclei, and substantia nigra of the rat. I. Cell differentiation. *J. Comp. Neurol.*, 155:469–482.
22. Lee, T., Seeman, P., Tourtellote, W. W., Farley, I. J., and Hornykeiwicz, O. (1978): Binding of ^{3}H-neuroleptics and ^{3}H-apomorphine in schizophrenic brains. *Nature*, 274:897.
23. Lewander, T. (1977): Effects of amphetamine in animals. Drug Addiction. In: *Handbook of Experimental Pharmacology*, edited by W. R. Martin. 45:33–246. Springer-Verlag, Berlin.
24. Lindvall, O., and Björklund, A. (1979): The organization of catecholamine neurons in the rat central nervous system. In: *Handbook of Psychopharmacology, Vol. 9: Chemical Pathways in Brain*, edited by L. L. Iverson and S. H. Snyder, pp. 139–231. Plenum Press, New York.
25. Maas, J. W. (1963): Neurochemical differences between two strains of mice. *Nature*, 197:255–257.
26. Mandel, P., Ebel, A., Mack, G., and Kempf, E. (1971): Neurochemical correlates of behavior. In: *Genetics of Behavior*, edited by J. H. F. Van Abeelen, pp. 397–415. North Holland, Amsterdam.
27. McGeer, P. L., and McGeer, E. G. (1974): Enzymes associated with the metabolism of catecholamines, acetylcholine, and GABA in human controls and patients with Parkinson's disease and Huntington's chorea. *J. Neurochem.*, 26:65–76.
28. McGeer, P. L., McGeer, E. G., Singh, V. K., and Chase, W. H. (1974): Choline acetyltransferase localization in the central nervous system by immunohistochemistry. *Brain Res.*, 81:373–379.
29. Moisset, B. (1977): Genetic analysis of the behavioral response to *d*-amphetamine in mice. *Psychopharmacology*, 53:263–267.
30. Moore, K. E., and Wuerthele, S. M. (1979): Regulation of nigrostriatal and tuberoinfundibular-hypophyseal dopaminergic neurons. *Prog. Neurobiol.*, 13:325–359.
31. Moore, R. Y., and Bloom, F. E. (1978): Central catecholamine neuron systems: Anatomy and physiology of the dopamine systems. *Ann. Rev. Neurosci.*, 1:129–170.
32. Niemegeers, C. J., and Janssen, P. A. (1979): A systematic study of the pharmacological activities of dopamine agonists. *Life Sci.*, 24:2,201–2,216.
33. Olson, L., and Seiger, A. (1972): Early prenatal ontogeny of central monoamine neurons in the rat: Fluorescence histochemical observations. *A. Anat. Entwickl.-Ges.*, 137:302–416.

34. Pickel, V. M., Sumal, K. K., Beckley, S., Miller, R., and Reis, D. J. (1980): Immunocytochemical localization of enkephalin in the neostriatum of rat brain: A light and electron microscopic study. *J. Comp. Neurol.*, 189:721–740.
35. Reis, D. J., Baker, H., and Fink, J. S. (1982): A genetic control of the number of dopamine neurons in mouse brain: Its relationship to brain morphology, chemistry, and behavior. In: *Genetic Strategies in Psychobiology and Psychiatry*, edited by E. S. Gerhson, S. Matthyse, X. O. Breakfield, and R. Ciaranello. Foxwood Press, Pacific Grove, California *(in press)*.
36. Reis, D. J., Baker, H., Fink, J. S., and Joh, T. H. (1979): A genetic control of the number of central dopamine neurons in relationship to brain organization, during responses and behavior. In: *Catecholamines: Basic and Clinical Frontiers*, Vol. 1, edited by E. Usdin, I. J. Kopin, and J. Barchas, pp. 23–33. Pergamon Press, New York.
37. Ross, R. A., Judd, A. B., Pickel, V. M., Joh, T. H., and Reis, D. J. (1976): Strain-dependent variations in number of midbrain dopaminergic neurons. *Nature*, 264:654–656.
38. Segal, D. S., Kuczenski, R. T., and Mandall, A. J. (1972): Strain differences in behavior and brain tyrosine hydroxylase activity. *Behav. Biol.*, 7:75–81.
39. Specht, L. A., Pickel, V. M., Joh, T. H., and Reis, D. J. (1978): Immunocytochemical localization of tyrosine hydroxylase in processes within the ventricular zone of prenatal rat brain. *Brain Res.*, 156:315–321.
40. Specht, L. A., Pickel, V. M., Joh, T. H., and Reis, D. J. (1982): Light microscopic immunocytochemical localization of tyrosine hydroxylase in prenatal rat brain. I. Early ontogeny. *J. Comp. Neurol. (in press)*.
41. Teitelman, G., Baker, H., Joh, T. H., and Reis, D. J. (1979): Appearance of catecholamine-synthesizing enzymes during development of rat sympathetic nervous system: Possible role of tissue environment. *Proc. Natl. Acad. Sci. USA*, 76:509–513.
42. Teitelman, G., Joh, T. H., and Reis, D. J. (1978): Transient expression of a noradrenergic phenotype in cells of the rat embryonic gut. *Brain Res.*, 158:229–234.
43. Teitelman, G., Reis, D. J., and Joh, T. H. (1982): Transformation of catecholamine precursors into glucagon (A) cells in the mouse embryonic pancreas. *Proc. Natl. Acad. Sci. USA (in press)*.
44. Tiplady, B., Killian, J. J., and Mandel, P. (1976): Tyrosine hydroxylase in various brain regions of three strains of mice differing in spontaneous activity, learning ability, and emotionality. *Life Sci.*, 18:1,065–1,070.
45. Tunnicliff, G., Wimer, C. C., and Wimer, R. E. (1973): Relationships between neurotransmitter metabolism and behavior in seven inbred strains of mice. *Brain Res.*, 61:428–434.
46. Ungerstedt, U. (1971): Sterotoxic mapping of the monoamine pathways in the rat brain. *Acta Physiol. Scand (Suppl.)*, 82:1–48.
47. Will, B. (1977): Neurochemical correlates of individual differences in animal learning capacity. *Behav. Biol.*, 19:143–171.
48. Zimmerman, E. A. (1976): Localization of hypothalamic hormones by immunocytochemical techniques. In: *Frontiers in Neuroendocrinology*, Vol. 4, edited by L. Martini, and W. Ganong, pp. 25–62. Raven Press, New York.

Genetics of Neuronal Form

Steven Matthysse and Roger Williams

Mailman Research Center, McLean Hospital, Belmont, Massachusetts 02178

Our studies have concerned methods for the description of the geometry of the dendritic arbor and formulation of branching rules, which we hope will contribute to future studies of genetic and environmental effects on neuronal form. As Dr. Levinthal has stated *(this volume)*, the interplay of genetic and environmental factors in the determination of the shape of plants has long been studied. The *Vochsyia* tree in Fig. 1 is a particularly good example. It is possible for field workers to tell that this tree has suffered damage because of its bayonet-shaped branches (2). When the growing tip of a plant (the apical meristem) is damaged, a lateral meristem takes over; it starts to grow at right angles to the axis of the plant, and then begins to curve upward in order to reach the light. The effect of environment on the form of trees has been documented ever since Theophrastus ("Enquiry Into Plants," c. 350 BC):

> Trees growing in a sunny or windy position are more branched, shorter and less straight, and in general mountain trees have more knots than those that grow in plains, and those that grow in dry spots more knots than those that grow in marshes.

We postulate that the shape of the dendritic tree is a permanent record of the life history of the cell, just as it is in the woody tree.

We set ourselves the goal of using the shape of the dendritic arbor, as revealed by the Golgi method, as an indicator of the developmental history of the neuron, and especially as a sign of developmental pathology. Environmental influences on the shape of Purkinje cells are particularly well known. Normal Purkinje cells are shown in Fig. 2. Figure 3 illustrates an abnormality in form caused by heterotopic location, in the Zellweger malformation. Figure 4 illustrates the "weeping willow" appearance of Purkinje cells in the Menke's steely hair (6) syndrome. In this disease the normal synaptic interaction with granule cells is lost. The weeping willow shape also occurs after radiation-induced granule cell destruction (1).

GEOMETRY OF MOUSE HIPPOCAMPAL GRANULE CELL

Our studies have begun with the mouse hippocampal granule cell, which has a characteristic "wine glass" shape (Fig. 5). With the aid of a computer microscope, we have studied the distributions of the branching angles characteristic of this cell.

78   GENETICS OF NEURONAL FORM

FIG. 1. Bayonet shape of *Vochsyia* tree branches after recovery from injury. (From Halle et al., ref. 2.)

In a quantitative analysis of dendritic trees, the angle that immediately comes to mind is the angle between the daughter branches at each bifurcation. We call this the daughter angle. In previous studies of pyramidal cells from adult rat neocortex, Lindsay and Scheibel (4) found the daughter angle to be highly variable. Our results on hippocampal granule cells in the mouse are in agreement (Fig. 6). Variability detracts from the usefulness of a geometric parameter in characterizing normal or abnormal developmental histories.

Certain other angles are more orderly in the normal hippocampal granule cell. In view of the cylindrical symmetry of the cell, an "ideal axis" can be defined by computing a unit vector in the direction of the vector sum of all the branches. This vector represents the general tendency of the outward growth of the cell. We define

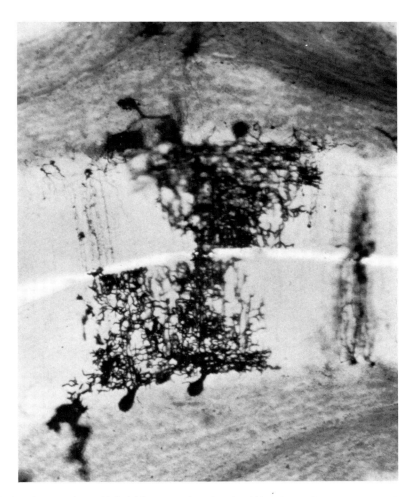

FIG. 2. As seen in rapid Golgi impregnation, the dendritic arbors of normal mature Purkinje cells are highly branched and oriented into a flat plane in the molecular layer. Terminal segments are richly invested with dendritic spines.

the planar angle as the angle of inclination of the plane of the daughter branches with respect to the main axis of the cell. As shown in Fig. 7, on the basis of data on 100 granule cells in the normal mouse hippocampus, it is strongly concentrated around 0. The bisector angle indicates whether, within their plane, the daughter branches emerge symmetrically or asymmetrically with respect to the central axis. Figure 8 indicates that the bisector angle is also concentrated around 0. These two regular features—the planar angle and the bisector angle—give rise to the characteristic wine glass shape of the hippocampal granule cell.

We noticed a feature of the geometry of these cells that may be a sensitive indicator of the effects of the neuronal environment. Imagine straight lines drawn

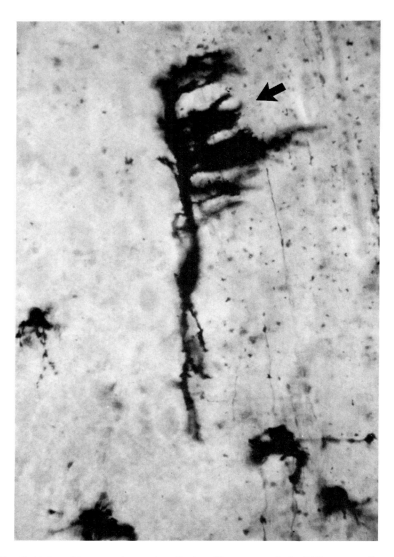

FIG. 3. Purkinje cell in anomalous subcortical position as seen in the Zellweger malformation. The number of dendritic branches is reduced and they are not oriented into a flat plane. Instead, the cell is bipolar and oriented parallel to the fibers passing into and out of the overlying cortex. Spine-rich terminal branches *(arrow)* are oriented away from the cell-rich heterotopia into the shell of investing fibers. Rapid Golgi.

from each branch point to the next, and this "skeletonized" arbor substituted for the actual dendritic tree. In this process, the detailed shape of the dendrite between the branch points is omitted. The curvature (rate of change of the tangent) and torsion (degree of twisting) of the dendritic segment between branch points are

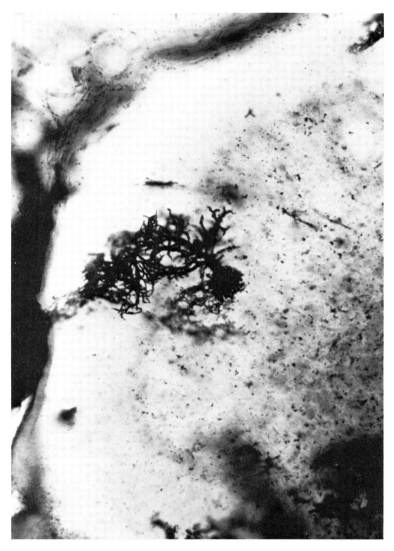

FIG. 4. Purkinje cell from Menke's disease. A stout primary dendrite courses over an abnormally long distance in the molecular layer before giving rise to reduced numbers of subordinate branches that recurve inward, accounting for the weeping willow appearance. Rapid Golgi.

interesting aspects of its geometry, which might also be sensitive to deviant environmental conditions, but the skeletonized arbor is particularly useful for studying the geometry at the branch points. Histograms of daughter, planar, and bisector angles can be computed for the skeletonized cell just as for the actual cell. The

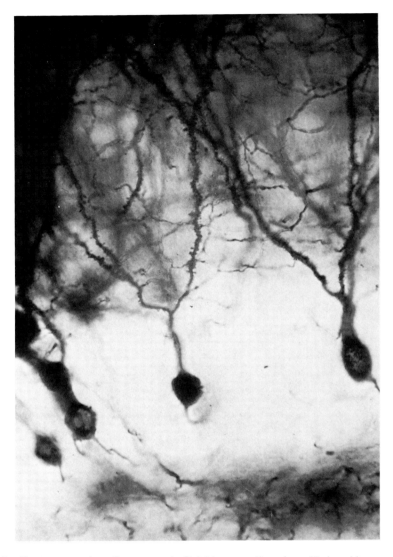

FIG. 5. Dentate granule cells as seen in Golgi impregnations from 60-day-old mouse. The geometric structure inscribed by the dendritic envelope is an inverted cone. Cells with one primary dendrite have a wine glass shape.

skeletonized cells have planar, daughter, and bisector angles closer to 0 than the actual branches, as shown by the means and standard deviations in Table 1. Our tentative interpretation of this difference is that there is an "internodal correction process" operating in the normal hippocampal granule cell. Branches that emerge at deviant planar, bisector, or daughter angles tend to be corrected before they reach

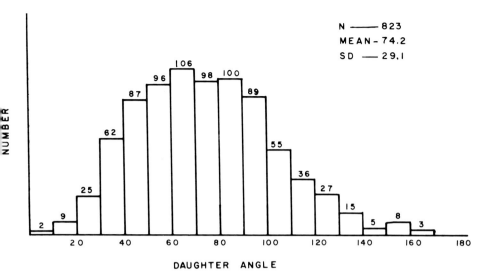

FIG. 6. Histogram of daughter angles from 100 dentate granule cells of 60-day-old mice.

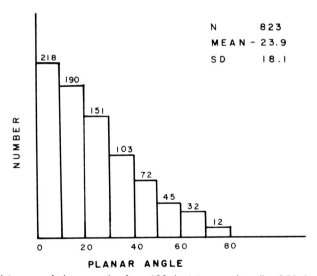

FIG. 7. Histogram of planar angles from 100 dentate granule cells of 60-day-old mice.

the next branch point. The correction is presumably caused by factors in the immediate environment of the growing dendritic arbor, perhaps by biochemical gradients. In a similar way, sunlight and gravity stabilize the arbor of the growing plant.

Preliminary data suggest that the internodal correction process is absent in the hippocampus of the *Reeler* mouse (Table 2). A plausible explanation is the mal-

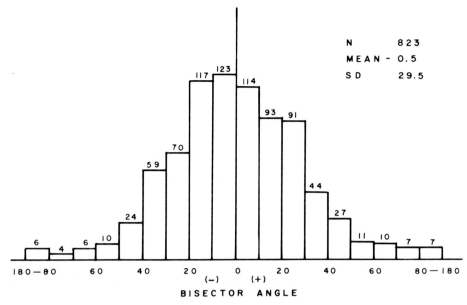

FIG. 8. Histogram of bisector angles from 100 dentate granule cells of 60-day-old mice.

TABLE 1. *Comparison of values for daughter, planar, and bisector angles, as determined from a sample point near the branch point (N), or between successive branch points (skeletonized, S)*

		Mean	SD
Daughter angle	N	74.2	29.1
	S	47.8	24.6
Planar angle	N	23.9	18.1
	S	14.9	11.4
Bisector angle	N	0.5	29.5
	S	0.1	19.8

$N = 823$.

position of many hippocampal granule cells in this mutant (5; see Fig. 9) which may deprive them of growth-stabilizing factors in the immediate environment.

GEOMETRY OF SYNTHETIC CELLS

We wondered if our histograms of angles characterizing the hippocampal granule cell provide a complete set of branching rules determining its form. In order to approach this problem, we artificially synthesized hippocampal granule cells according to the branching rules derived from the data. If the synthetic arbors do not convincingly resemble actual branching patterns, we can infer what is lacking in

TABLE 2. *Internodal correction in normal and reeler cells*

	Internodal correction	
	Planar angle ≤ 20° (%)	Bisector angle ≤ 20° (%)
Near		
Normal	49.6	54.3
Reeler	46.8	53.2
Skeleton		
Normal ($N = 823$)	71.9	74.0
Reeler ($N = 47$)	59.6	63.8

When planar and bisector angles are computed using a reference point near the branch point, about 50 % of the values fall between 0–20 degrees in both *Reeler* and normal mice. When these angles are computed between branch points (skeletonized), the number that fall within this narrow range in the normal mice increases, indicating internodal correction. The tendency to internodal correction is substantially less in *Reeler* mice.

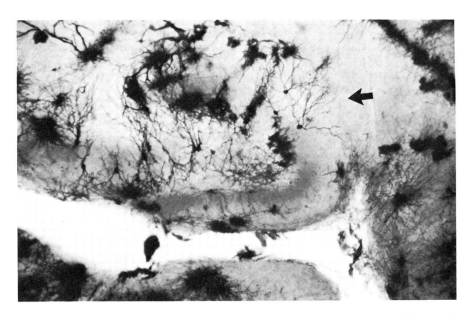

FIG. 9. Golgi impregnations of *Reeler* mouse dentate gyrus, illustrating more variable granule cell position and dendrite orientation *(arrow)*.

the quantitative description and add to our set of branching rules. This procedure has been used effectively in formalizing the shapes of trees. Figures 10 and 11 show syntheses, which look remarkably realistic, constructed by Honda (3) on the basis of a very simple set of branching rules.

Synthetic cells were constructed by randomizing the lists of angles and branch lengths derived from actual cells, and then computing the shape that would emerge

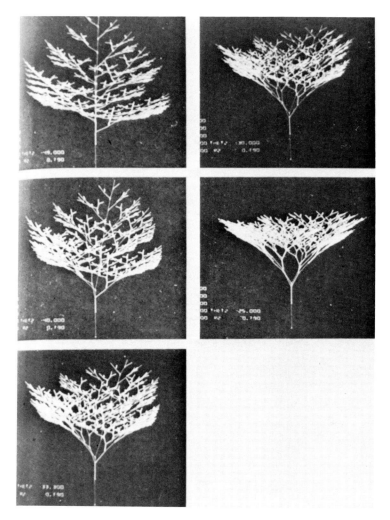

FIG. 10. Tree-like shapes generated by computer models using simple sets of branching rules. (From Honda, ref. 3.)

if an actual neuron had these randomized angles and lengths at each of its branch points. The most natural-appearing synthetic cells were obtained if randomization for branch length was carried out on a laminar basis. The dendritic arbor was divided into four laminae, in terms of increasing distance from the cell body along the ideal axis, and separate length histograms were computed for each lamina. This procedure was necessary because branches tend to be short in the outer laminae. Figure 12 shows a typical skeletonized hippocampal granule cell. Figures 13 and 14 are representative examples of the synthetic skeletonized cells obtained by ran-

FIG. 11. Tree-like shapes generated by computer models using simple sets of branching rules. (From Honda, ref. 3.)

domizing the angle and branch length histograms. They differ from the actual cells in that there is too much branching in the intermediate zone, with many short branches that do not reach upward to the pial surface.

The synthetic dendritic arbors resemble immature (postnatal day 7) rather than adult hippocampal granule cells. A typical skeletonized immature cell is shown in Fig. 15, and the Golgi impregnation of an immature cell is shown in Fig. 16. In addition to their thin, irregular dendrites that terminate short of the pial surface,

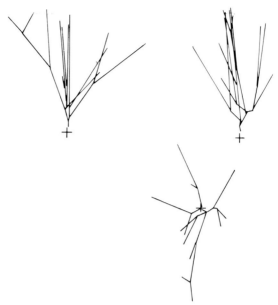

FIG. 12. Computer reconstruction of dentate granule cell from 60-day-old mouse after sample points are extracted and branch points are connected by a straight line (skeletonized). Views in three planes.

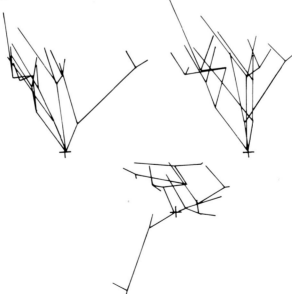

FIG. 13. Tree-like structures generated by random synthesis programs using data derived from actual cells.

GENETICS OF NEURONAL FORM 89

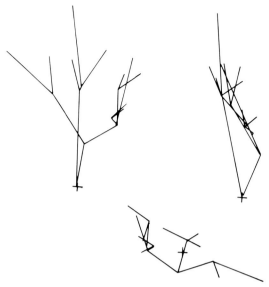

FIG. 14. Tree-like structures generated by random synthesis programs using data derived from actual cells.

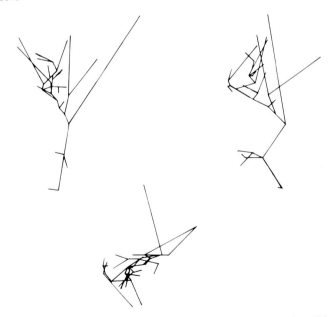

FIG. 15. Computer reconstruction of typical skeletonized dentate granule cell from 7-day-old mouse.

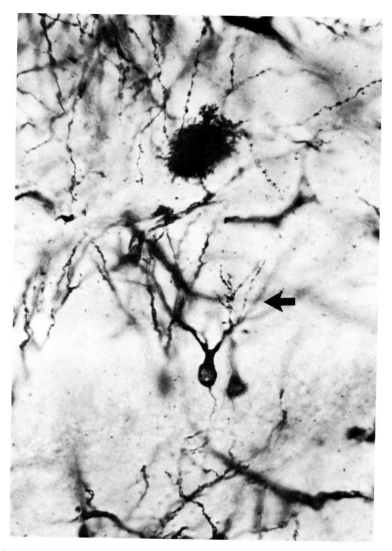

FIG. 16. Golgi-impregnated dentate granule cell from 7-day-old mouse. At this early age, some dendrites *(arrow)* appear extensively branched and less symmetrical.

these cells can be recognized by their small soma and paucity of mature spines. Our branching rules seem more adequate than we first thought, except that they must be supplemented by a "pruning" rule in order to account for adult, rather than immature, shape.

GENETIC CONTROL OF NEURONAL FORM

How a relatively small number of genes can control the complexity of form of the mammalian nervous system is a long-standing problem in developmental neurobiology. Our syntheses suggest that the number of rules required to determine neuronal form may not be large. It is likely that the class-characteristic shapes of most neuronal phenotypes result largely from environmental interactions to which cells of that class are exposed during the course of dendritic development. Therefore, the shape of the dendritic arbor may also be a sensitive indicator of abnormalities in the neuronal environment. We hope that these beginnings of quantitative description and axiomatic formulation of neuronal geometry will contribute to the foundation for understanding genetic control and gene-environment interaction in neuronal morphology.

ACKNOWLEDGMENTS

We are grateful to Dr. Richard Sidman for making the computer microscope in his laboratory available to us, and to Drs. Verne Caviness and Alfred Pope for helpful advice.

Supported in part by the Schizophrenia Research Foundation of the Scottish Rite, N.M.J.; U.S.P.H.S. grants R01-NS-12005, R01-MH-34079, 3-P01-MH-30511, 5-P01-MH-31154; and Research Scientist Development Award 5-K02-MH-00108 to S. M.

REFERENCES

1. Bradley, P., and Berry, M. (1976): The effects of reduced climbing and parallel fiber input on Purkinje cell dendritic growth. *Brain Res.*, 109:133–151.
2. Halle, F., Oldeman, R. A. A., and Tomlinson, P. B. (1978): *Tropical Trees and Forests: An Architectural Analysis*, p. 285. Springer-Verlag, Berlin.
3. Honda, H. (1971): Description of the form of trees by the parameters of the tree-like body: Effects of the branching angle and branch length on the shape of the tree-like body. *J. Theor. Biol.*, 31:331–338.
4. Lindsay, R. D., and Scheibel, A. B. (1974): Quantitative analysis of the dendritic branching pattern of small pyramidal cells from adult rat somesthetic and visual cortex. *Exp. Neurol.*, 45:424–434.
5. Stanfield, B., and Cowan, W. M. (1979): The morphology of the hippocampus and dentate gyrus in normal and reeler mice. *J. Comp. Neurol.*, 185:393–422.
6. Williams, R. S., Marshall, P. C., Lott, I. T., and Caviness, V. S., Jr., (1978): The cellular pathology of Menke's steely hair syndrome. *Neurology*, 28:575–583.

> *Genetics of Neurological and Psychiatric Disorders*, edited by Seymour S. Kety, Lewis P. Rowland, Richard L. Sidman, and Steven W. Matthysse. Raven Press, New York © 1983.

Some Perspectives from Population Genetics

James F. Crow

Department of Medical Genetics, University of Wisconsin, Madison, Wisconsin 53076

> That was the secret of Darwin's popularity. He never puzzled anybody. If very few of us have read the *Origin of Species* from end to end, it is not because it overtaxes our mind, but because we take in the whole case and are prepared to accept it long before we have come to the end of the innumerable instances and illustrations of which the book mainly consists. Darwin becomes tedious in the manner of a man who insists on continuing to prove his innocence after he has been acquitted.
>
> G. B. Shaw
Back to Methuselah

It would be appropriate in this volume to discuss specifically the population genetics of human behavioral traits. Actually, the population genetics of genes affecting the nervous system is much like that of genes in general, and behavioral population genetics is a subject that is in its infancy. Therefore, I concentrate mainly on population genetics in general and have chosen particularly to consider the manifold ways in which natural selection acts. For the most part, I am reviewing the ideas and observations of others; in some cases, the ideas are so familiar that it is no longer clear who first thought of them.

Biological inquiry, unless it is mainly observational or classificatory, usually involves one of two kinds of issues. One type includes such questions as: What is the mechanism? What is the underlying chemistry? What is the mode of inheritance? What is the developmental process? In short: How does it work?

Yet a biologist is typically unsatisfied with having an answer to only the first kind of question. There is also interest in the historical origin. How did this structure or process come about? How did natural selection and other evolutionary forces produce this result? What function does this structure or process serve, and what was its "purpose" in its evolutionary past? In short: How did it get this way?

Population genetics is concerned with such evolutionary processes. It studies the evolutionary history but also asks about the mechanisms responsible for maintaining

---

This chapter is paper number 2612 from the Laboratory of Genetics, University of Wisconsin.

the current level of variability in the population. In the past, it could study the frequencies of genes only by studying the traits determined by these genes, but molecular biology has at last made possible the study of the evolution of the genes themselves.

Since the beginning of the 20th century, we have known that the mechanism of inheritance in multicellular, sexually reproducing species is, with minor but interesting exceptions, Mendelian. The underlying mechanism is provided by the behavior of chromosomes in meiosis and fertilization. Since Darwin, biologists have known that the molding force of evolution is natural selection acting on variations that are randon within the constraints of mutational possibility. Since the middle of the 20th century, the chemical nature of the gene, of its replication process, and of mutation have been known. This knowledge has been greatly deepened and enriched in its details in recent years.

Mendelism is exceedingly simple. It would be hard to devise a workable system of biparental inheritance with simpler rules. Mendelian segregation and recombination insure that each gene gets a completely unbiased test in many combinations with other genes, as these are scrambled every sexual generation. Likewise, Darwinian natural selection is so self-evident, at least in retrospect, that one is puzzled as to why it was so slow to be discovered and accepted. Many biologists must have felt somewhat the way Shaw did in the quotation at the beginning of this chapter.

Yet, the neo-Darwinian theory goes much farther. It considers, in rich observational, experimental, and mathematical detail, the joint action of mutation, migration, natural selection, and random processes. It provides answers to a number of questions that Darwin found insoluble—the evolution of the sex ratio, for example (8). The mechanistic basis and symmetry of the Mendelian mechanism provides an inviting model for developing a mathematical theory, and the foundations were laid by R. A. Fisher (8), J. B. S. Haldane (9), and Sewall Wright (16). More recently, some very gifted mathematicians have entered the field, with a resulting increase in rigor and precision (7,13). This, plus the recent advances, have made the field quite different from what it was only a few years ago (11).

SELECTION AT VARIOUS LEVELS

The most significant natural selection, as Darwin said, is between individual animals and plants. Yet there is the opportunity for selection at many levels of organization.

Moving downward, there can be selection among tissues and cells. There is competition for nutrients and energy between, for example, the left and right kidney. Yet, this is somehow kept in control by coordinating mechanisms within the developing animal. We realize the effectiveness of these mechanisms by observing the tendency for one of a pair of organs to enlarge when the other is removed.

There is a striking example of intercellular selection in the X chromosomes of human females, where in each cell only one of the two X chromosomes is functional. If one of the two X chromosomes is abnormal, for example, by having a large

DNA segment deleted, the cells in which this X is turned on lose out in competition with cells in which the normal X is active. This is often observed by finding only one class of white blood cells when two would have been expected. No doubt, there is a great deal of cleansing of the organism by selective death of the most grossly abnormal cells, those that are the result of a mistake in chromosome distribution or a mutation.

Yet, it is important to the organism that such processes do not get out of control. The organism as a whole has to develop in a coordinated way following instructions programmed in the inherited genotype. It can not let one organ or one cell type grow disproportionately. The most obvious example of the destructive effect of a cell that competes individualistically and gets out of hand is malignancy.

One way that the organism keeps such selection in check is by the process of mitosis. This insures that all cells, except when there are errors or mutations, have the same genotype. If there were some sort of process that led to cells of different genotypes, there would undoubtedly be differences in growth rates among the different genotypes, and the organism would lose control. From this viewpoint, it is not difficult to understand why differentiation takes place by epigenetic mechanisms rather than by segregation of genetic material. Cellular changes, such as malignancy, upset this orderly system, and there are other mechanisms, such as immune surveillance, that remove such noncooperative cells.

The one place where there is regular segregation leading to cells of different genotypes is in meiosis. This leads to sperm of different genotypes. There is, of course, intense competition among different sperm in that only one succeeds in fertilizing the egg. Yet, this competition is not based on genetic differences among sperm cells produced by a single male. It is well established in *Drosophila* that the function of a sperm does not depend on its gene content. In fact, a *Drosophila* sperm can function normally with no chromosomes at all, as can be demonstrated by contriving to have a diploid egg fertilized by such a sperm. Many similar experiments, similar in principle by combining two complementary wrongs to make a right, have demonstrated in several species that spermatozoa with grossly abnormal chromosome content can compete successfully with other sperm.

I think this is no accident. If there were competition among sperm based on their genotypes, then genes causing a sperm to be particularly effective would have an enormous selective advantage and would increase rapidly in the population. But, it is typical for genes to affect more than one structure or process. If a gene that enhanced the motility of the sperm carrying it also caused abnormal liver or brain function, the population would be harmed, for natural selection among sperm would be much more intense and effective than selection for bodily functions. So, I believe that one reason—possibly *the* reason—for cessation of gene transcription in post-meiotic cells is to protect the organism from inappropriate and potentially harmful intersperm competition. Of course, this does not limit competition between sperm produced by different males.

In *Drosophila* there is a gene—really a cluster of closely linked genes—called *segregation distorter (SD)* which affects sperm function. A male that is heterozygous for an *SD* chromosome and a normal chromosome produces mainly only one kind of sperm, those containing an SD chromosome. Somehow, at the time of meiotic pairing, the SD chromosome tells its partner chromosome to self-destruct. The sperm carrying the partner chromosome fails to mature properly (4).

This process is striking in two regards. First, the *SD* chromosome somehow communicates with its homolog and persuades it to behave abnormally during spermiogenesis, while it itself behaves normally. Second, the partner chromosome cannot simply fail to carry out its normal function, for as previously indicated, the chromosomes of sperm play no role in sperm function; there is no normal function to carry out. There must be a positive act of damage, not simply a failure to do the right thing. It is of course much easier physiologically to fail to carry out a normal process than to do something that was not done before. If the chromosomes in sperm had a job to do, it would be much easier to interfere with this. So the nonfunctioning of genes in sperms makes it much harder for harmful systems like SD to arise.

What happens to the population? At present, many of the chromosomes in the population have evolved to become insensitive to the *SD* effect. Therefore, the *SD* chromosomes remain at low frequency in the population and do little harm. There was surely a time in the past, however, when the gene did a great deal of harm, since it causes lethal or sterilizing effects when homozygous. If a resistant chromosome had not arisen, the fly population might now be in a sad state (4).

Something very similar happens in the mouse. The *T* genes that Dorothea Bennett has discussed (D. Bennett, *this volume*) have an analogous effect in distorting the normal Mendelian mechanism.

These are examples of ultraselfish or cheating genes. By subverting the Mendelian mechanism, which ordinarily is precise in giving each gene an equal chance of being represented in a functional sperm, the *SD* and *T* genes gain an unfair advantage. In the long run, the species suffers by this distortion of the usually fair Mendelian process.

At another level, there are organelles inside the cell, such as mitochondria and plastids, which play important roles. For example, many energy-producing cell processes are in the mitochondria. These organelles carry their own DNA which encodes for their own proteins. In many ways, they are small organisms within the cell. At the same time, some of the organelle proteins are encoded by chromosomal genes. This means that the rate of multiplication of the organelle can be controlled by the chromosomes, and the number of organelles kept in line.

If this were not so there would be a great deal of intracellular competition between different organelles. The animal or plant would suffer, for in general, the organelle that can multiply the most rapidly is not necessarily the one that is best for the organism. I think that the nuclear control over some organelle proteins is evolution's way of bringing the function of these organelles under control of the well-behaved chromosomes. This subject has recently been reviewed by Eberhard (6).

One of the most interesting findings of molecular biology in recent years has been that only a small minority of the DNA is used to encode proteins. Some of the rest play a regulatory role. Some DNA is highly repetitive in its nucleotide content, and for some of this, the function is known. But there is a large amount of chromosomal DNA whose function is unknown and which, at least superficially, seems unnecessary.

There are many ways in which the amount of DNA in the cell can change. Some of these involve accidents in the distribution of chromosomes during cell division. Usually, these produce harmful if not lethal effects. But sometimes, this is not so, and the species is changed. This is especially striking when the entire set of chromosomes is duplicated, leading to polyploidy. This has been especially important in the evolution of plants.

There are also mechanisms by which the amount of DNA in the chromosomes can change by small increments. One way is by inaccurate exchange of parts during crossing over, so that one partner gets too much and the other too little material. The frequent finding of closely linked genes with identical or similar effects suggests that this is an important way for the number of genes to increase. This has an important evolutionary consequence, as first pointed out by Calvin Bridges in the early years of *Drosophila* genetics. Duplicating a gene gives the organism a chance to eat its cake and still have it; for with duplicate genes, one is unnecessary and is free to mutate to a new function while the old function is retained by the other. Of course, it may be that one of the extra genes is simply excess baggage. In this case, it may become inactive by the accumulation of mutations that are sheltered by the other, normal genes. Such nonfunctional pseudogenes have been found in DNA clones; they could never have been found by conventional genetics, for they do not do anything to make them detectable. They are presumably a dead relic of an earlier gene that, perhaps because its function was duplicated elsewhere, simply lost its function over millions of years of evolutionary time.

There is another mechanism that may be responsible for the accumulation of a great deal of extra DNA. There are many examples of microorganisms of transposons—elements that are able to move from place to place in the chromosome. These are pieces of DNA that become inserted into the chromosome by a process analogous to crossing over and can be removed from the chromosome by reversing the process. Hence, they can move from place to place. In some instances, the inserted DNA can produce an extra copy that moves elsewhere while the original stays behind, thus producing a small increase in the total chromosomal DNA. This provides a way for the DNA of the cell to increase. Transposons are very much the center of attention in genetics now, and the 1980 Cold Spring Harbor Symposium was devoted to these and other movable genetic elements (1).

If the piece of DNA is inserted into a gene, it is quite likely to cause an impairment of the gene function, so it is disadvantageous and removed by natural selection. On the other hand, if it settles into a nonessential part of the DNA, it may do no harm. The replication of DNA is rapid and energetically cheap for the organism, so there may be so little selective disadvantage that more and more extra material accumulates. Very likely, such "selfish DNA" (5,14) is one reason, perhaps the main reason, why there is so much seemingly useless DNA in the cell.

Transposons can also move from one organism to another in the manner of virus infections. There are strong molecular similarities between transposons and RNA tumor viruses. Since viruses can move from species to species, transposons need not have originated in the species where they are found. In some cases, the transposon is useful to its carrier, for example, those associated with a drug resistance factor.

What about higher organisms? Transposons are well established in maize and *Drosophila*; very likely they are in mammals as well but have not yet been identified. One of the most studied examples is "hybrid dysgenesis" in *Drosophila*, now being actively studied in several laboratories including mine. It is characterized by production of a number of disastrous effects, sterility, chromosome breakage, and high mutability among others. But such effects occur only in certain types of hybrids between different strains. We should not conclude from this, however, that transposons in general will have grossly harmful effects. This case was discovered precisely because it produced such effects. Very likely, there are movable elements that move around the DNA, doing no harm until they happen to insert into a crucial gene. I do not mean to imply that such movement need be frequent. A transposition every hundred or thousand generations would still be important in evolutionary time.

We see, then, that there is selection, or at least the possibility thereof, at many different levels. For the welfare of the individuals of a species, and therefore of the species itself, it is important that the bulk of the selection be between organisms. The most important insurance of this is the orderly process of cell division by mitosis. This insures that each cell in the body has the same genetic makeup and, therefore, that there is no genetic competition among developing body cells. At the same time, meiosis insures that each gene gets the fairest possible test by scrambling the genes every sexual generation and making sure that each individual gene has an equal chance of being included in a sperm or egg. It goes without saying that it is important for these processes to be as accurate as possible, and natural selection has kept the error rate quite low. Control over interorganelle competition is exercised by having part of the essential proteins of the organelle controlled by the chromosomes. Control over inappropriate competition between sperm of different haploid genotypes is exercised by shutting off postmeiotic transcription and translation. Yet, the process is not quite perfect, for there are always selfish or cheating entities that beat the system despite built-in mechanisms to prevent them.

The tissue of the nervous system is particularly important in higher organisms. It is therefore particularly important that the genetic program be carried out correctly, for the system is especially delicate as regards genic balance. For example, one of the most frequent manifestations of even a slight degree of chromosomal unbalance is mental retardation or behavioral change. Perhaps nervous tissue, by having a minimum number of cell divisions, is particularly well buffered against possible effects of mitotic mistakes or such subverting entities as mentioned above.

SELECTION AT LEVELS HIGHER THAN THE INDIVIDUAL

Just as there can be selection between cells or smaller entities within the organism, there can be selection involving entities larger than the individual organism. This can be particularly important in behavioral traits, because interactions, ranging from criminally selfish to outright altruistic, are much more likely to be behavioral than to be based on other kinds of processes.

One example of selection in the human population at higher than the individual level is the indirect selection on other members of the family when one member is affected with a debilitating disease. Several conditions including Huntington's disease, schizophrenia, and mental retardation have been studied quantitatively. The presence of one affected person in the sibship depresses the reproductive rate of normal sibs (S. Yokoyama, *personal communication*). If widespread, this could cause difficulties with the standard method of using a sib control in the study of fitness of persons with various conditions. It also means that, unless the trait is dominant and fully penetrant, selection is more effective in eliminating the causative genes than simple, individualistic theory would predict.

Another way in which natural selection can act through relatives is kin selection. This is an old idea, mentioned by both Haldane and Fisher but developed especially by Hamilton (10). The idea is easily grasped. For example, sibs share half their genes. Therefore, genes that cause a person to protect or aid his brother will tend to increase in the population because of the greater chance of the brother's reproducing. Since his brother shares half of his helper's genes, these genes will increase. Thus, a genetic tendency to protect or aid near relatives would be expected to increase in the population and eventually become prevalent.

This should happen even if aiding the relative is done at some expense to the helper. In general, one would expect behavior-affecting genes to increase in the population if they cause a benefit to the recipient that, when weighted by the fraction of shared genes, is greater than the cost to the donor. Both cost and benefit are measured in terms of survival and fertility, that is to say, Darwinian fitness. Specifically, if c is the cost of an altruistic act to the donor, b is the benefit to the recipient, and r is the proportion of shared genes, then genes causing altruistic behavior will increase if $c/b < r$. For relatives that are not inbred, r is Wright's coefficient of relationship (16).

Of course, the effectiveness of such selection depends on whether there is genetically based altruistic behavior and whether there are also genes that affect the ability of an individual to be selectively atruistic toward relatives. We would not expect any such principle to work precisely, and there would surely be frequent miscalculations, such as confusing a close neighbor with a relative. But such a system might be expected, especially for close relatives such as sibs or parent and child. I do not mean to imply that animals, including humans, have genes that enable them to compute relationship coefficients. Yet, it is clear that if there were a heritable tendency to behave as if such a calculation were being made, the genes causing such behavior would tend to increase in the population.

Furthermore, it would be expected that altruism would be greatest when the altruist is at the end of the reproductive period and the recipient is at the beginning. More specifically, the altruist should have minimum reproductive value, that is, expectation of future progeny, and the recipient maximum. Hence it is no surprise that in all species parents are more unselfishly protective of their children than vice versa.

H. J. Muller (12) has emphasized the importance of a tribal structure in the human past for the evolution of cooperative and altruistic behavior. In the human ancestry, the population structure was that of small, nearly isolated units. Within such a unit, all the individuals would tend to be related to each other, so the kin selection principle should operate, leading to cooperation and even altruism within the group. In Muller's view, much of human nobility—to the extent that it is genetic—is attributable to a tribal population structure in the past.

There has been considerable development of the theory of kin selection beyond the simple Haldane–Hamilton equation given here. Different forms of inheritance and different patterns of behavior have been considered in detail. As an example of a theoretical advance, one can ask the question: At what c/b value will the population equilibrate if there is a wide range of possible altruistic behavior patterns to choose from. The somewhat surprising answer obtained by my colleague William Engels *(personal communication)* is that the equilibrium value is one-half the coefficient of relationship, not the value itself as one might naively think from the Hamilton equation.

The greatest success of kin selection theory has been in understanding the evolution of social hymenoptera such as honey bees. How applicable it is to human behavior remains to be demonstrated. The whole subject of sociobiology depends heavily on kin selection. Unfortunately, the subject has been one of debate and frequent vituperation, with political overtones.

The evolutionary origin of behavior in the mammals, including ourselves, constitutes a biological question of enormous interest. The sociobiological approach is only one among many that can be used to unravel the various causal threads. It is a theory of heuristic value, leading to testable hypotheses. I hope that these ideas can be developed and tested in an atmosphere free of any tendency to judge the correctness of a scientific theory by its perceived political or social consequences rather than by evidence for or against its correctness. We need more research and less dialectic.

At the still higher level is intergroup selection. This is particularly difficult to assess. Group selection depends on the survival and extinction of populations. It is easy to fall into the facile conclusion that group selection can be effective in the evolution of traits that appear to benefit the group rather than the individual. Yet, there are difficulties, as first pointed out by Fisher (8). In selection between groups, the number of groups is always far less than the number of individuals, with consequent lower efficiency of selection. Furthermore, the lifetime of a group is not that of the individuals composing it but the life of the group, which is many more generations. Hence, we would expect individual selection to predominate.

At the moment there is a great deal of discussion of the possible evolutionary importance of group selection. It has been hard to find really convincing examples of group selection. The general working assumption is to invoke group selection only where individual selection or kin selection seems inadequate. That group selection is of minor importance compared to individual selection is clearly the consensus view (15).

THE PRIMACY OF ORGANISMIC SELECTION

All of this discussion leads to the conclusion that the most important force molding evolution is selection among individual organisms. Selection below this level occurs. So does selection at higher levels, in particular, kin selection. But Darwin was correct in emphasizing individual selection.

In developing a theory of evolution, it is necessary to build models and deal with abstractions. Such an abstraction is to treat a gene, or a region of the chromosome, as a unit of selection. In this sense, we can speak of the fitness of a gene as being nothing more or less than the average fitness of the individuals that carry this gene. Much of our present theory derived from Fisher, Haldane, and Wright is genic in this sense. It leads to great mathematical simplification, and therefore to intuitive understanding. But more complete and exact formulations can be made, although they greatly complicate the mathematics; present trends are in this direction.

It goes without saying that in human genetics and in behavioral genetics, and especially in human behavioral genetics, separation of genetic from environmental effects is particularly difficult.

Consider the almost universal taboo against incest. There is a clear and obvious selectionist explanation: those individuals who avoided mating with first-degree relatives produced a larger fraction of surviving children by avoiding the deleterious effects of inbreeding. So one would expect that genes favoring such behavior would tend to increase in the population. Of course, the psychological or social basis of the behavior is irrelevant; natural selection cares only that by avoiding incestuous mating for whatever reason, the number of surviving progeny is increased. So it may be that psychology or anthropology may provide part of the answer and natural selection theory other parts. Of course, there may be totally different reasons for incest avoidance, although the genetic explanation gains some support from the widespread occurrence of inbreeding avoidance systems in plants and animals. Furthermore, regardless of how the behavior pattern of incest avoidance got started, the social mechanisms that perpetuate the custom may be of a totally different sort, having little to do with what brought it about in the first place.

Let me emphasize the frequently made point that natural selection has only one criterion for merit—the capacity to survive to the reproductive ages and reproduce. In organisms that care for their young, there will also be some selection, through caring for children until they can fend for themselves, for survival of the caring parent. Selection for longer survival of parents is kinetically similar to selection for altruism; from an evolutionary viewpoint, survival of a postreproductive mother is important only to the extent that this enhances the survival and fertility of her children.

This means that natural selection has done little to preserve life and health much beyond the child-caring period. There is considerable variability in the length of life-span, and studies of pedigrees and of identical twins suggest that a considerable fraction of this variability is genetic. Nevertheless, since there is no mechanism for natural selection to favor those who live to a ripe old age, we would not expect to find extremes of longevity.

The animal breeder can select for longevity by breeding the relatives of those who die at a very old age. In principle, human society could do the same. Yet, such selection experiments have usually failed. The main difficulty is simply time and logistics. It is necessary to save the progeny of a large number of oldsters while waiting for them to die. Then one must breed from the progeny of those who live the longest; by this time some of the progeny may be past the best breeding age. Actually, the only experiments that have produced any substantial change in longevity have been done in *Drosophila* by selecting instead for a long reproductive period. By selecting those strains that reproduce until the oldest ages, the onset of death was postponed along with the onset of sterility. In *Drosophila*, the age of sterility and the age of death are strongly coupled.

Senility and death are then to be expected as a natural consequence of evolution. This does not mean that we can do nothing to extend human life, but it does mean that we can not expect much help from natural selection.

RELAXED SELECTION IN CONTEMPORARY SOCIETY

We can expect, as a consequence of the development of civilization, that the directions of natural selection have greatly changed. Genotypes that favored survival in a hunting and gathering society may differ in many ways from those that are optimal in the contemporary world. In the middle ages, the most important genotypes to have, I suppose, would have been those conferring some measure of resistance to smallpox and plague. Such genes are now almost irrelevant. Selection by differential mortality is considerably reduced in modern times. Human fertility is now determined by psychological and social factors more than biological.

It is easy to quantify the extent to which changing demographic patterns have lessened the effectiveness of natural selection. We can define a quantity called the Index of Opportunity for Selection (2). Examination of demographic patterns shows that the opportunity for natural selection from differential mortality has greatly reduced, to nobody's surprise. But the effect of changing fertility in the United States has varied. The opportunity for selection actually increased, despite decreased average family sizes, in the first half of this century as contraceptive practice was stratified by educational, economic, and religious factors. More recently, it has gone down, as there is now a more uniform tendency to limit family size.

As natural selection becomes less important, one factor clearly becomes more important. The mechanism by which deleterious mutant genes are eliminated from the population is steadily less effective. Hence, mutant genes are accumulating to a higher level in the population, although their individual degree of detriment is lessened by environmental improvements. The process of arriving at a new mutational equilibrium is very slow, so at present, the population is far from equilibrium. We are improving the environment through higher living standards, better sanitation, and medical advances faster than new mutants are accumulating. It is not clear how long such enviromental improvements can continue. The worst thing would be for us to return to the harsher environments of the past, such as might happen with overpopulation or nuclear war—the twin problems of overpopulation and no population. We would then have all the problems with mutant genes that our ancestors had plus whatever mutants have accumulated during the period of relaxed selection since that time (3).

It is important that we make sure that the human mutation rate does not get higher. Possibly, molecular biology will find ways of lowering the spontaneous mutation rate.

FINAL REMARKS

I hope it is clear from this chapter that a great deal has been learned about evolution in the period since the fusion of the results of Darwin and Mendel. This includes increased knowledge of the genetics and evolution of behavior in many species, including our own. We can expect much more rapid increase of knowledge in the future, I think.

Behavior genetics is a relatively new field. So is the evolution of behavior. But increases are rapid, as they are in basic neurology and molecular biology. The possibility of a real understanding, at many levels, of the genetics and evolution of behavior is highly promising. The better we understand our past, the better we can understand ourselves.

REFERENCES

1. *Cold Spring Harbor Symposium on Quantitative Genetics* (1981) Vol. 45: Cold Spring Harbor Laboratory, New York *(in press)*.
2. Crow, J. F. (1958): Some possibilities for measuring selection intensities in man. *Hum. Biol.*, 30:1–13.
3. Crow, J. F. (1973): Population perspective. In: *Ethical Issues in Human Genetics*, edited by B. Hilton, D. Callahan, M. Harris, and B. Berkley, pp. 73–81. Plenum Press, New York.
4. Crow, J. F. (1979): Genes that violate Mendel's rules. *Sci. Am.*, 240(2):134–146.
5. Doolittle, W. F., and Sapienza, C. (1980): Selfish genes, the phenotype paradigm and genome evolution. *Nature*, 284:601–603.
6. Eberhard, W. G. (1980): Evolutionary consequences of intracellular organelle competition. *Q. Rev. Biol.*, 55:231–249.
7. Ewens, W. J. (1979): *Mathematical Population Genetics*. Springer-Verlag, Berlin, Heidelberg, New York.
8. Fisher, R. A. (1930): *The Genetical Theory of Natural Selection*. Clarendon Press, Oxford. Reprinted 1958, Dover Publications, New York.
9. Haldane, J. B. S. (1932): *The Causes of Evolution*. Harper, New York. Reprinted 1966, Cornell University Press, Ithaca, New York.
10. Hamilton, W. D. (1964): The genetical evolution of social behavior. *J. Theor. Biol.*, 7:1–16,17–52.

11. Hartl, D. L. (1980): *Principles of Population Genetics*. Sinauer Associates, Sunderland, Massachusetts.
12. Muller, H. J. (1968): What genetic course will man steer? In: *Proceedings of the Third International Congress on Human Genetics*, pp. 521–543. Johns Hopkins University Press, Baltimore.
13. Nagylaki, T. (1977): Selection in one-and-two-locus systems. In: *Lecture Notes in Biomathematics*, Springer-Verlag, Berlin, Heidelberg, New York.
14. Orgel, L. E., and Crick, F. H. C. (1980): Selfish DNA: The ultimate parasite, *Nature*, 284:604–607.
15. Williams, G. C. (1966): *Adaptaion and Natural Selection*. Princeton University Press, Princeton.
16. Wright, S. (1968),1969,1977,1978): *Evolution and the Genetics of Populations, Vols. 1–4*, University of Chicago Press, Chicago.

Genetics of Neurological and Psychiatric Disorders, edited by Seymour S. Kety, Lewis P. Rowland, Richard L. Sidman, and Steven W. Matthysse. Raven Press, New York © 1983.

Observations on Genetic and Environmental Influences in the Etiology of Mental Disorder from Studies on Adoptees and Their Relatives

Seymour S. Kety

Laboratories for Psychiatric Research, Mailman Research Center, Belmont, Massachusetts 02178 and Department of Psychiatry, Harvard Medical School, Boston, Massachusetts 02115

There are a number of problems in the identification and differentiation of genetic and environmental factors in human disease, and these are aggravated in psychiatry because of the subjective nature of many of the symptoms and diagnostic criteria, the absence of pathognomonic and objective manifestations, and the lack of stability and consistency in nosology. The tendency displayed by several of the psychiatric syndromes to occur in other members of the patient's family has been recognized since these syndromes were first described and has been used variously to support genetic or environmental inferences of etiology. To what extent the symptoms, syndromes, and diagnoses in the natural relatives of patients reared together are affected by genetic or environmental factors, enhanced by ascertainment and subjective bias, and molded or confounded by the relatives' close association with the patient is difficult to determine. Where adoption has segregated the genetic and familial environmental influences, they can more readily be differentiated and evaluated. Mental disorders, should they occur in the adoptee and certain of the biological relatives, will have developed independently and can be ascertained and diagnosed more objectively. An increased prevalence of the disorder or any of the characteristics that comprise it in the biological or the adoptive relatives, when compared with suitable matched control groups, would be compatible with the operation of genetic and family-associated environmental factors, respectively. A number of studies using the adoption strategy have been conducted over the past 20 years in Denmark (5,7–9,11,13–16,19–22,24,26,27), the United States (2–4,6,10,25,27,28), Sweden (1), and Belgium (18). The results of these indicate the operation of genetic factors to a significant extent in schizophrenia, sociopathy, alcoholism, affective disorders, and suicide.

In addition to the obvious advantages that studies of adopted individuals and their two families provide, a number of circumstances and characteristics differentiate such individuals and families from the general population and must be taken into account, since they may augment or diminish the apparent effects of genetic or rearing influences.

Differences in demographic and other characteristics may occur in the case of adoptees and their two types of relatives which would tend to distinguish them from each other and from their counterparts in the general population. Biological parents tend to be younger than adoptive parents, and there are more biological siblings and half siblings of adoptees than the adoptive counterparts (2). Biological parents also tend to have a somewhat lower socioeconomic status than adoptive parents, although the range may be quite broad in both groups, and the difference in the means not large (24). There are reports suggesting a greater prevalence of various types of psychopathology in samples of adoptees than in the general population (17), but these have rarely been carried out in a way to minimize selective and subjective bias and to match the samples demographically. If adoptive parents are more likely than natural parents to bring deviant behavior in their child to the attention of a psychiatrist, the selective bias could account for much of the difference. To evaluate that question more systematically and objectively, it is necessary to compare large samples of adoptees and nonadopted individuals, matched on variables that affect the risk for mental illness and studied simultaneously with the same techniques of ascertainment and diagnosis. In connection with our studies on the 5,483 adoptees in the Copenhagen sample and their families (14,20), we identified an equal number of individuals reared by their natural parents and matched with a corresponding adoptee on age, sex, and neighborhood of rearing, whose records were searched for mental illness in the same procedure and diagnosed by the same raters blind to the basis of selection. We did not find an increased prevalence of schizophrenia or affective disorder or more admissions to mental hospitals in the Copenhagen adoptees than in the matched sample of nonadoptees. At least for mental disorders serious enough to require hospitalization, adoptees in Copenhagen do not have a higher risk than their counterparts in the general population.

In adoptions that have had legal recognition, there is a likelihood that some selection has occurred that would reduce the incidence of recognizable mental disorder in adoptive parents at the time of adoption. This could render the sample less representative of mental illness attributable to parental deviance and diminish the apparent influence of rearing factors in mental disorders that later developed among the adoptees. Another factor of at least equal significance operates to reduce the apparent influence of genetic factors for serious mental disorder among adoptees. This factor is represented by the selective processes that act to prevent the birth or the adoption of offspring of biological parents with obvious mental disorder (14). In Denmark, such infants were usually not put out for adoption by the social agencies but were reared in child-care institutions. Moreover, the liberal abortion laws markedly reduced the number of such pregnancies that were carried through to term. Among the biological parents of our Copenhagen sample of adoptees, 84% of those who were hospitalized for mental disorder had their first hospital admission after the adoptee was born, more than 11 years later on the average (20). This would produce a significant reduction in our adoptee sample of offspring from parents with severe schizophrenia of early onset, the type of illness that is likely to be most loaded genetically. Thus, the Copenhagen sample of adoptees would have a considerably lower genetic risk for schizophrenia than the sample studied by Heston (6), all of whom were born of schizophrenic mothers confined to a mental institution. This probably accounts for the lower prevalence of definite schizophrenia in the Copenhagen adoptees (20) and for the reduced prevalence of schizophrenia in the biological parents of the adoptees who became schizophrenic in comparison with the incidence in their biological siblings and half siblings (14,15).

The factors described in the preceding paragraphs would act to diminish the likelihood of a significant difference between experimental and control groups. Moreover, the biological and adoptive relatives of matched groups of adoptees without known mental illness, which are simultaneously examined in most of the adoption studies, should go far toward minimizing such effects.

The adoption strategy depends on the ability of that approach to dissociate and randomize the genetic and family-dependent rearing variables in the transmission of the disorders studied. There are circumstances, however, that could make that dissociation imperfect. An adopted child has spent 9 months in the uterus of its biological mother where it shares with her a number of environmental influences and has had a certain amount of early mothering at her hands. It is possible to circumvent that problem by the study of biological paternal half siblings of the adoptees. Since the biological parents are usually not married to each other and have children with another spouse or partner, and since males can produce more children than females, there is among the biological relatives of adoptees a substantial number of paternal half siblings who shared neither the same uterus nor the same early mothering with the adoptee. In the Copenhagen study of the two familes of schizophrenic adoptees, there was no higher prevalence of schizophrenia spectrum disorders in the maternal than in the paternal biological half siblings, and both showed a significantly higher prevalence than the corresponding control half siblings (15).

Another circumstance that could operate to diminish the complete separation or randomization of genetic and environmental variables is the possibility of selective placement. Since only a very few environmental variables have as yet been shown to be associated with a higher risk for schizophrenia, it is not likely that a social agency, even if it set about doing so deliberately, could find sufficient of the unknown variables in the prospective adoptive parents materially to affect the risk of illness in the adoptee. One environmental variable, however, that appears to be correlated with the prevalence of schizophrenia and on which selective placement could be based to some extent is socioeconomic status. Where the correlations between biological and adoptive parents have been measured with respect to socioeconomic class, the correlations have been low. In one study of 206 adoptions in Colorado (10), average weighted correlations between biological and adoptive parents for education and occupation were 0.19 and 0.13, respectively. Teasdale (24) found a correlation coefficient of 0.15 between socioeconomic status of biological and adoptive fathers in 11,000 adoptions in Denmark. These correlations, although statistically significant, are both small and indicate that selective placement on the basis of socioeconomic status would not account for more than 3% of the variance. In the Danish sample of adoptees, selective placement on the basis of risk for mental illness in the child could hardly have occurred to any significant extent, since in very few, if any, instances was there an indication of mental illness in the biological parents at the time the adoptive parents were selected. More important, perhaps, than selective placement is the possibility of knowledge on the part of adoptive parents, adoptees, and psychiatrists regarding the presence of mental illness in the biological parents and the effect such knowledge could have on the occurrence, perception, and diagnosis of mental illness in the adoptee. Although this problem could be of some significance in studies in which the adoptee sample represented children born of mentally ill mothers, it would be negligible in the Danish adoptee sample for reasons that have been discussed.

The problem of assortative mating is one that complicates human family studies generally and adoption studies in particular. Like the problem of selective placement it makes imperfect the randomization of etiological factors, and the effect of assortative mating in the case of adoption studies is more likely to enhance than diminish genetic expression of mental disorder. An interesting example of this effect was found in the study of adoptees of whom at least one biological parent became mentally ill. The prevalence of schizophrenia spectrum disorders in adoptees was enhanced in those cases where the other biological parent was diagnosed schizoid or inadequate personality disorder in comparison with adoptees whose other parent was psychiatrically normal (19). It is worth pointing out that although assortative mating will enhance genetic transmission where it exists, it will not serve as a false substitute for it.

The studies of biological and adoptive relatives of schizophrenic adoptees in Denmark took place in two distinct stages separated by several years, so that the second study in the rest of Denmark (16) will serve as a replication of the first which was limited to the city and county of Copenhagen. The results of the second study, which only recently have been tabulated, have confirmed the first in finding a significantly higher prevalence of schizophrenia spectrum disorders in the biological relatives of the schizophrenic adoptees (Table 1). There are some striking differences, however, in the distribution of schizophrenic illness in biological full and half siblings. Whereas the 1968 Copenhagen study found only two full biological index siblings of whom one was schizophrenic, there were 29 in the second study, with a significant prevalence of schizophrenia among them. Schizophrenia spectrum disorders were significantly more prevalent in the biological half siblings of the index adoptees in Copenhagen than in the rest of Denmark. The most reasonable explanation of these differences appears to be the difference in life style and mating behavior in Copenhagen in contrast to the rural areas of Denmark with a resultant increase in assortative mating and a decrease in stable monogamous relationships in the case of biological parents of adoptees in Copenhagen.

TABLE 1. Schizophrenia spectrum disorders[a] in the biological and adoptive relatives of adoptees who became schizophrenic

Probands	Copenhagen study			Provincial study[c]			Total patient sample[b]		
	No. of probands	Relatives		No. of probands	Relatives		No. of probands	Relatives	
		Biological	Adoptive		Biological	Adoptive		Biological	Adoptive
Index	34	13/150	2/74	42	11/238	0/99	76	24/396	2/172
Control	34	3/156	3/83	42	2/206	2/105	76	5/358	5/192
p		0.007			0.019			0.0006	

[a] Comprising chronic, latent, acute, and uncertain schizophrenia, schizoid or inadequate personality as described in DSM-II on the basis of global consensus diagnoses by three raters from comprehensive abstracts of hospital records (14).
[b] The denominators in each fraction represent the number of identified relatives at risk (lived to age 15 or more in Denmark). The denominators for the national sample represent the relatives identified and at risk at the most recent updating.
[c] Although this study confirms the finding in the Copenhagen study of a significantly increased prevalence of schizophrenia spectrum disorders restricted to the relatives genetically related to the schizophrenic adoptees, the prevalence of these disorders is considerably less than that in the Copenhagen sample. This is compatible with the lower prevalence of hospitalized mental illness generally in rural as opposed to urban populations.
[d] p values are Fischer one-tailed exact probabilities between biological index and control relatives. Differences between adoptive relatives are not significant.

ADOPTION STUDIES ON SCHIZOPHRENIA

Since the chapters by Gershon and by Cloninger and Reich that follow will review the findings of family, twin, and adoption studies in affective disorder, alcoholism, and sociopathy, this discussion centers on schizophrenia. Here, the results have had several implications, shedding some light on the strength and nature of genetic and environmental contributions to etiology, enhancing the validity of family and pedigree studies in that syndrome, and providing further opportunity for defining the limits of the syndrome and examining the genetic relationships of putative subgroups to the original syndrome.

There have been three studies of the prevalence of schizophrenia and other mental disorders in the reared-apart offspring of a parent who was or became schizophrenic (6,7,20). In two of the studies, the mentally ill parent was a mother hospitalized with chronic schizophrenia, and the prevalence of schizophrenia in the offspring was approximately 10%. In one study (20), the offspring were adopted away in all but a few instances before either the biological mother or father were overtly psychotic. Although the mental illness in these offspring was less severe, a higher prevalence of chronic and latent schizophrenia and schizoid personality were found in them than in their matched controls.

An alternative approach has been the examination of the biological and adoptive relatives of adoptees who became schizophrenic, and the national sample of adoptees in Denmark has provided two independent samples, the results of which have been discussed briefly in the preceding section and are summarized in Table 1. The conclusion that seems to be permissible from both types of approach is that genetic factors operate significantly in a major segment of schizophrenic illness.

Since the prevalence of schizophrenia in the offspring, parents, siblings, and half siblings of schizophrenic individuals separated from them and reared apart is not significantly different from that found in the families of schizophrenics reared together, it appears that the well-known and repeatedly documented familial tendency of schizophrenia is an expression of genetic factors which family members share. This should enhance the value of studies of unseparated families and pedigrees in the resolution of genetic heterogeneity and in defining the modes of genetic transmission.

The adoption studies in schizophrenia have also provided a means of validating hypothetical relationships within the schizophrenia spectrum of disorders. Although the original diagnoses on relatives in the 1968 Copenhagen schizophrenia study (14) were made on the basis of global consensus judgments of three raters based on hospital records, there has been an opportunity more recently to derive specified diagnostic criteria from exhaustive psychiatric interviews of the relatives (15) and to compare diagnoses thus made with the global diagnoses utilizing the same interviews (12,13,23). In 24 probands and relatives in whom a global consensus diagnosis of chronic schizophrenia had been made, the independent application of Research Diagnostic Criteria confirmed that diagnosis in 22 (23). From all of the psychiatric interviews on the basis of which global diagnoses of latent or uncertain schizophrenia had been made, including a few examples of schizoid personality, eight features have been derived which constitute the criteria for schizotypal personality in DSM-III.

Recently Kendler, Gruenberg, and Strauss (12) independently reviewed the complete interviews on biological and adoptive relatives of the schizophrenic and nonschizophrenic control adoptees in the Copenhagen sample, making consensus diagnoses based on DSM-III criteria. They made a significantly larger number of diagnoses of schizophrenia and schizotypal personality as well as paranoid personality disorder, but not anxiety disorder or major affective disorder, in the biological relatives of schizophrenic adoptees in comparison with their control relatives. Their diagnoses of schizotypal personality disorder had a greater specificity but a somewhat lessened sensitivity in discriminating biological relatives of schizophrenic adoptees from controls than did the original global diagnoses of latent and uncertain latent schizophrenia (Table 2).

The adoption studies in schizophrenia have also permitted the examination of certain hypotheses with respect to the type and significance of specific environmental influences on etiology. Failure to demonstrate a higher prevalence of schizophrenia in the adoptive relatives of schizophrenic adoptees suggests that the presence or absence of schizophrenia in the rearing environment has little effect on morbid risk. A high prevalence of schizophrenia in the biological siblings of schizophrenic adoptees and its absence in their adoptive siblings argues against an exclusive role for family-associated environmental influences but does not rule out the possibility that rearing factors may potentiate or diminish the effects of genetic vulnerability or predisposition. Two studies, however, that compared the effects of strikingly

TABLE 2. *Consensus diagnosis[a] by two independent groups based on psychiatric interviews (Copenhagen study)*

Probands	Relatives[b]		Relatives[c]	
	Biological	Adoptive	Biological	Adoptive
34 Schizophrenic adoptees	18/113	1/38	11/105	0/38
34 Control adoptees	6/138	4/50	2/138	1/48
p[d]	0.002		0.002	
23 Screened[e] controls	1/86	1/36	0/88	1/35
p	0.0002		0.001	

[a]Based on comprehensive psychiatric interviews in more than 90% of the relatives and control adoptees alive and residing in Scandinavia.

[b]Data from Kety et al. (15); diagnosis of latent or ? latent schizophrenia based on DSM-II.

[c]Data from Kendler et al. (12); diagnosis of schizotypal personality disorder based on DSM-III.

[d]p values are Fischer one-tailed exact probabilities between biological index and control relatives; differences between adoptive relatives are not significant.

[e]Excluding eight of the 34 control adoptees who had died, emigrated, or refused interview and three who received blind consensus diagnoses of schizoid or inadequate personality (DSM-II).

different types of rearing (with a psychotic parent, in a normal adoptive or foster family, or in a child-care institution) failed to reveal any differential effect on the risk for schizophrenia in genetically vulnerable offspring (6–8). Furthermore, a cross-fostering study (26) which compared the incidence of schizophrenia in adoptees with high or low genetic risk reared by psychotic or normal parents found no evidence for an effect of rearing on the risk for schizophrenia.

Effects of certain environmental variables not closely associated with rearing have been found in studies of the Danish adoption sample (11) that confirm earlier observations on a relationship between birth injury or season of birth on the prevalence of schizophrenia.

The observation that parents who have reared an adoptee or their natural child who eventually became schizophrenic show characteristic signs of communication deviance elicited by the Rorschach test is compatible with a role of that rearing variable in etiology (28), although alternative hypotheses remain to be ruled out, and one attempt to replicate that observation was unsuccessful (27).

Psychiatric interviews and a battery of tests for thought disorder, communication deviance, and other rearing variables are presently being applied in the remaining sample of relatives of schizophrenic adoptees in Denmark outside of Copenhagen in a continuing search for rearing and other environmental factors other than genetic that may operate in the transmission of schizophrenia.

Analyses of the causes of death of the relatives in the study on schizophrenia (14) had revealed a high incidence of suicide restricted to the biological relatives of the schizophrenic adoptees. The results in the study of the relatives of mood-disordered adoptees, as yet unpublished, found an even more striking concentration of suicide almost exclusively in their biological relatives (Table 3). A further study of relatives of all adoptees who had committed suicide (22) found a similar high incidence of suicide restricted to their biological relatives. It is clear that genetic factors play a significant role in suicide. It is equally clear, however, that environmental influences are at least as important; witness the considerable variance

TABLE 3. *Suicide in the biological and adoptive relatives of adoptees who developed affective disorder[a] or who themselves committed suicide[b]*

Probands	Relatives		Probands	Relatives	
	Biological	Adoptive		Biological	Adoptive
71 Adoptees with affective disorder[c]	15/388	1/181	57 Adoptees who had committed suicide	12/269	0/148
71 Control adoptees	1/346	2/169	57 Control adoptees	2/269	0/150
p[d]	0.005		p	0.006	

[a]From S.S. Kety, P.H. Wender, D. Rosenthal, and F. Schulsinger *(unpublished data)*.
[b]From Schulsinger et al. (22).
[c]Comprising affect reaction, neurotic depression, bipolar and unipolar depression. The incidence of suicide was higher than that of controls in the biological relatives of each type of proband.
[d]p are Fischer one-tailed exact probabilities between biological index and control relatives; differences between adoptive relatives are not significant.

from one country or one religion to another and for different time periods in the same country, and the observation that suicide is usually preceded by some seriously disturbing life event. Suicide may be a striking example of the major conclusion that may be drawn from all of these studies—that genetic factors contribute significantly but not exclusively to most of the serious mental disorders. Much remains to be done to delineate these disorders and the mode of their genetic contributions more clearly and to identify more definitively the environmental factors such as infections, diet, birth injury, as well as rearing influences that operate in conjunction with those of genetic origin to precipitate, amplify, or diminish their ultimate expression.

REFERENCES

1. Bohman, M. (1978): Some genetic aspects of alcoholism and criminality: A population of adoptees. *Arch. Gen. Psychiatry*, 35:269–276.
2. Cadoret, R. J. (1978): Psychopathology in adopted-away offspring of biological parents with antisocial behavior. *Arch. Gen. Psychiatry*, 35:176–184.
3. Cadoret, R. J., and Gath, A. (1978): Inheritance of alcoholism in adoptees. *Br. J. Psychiatry*, 132:252–258.
4. Crowe, R. D. (1974): An adoption study of antisocial personality. *Arch. Gen. Psychiatry*, 31:785–791.
5. Goodwin, D. W., Schulsinger, F., Hermansen, L., Guze, S. B., and Winokur, G. (1973): Alcohol problems in adoptees raised apart from alcoholic biological parents. *Arch. Gen. Psychiatry*, 28:238–243.
6. Heston, L. L. (1966): Psychiatric disorders in foster home reared children of schizophrenic mothers. *Br. J. Psychiatry*, 112:819–825.
7. Higgins, J. (1966): Effective child rearing by schizophrenic mothers. *J. Psychiatr. Res.*, 4:153–167.
8. Higgins, J. (1976): Effects of child rearing by schizophrenic mothers: A follow up. *J. Psychiatr. Res.*, 13:1–9.
9. Hutchings, B., and Mednick, S. A. (1975): Registered criminality in the adoptive and biological parents of registered male criminal adoptees. In: *Genetic Research in Psychiatry*, edited by R. R. Fieve, D. Rosenthal, and H. Brill, pp. 105–116. Johns Hopkins University Press, Baltimore.
10. Ho, H., Plomin, R., and DeFries, J. C. (1979): Selective placement in adoption. *Soc. Biol.*, 26:1–6.
11. Jacobsen, B., and Kinney, D. K. (1980): Perinatal complications in adopted and nonadopted schizophrenics and their controls: Preliminary results. *Acta Psychiatr. Scand. [Suppl.]*, 285:337–346.
12. Kendler, K. S., Gruenberg, A. M., and Strauss, J. S. (1981): An independent analysis of the Copenhagen sample of the Danish adoption study of schizophrenia. *Arch. Gen. Psychiatry*, 38:973–987.
13. Kety, S. S., Rosenthal, D., and Wender, P. H. (1978): Genetic relationships within the schizophrenia spectrum: Evidence from adoption studies. In: *Critical Issues in Psychiatric Diagnosis*, edited by R. L. Spitzer and D. F. Klein, pp. 213–223. Raven Press, New York.
14. Kety, S. S., Rosenthal, D., Wender, P. H., and Schulsinger, F. (1968): The types and prevalence of mental illness in the biological and adoptive families of adopted schizophrenics. In: *The Transmission of Schizophrenia*, edited by D. Rosenthal and S. S. Kety, pp. 345–362. Pergamon Press, Oxford.
15. Kety, S. S., Rosenthal, D., Wender, P. H., Schulsinger, F., and Jacobsen, B. (1975): Mental illness in the biological and adoptive families of adopted individuals who have become schizophrenic: A preliminary report based on psychiatric interviews. In: *Genetic Research in Psychiatry*, edited by R. R. Fieve, D. Rosenthal, and H. Brill, pp. 147–165. Johns Hopkins University Press, Baltimore.
16. Kety, S. S., Rosenthal, D., Wender, P. H., Schulsinger, F., and Jacobsen, B. (1978): The biological and adoptive families of adopted individuals who became schizophrenic: Prevalence of

mental illness and other characteristics. In: *The Nature of Schizophrenia*, edited by L. C. Wynne, R. L. Cromwell, and S. Matthysse, pp. 25–37. John Wiley & Sons, New York.
17. Mech, E. V. (1973): Adoption: A policy perspective. In: *Review of Child Development Research, Vol. 3, Child Development and Social Policy*, edited by B. M. Caldwell and H. N. Ricciuti, pp. 467–508. University of Chicago Press, Chicago.
18. Mendlewicz, J., and Rainer, J. D. (1977): Adoption study supporting genetic transmission in manic depressive illness. *Nature*, 268:327–329.
19. Rosenthal, D. (1975): Discussion: The concept of subschizophrenic disorders. In: *Genetic Research in Psychiatry*, edited by R. R. Fieve, D. Rosenthal, and H. Brill, pp. 199–215. Johns Hopkins University Press, Baltimore.
20. Rosenthal, D., Wender, P. H., Kety, S. S., Schulsinger, F., Welner, J., and Ostergaard, L. (1968): Schizophrenics' offspring reared in adoptive homes. In: *The Transmission of Schizophrenia*, edited by D. Rosenthal and S. S. Kety, pp. 377–391. Pergamon Press, Oxford.
21. Schulsinger, F. (1972): Psychopathy: Heredity and environment. *Int. J. Ment. Health*, 1:190–206.
22. Schulsinger, F., Kety, S. S., Rosenthal, D., and Wender, P. H. (1979): A family study of suicide. In: *Origin, Prevention and Treatment of Affective Disorders*, edited by M. Schou and E. Stromgren, pp. 277–287. Academic Press, New York.
23. Spitzer, R. L., Endicott, J., and Gibbon, M. (1979): Crossing the border into borderline personality and borderline schizophrenia: The development of criteria. *Arch. Gen. Psychiatry*, 36:17–24.
24. Teasdale, T. W. (1979): Social class correlations among adoptees and their biological and adoptive parents. *Behav. Genet.*, 9:103–114.
25. Wender, P. H., Rosenthal, D., and Kety, S. S. (1968): A psychiatric assessment of the adoptive parents of schizophrenics. In: *The Transmission of Schizophrenia*, edited by D. Rosenthal and S. S. Kety, pp. 235–250. Pergamon Press, Oxford.
26. Wender, P. H., Rosenthal, D., Kety, S. S., Schulsinger, F., and Welner, J. (1974): Cross-fostering: A research strategy for clarifying the role of genetic and experiential factors in the etiology of schizophrenia. *Arch. Gen. Psychiatry*, 30:121–128.
27. Wender, P. H., Rosenthal, D., Rainer, J. D., Greenhill, L., and Sarlin, B. (1977): Schizophrenics' adopting parents: Psychiatric status. *Arch. Gen. Psychiatry*, 34:777–784.
28. Wynne, L. C., Singer, M. T., and Toohey, M. L. (1976): Communication of the adoptive parents of schizophrenics. In: *Schizophrenia 75: Psychotherapy, Family Studies, Research*, edited by J. Jorstad and E. Ugelstad, pp. 413–451. University of Oslo Press, Oslo.

Use of the Danish Adoption Register for the Study of Obesity and Thinness

*Albert J. Stunkard, **Thorkild Sørensen, and **Fini Schulsinger

*Department of Psychiatry, University of Pennsylvania, Philadelphia, Pennsylvania 19104; and **Psykologisk Institut, Kommunehospitalet, Copenhagen, Denmark

During the past 3 years, we have extended the use of the Danish Adoption Register beyond its traditional application to mental illness and applied it to the study of somatic disease, specifically, the genetics of human obesity and thinness.

The opportunity of using the Adoption Register for the study of the genetics of human obesity is a particularly timely one, for our understanding of this topic is, unfortunately, very limited, and the available data are contradictory. Clearly, obesity can be highly heritable. Millennia of experience have shown that the fat content of farm animals can be genetically determined. More recently, several forms of genetic obesity in rodents have indicated how genetic mechanisms of obesity may be expressed. But when we turn to human obesity, we enter an area of vast ignorance.

Three twin studies have suggested that human obesity may be highly heritable (6), but recent criticism of the twin study method renders these suggestions less compelling. Two adoption studies, both of small size and limited to children, have yielded diametrically opposite results. The data of Garn et al. suggest that human obesity is almost exclusively of environmental origin (7–9). The data of Biron et al., on the other hand, indicate that heredity accounts for almost all of the variance (1,2). Brooks et al. have concluded that genetics plays the major part in determining human obesity; reanalysis of their data suggests the opposite (3).

This rampant confusion suggests that the Danish Adoption Register can play an invaluable part in bringing order out of chaos, and we are trying to do just that. Furthermore, data collected for the study of the genetics of human obesity should also make it possible to study the heritability of human thinness, a heretofore totally unexplored area with profound implications for our understanding of the regulation of body weight. There are, however, problems in this study.

The first problem that we encountered in use of the Register for the study of obesity was that of ascertainment of obese (and thin) probands. It was not possible to utilize the method of ascertainment of the earlier studies of schizophrenia, alcoholism, and psychopathy for which registers were available. There is no Obesity Register in Denmark, and efforts to ascertain obese probands by means of written records were unsuccessful. Inpatient hospital records were perfectly adequate but did not permit complete ascertainment; outpatient records permitted complete ascertainment but were insufficiently accurate for research purposes.

To deal with this problem, we adopted a totally new strategy of ascertainment of the Danish adoptees, and this strategy has provided information about a number of other chronic conditions. The strategy was a mail questionnaire directed to 4,621 adoptees living in Copenhagen. In addition to questions about height and current and maximum weight, we also asked about the presence of coronary heart disease, high blood pressure, diabetes, ulcer, asthma, hay fever, and history of smoking. One of the benefits of research in Denmark is the cooperativeness of the population, and the response to our mail survey was no exception. A response rate of 79% has identified 3,651 adoptees with the chronic conditions described above. These adoptees are aged from 34 to 57 years and thus are entering the age range when these disorders assume great significance. It is our hope that this survey may provide investigators of the genetics of these disorders with a resource as useful in its own way as was the construction of the original Register for investigation of the genetics of psychiatric disorders.

Although the problem of ascertainment was solved by the use of a mail survey, the survey presented a second problem. The most prominent question was just how accurate are self-reports of body weights and whether the inaccuracy may not be greatest among precisely those adoptees in whom we are most interested—the obese. This problem led us to undertake a study of the accuracy of self-reported weights which has left us with the gratifying conclusion that self-reported weights are surprisingly accurate and that obesity does not unfavorably influence the accuracy of self-reports (11). Since the validity of the study of the heritability of human obesity and thinness depends on the accuracy of self-reported weights, this study will be described in some detail.

Data were collected in the United States and in Denmark. American subjects were 550 men and women from seven different sites; Danish subjects were 750 men and women from one site—applicants for medical insurance in Copenhagen. The American sites were selected to meet three criteria: the inclusion of a broad range of persons, the inclusion of a broad range of reasons for weighing them, and the fact that they did not know that they would be weighed. A major source of bias in self-reports could be the knowledge that one was to be weighed shortly after the report. Accordingly, sites were selected to minimize such indications. The three criteria were met at all sites.

Analysis of the data revealed that the association between reported and measured weights was surprisingly strong. Figure 1 shows this relationship, for all American sites combined in Fig. 1A, and for a single site (Fig. 1B), a union treatment program, chosen as representative of all sites. Figure 1B gives a clearer impression of the degree of association because there is less multiple counting of each point on the figure than was the case in Fig. 1A. The correlation coefficient for all sites combined was 0.992; that for the union site was 0.995.

Another method of assessing the accuracy of reported weights is the distribution of the error among subjects. The reported weights of 51% of subjects were within

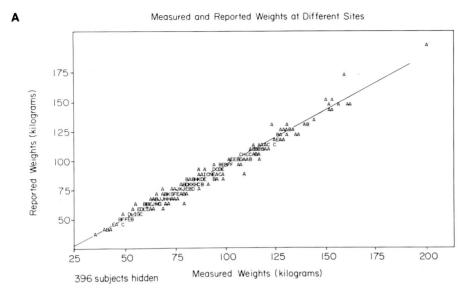

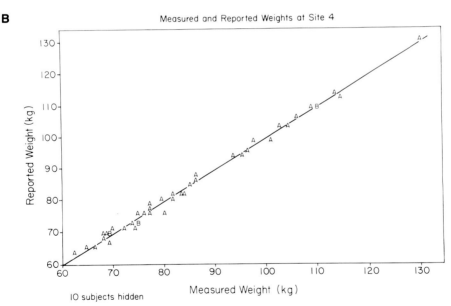

FIG. 1. Relationship between measured weights *(horizontal axis)* and reported weights *(vertical axis)* for all 550 subjects **(A)** and for the 66 subjects at site four **(B)**. *Letters* are used to represent data points: "A" represents one subject, "B" represents two subjects, and so forth. As a result, 72% of the data points in **A**, primarily the more highly correlated ones, are hidden, resulting in the appearance of greater variability than actually exists. In **B**, only three data points are hidden, and the diagram shows more accurately the strong association between measured and reported weights. Note that the slope of the regression line is close to 1.

1 kg of their measured weight, and the reported weights of 68% were within 2 kg of their reported weights.

A more detailed statistical analysis was carried out with the use of a simple linear regression model. It showed that age, height, sex, and interaction among them did not influence the accuracy of reports. The one factor that did influence accuracy was the absolute weight (but not degree of overweight) which was associated with a slight underestimation, averaging 1.2 kg with a standard deviation of 3.1 kg. This means that a weight of 80 kg was underestimated to the same extent by a tall man of normal weight and by a short woman who was severely overweight, a finding that provides assurance that the variable under consideration—obesity—did not influence the accuracy of estimation.

The association between reported and measured weights was not quite as strong in the Danish data as it was in the American data. This difference was a result of a somewhat greater degree of underestimation of body weight that was confined to older Danish women.

These reports confirm and extend the findings of four other investigations of the accuracy of self-reported weights (4,5,10,12) and provided encouragement to press ahead with the use of the Adoption Register for studying the heritability of obesity and thinness.

A third methodological problem that we faced in this study was determining the weight status of the adoptive and biological parents of the adoptees. At first, this problem appeared to be almost insurmountable. Only half of the parents are still alive, and some of them, aged about 54 to 77 years, were presumably no longer able to provide accurate information about their past weights. The limitations of both inpatient and outpatient records for verifying body weights had already been revealed by our earlier search for information about the adoptees.

After considerable exploration, we devised a novel method of determining the weight status of parents of the adoptees. It consisted of the use of a series of nine silhouettes ranging from very thin to very fat. Figures 2A and 2B depict the silhouettes of men and women that are being used. One method of use is for children to select the silhouette that most clearly resembles the body build of their parents. This method provides information about all adoptive parents and about those biological parents who have children who are siblings or half siblings of the adoptees, a total of over 50% of the biological parents. It does not provide information about biological parents who had no subsequent children.

The second method of determining the weight status of the biological parents is by asking them to select the silhouette that most closely resembles their usual weight and that of their spouse. This status is then compared with that of the adoptive parents determined in the same manner, that is, by their selection of their own silhouette. Although only 50% of parents are still living, mortality has been higher among males, and about 65% of the mothers are still alive. When their reports are added to those of the husbands whose wives are no longer living, we expect that information can be obtained from about 70% of the parents themselves.

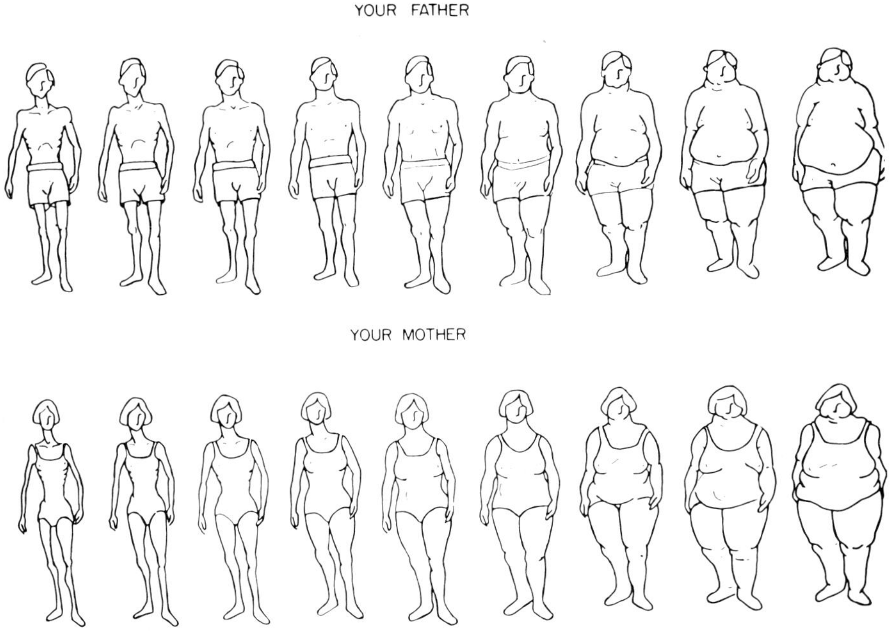

FIG. 2. Nine silhouettes of men and women ranging from very thin to very fat. Adult offspring pick the silhouette that most closely corresponds to the usual and to the most obese silhouette of their fathers and mothers. Validation studies show a high degree of correspondence between these choices and percentage overweight of the parents as determined by measurements of height and weight.

The fourth problem in determining the weight status of the parents was assessing the validity of the silhouette method. After many unsuccessful efforts, we finally succeeded by the use of a unique resource—the study of the health status of the population of Tecumseh, Michigan. The health status of the Tecumseh population has been measured by surveys carried out every 3 years for a period of more than two decades. Drs. Victor Hawthorne and Millicent Higgins of the Department of Epidemiology of the School of Public Health of the University of Michigan made it possible for us to have access to a sample of 1,000 adult residents of Tecumseh at the time of their most recent survey. The sample contained both men and women of approximately the same age as the adoptees. Members of the sample were asked to pick silhouettes that corresponded most closely to those of their mothers and fathers. Then, the Tecumseh data were searched for parents of members of the samples. Records of a total of 350 parents of members of the sample were identified, and heights and weights of these parents were determined. Analysis of these data is still underway, but it is already apparent that the silhouette method is surprisingly accurate (12a). For example, there is a monotonic increase in percentage overweight from the first to the ninth silhouette for both men and women. It may prove desirable to collapse some of the nine weight categories, such as the two extreme categories which were used only infrequently. But even the availability of only four or five categories of parental weight status would still permit a greater degree of discrimination than has been available for studies of psychopathology that have been forced to use categories of illness versus no illness.

After resolving these four problems, we have proceeded with the study of the heritability of obesity and thinness. A sample of 800 adoptees was drawn, half men and half women. Within each sex were 100 persons constituting the most obese, the next most obese, the thinnest, and those about the median in percentage overweight. Our prediction is that the most obese persons will have stronger genetic determinants than those who are less obese, in whom environmental determinants will play a more important part. These probands have been resurveyed to try to achieve a greater measure of accuracy of the self-reports. We are now proceeding to study the adoptive and biological parents of adoptees according to the procedures described above. The final statistical analysis will be under the direction of Dr. William J. Schull, Professor of Population Genetics at the University of Texas Health Sciences Center at Houston. It will consist of traditional analyses of variance, path analytic techniques, and the family set method developed by Dr. Schull.

ACKNOWLEDGMENT

The authors acknowledge with thanks Seymour Kety, M.D., David Rosenthal, Ph.D., and Paul Wender, M.D., for access to the data of the Danish Adoption Register.

REFERENCES

1. Biron, P., and Mongeau, J. G. (1978): Familial aggregation of blood pressure in the young and its components. *Pediatr. Clin. North Am.*, 25:29–33.
2. Biron, P., Mongeau, J. G., and Bertrand, D. (1977): Familial resemblance of body weight and weight/height in 374 homes with adopted children. *J. Pediatr.*, 91:555–558.
3. Brooks, C. G. D., Huntley, R. M. C., and Slack, J. (1975): Influence of heredity and environment in determination of skinfold thickness in children. *Br. Med. J.*, 2:719–721.
4. Charney, E., Goodman, H. C., and McBride, M., et al. (1976): Childhood antecedents of adult obesity. Do chubby infants become obese adults? *N. Engl. J. Med.*, 295:6–9.
5. Coates, T. J., Jeffery, R. W., and Wing, R. R. (1978): The relationship between persons' relative body weights and the quality of food stored in their homes. *Addict. Behav.*, 3:179.
6. Foch, T. T., and McClearn, G. (1980): Genetics, body weight and obesity. In: *Obesity*, edited by A. J. Stunkard, pp. 48–71. W.B. Saunders, Philadelphia.
7. Garn, S. M. (1976): The origins of obesity. *Am. J. Dis. Child.*, 130:465–467.
8. Garn, S. M., and Clark, D. C. (1975): Nutrition, growth, development and maturation: Findings from the ten-state nutrition survey of 1968–1970. *Pediatrics*, 56:306–319.
9. Garn, S. M., and Clark, D. C. (1976): Trends in fatness and the origin of obesity. *Pediatrics*, 57:443–456.
10. Perry, L., and Leonard, B. (1963): Letter to the editor. *J.A.M.A.*, 183:808.
11. Stunkard, A. J., and Albaum, J. M. (1981): The accuracy of self-reported weights. *Am. J. Clin. Nutr.*, 34:1593–1600.
12. Wing, R. R., Epstein, L. H., and Ossip, D. J., et al. (1979): Reliability and validity of self-report and observers estimates of relative weight. *Addict. Behav.*, 4:133.
12a. Sørensen, T. I. A., Stunkard, A. J., Teasdale, T. W., and Higgins, M. (1982): The accuracy of reports of weight: childrens' recall of their parents' weight 15 years earlier. *Int. J. Obesity. (in press)*.

Genetics of the Major Psychoses

Elliot S. Gershon

Section on Psychogenetics, Biological Psychiatry Branch, National Institute of Mental Health, Bethesda, Maryland 20205

The most powerful evidence that genetic transmission is taking place in psychiatric illness is offered by adoption studies in which psychiatric illness occurs more often in adoptees who have biological relatives with the same illness than in those who do not. A second type of evidence is found in twin and family studies in which more closely related individuals (to the proband) are more likely to show illness, particularly monozygotic twins who are expected to show greater concordance than dizygotic twins. For schizophrenia and the affective disorders, these two types of evidence support genetic transmission (Table 1). Furthermore, there is little crossover between the major diagnostic groupings (affective disorder and schizophrenia) in these studies, implying genetic specificity of each disorder. This evidence is strongly supportive of genetic transmission, although other events occurring before a few months of age (the latest age when adoption occurred) could conceivably be confounded with genetic influences.

The desirable next steps in genetic exploration are to identify the mode of genetic transmission (which requires an answer to the question of whether we are dealing with one or several genetic entities within each illness) and to identify the transmitted pathophysiologic abnormality. Methods currently in use to resolve these questions include genetic analyses of pedigree data, linkage marker studies, and biological investigations of patients and their relatives. The current status of these studies in affective disorders and schizophrenia and some general issues in psychiatric genetic studies are reviewed here.

METHODOLOGIC ISSUES

Diagnostic Reliability

The development of explicit diagnostic criteria for psychiatric disorders has led to acceptable reproducibility of diagnosis when the same individual is blindly examined by different interviewers/diagnosers (70,71,94,155). The same studies have shown that establishing a lifetime diagnosis from interview data is feasible in the affective disorders. Earlier studies based on family history taken only from the

TABLE 1. *Psychiatric disorders with family transmission data*

	Adoption transmission	Twin concordance MZ > DZ	Family (first-degree relative) prevalence exceeds population prevalence	References or review	Comments
Affective illness	Yes	Yes	Yes	Nurnberger and Gershon (120)	More severe forms heritable
Schizophrenia (chronic)	Yes	Yes	Yes	Kinney and Matthysse (88)	
Panic disorder/agoraphobia	Not studied	Yes	Yes	Pauls et al. (125,126) Crowe et al. (32) Carey and Gottesman (24)	
Alcoholism	Yes (males only)	Yes	Yes	Bohman (17) Goodwin et al. (65,66) Schukit (144) Seixas et al. (146) Slater and Cowie (150) Winokur et al. (180)	
Antisocial personality/criminality	Yes (2/3 studies)	Yes (most studies)	Yes	Crowe (31) Bohman (17) Hutchings and Mednick (76) Witkin et al. (182) Hook (75) Guze et al. (69) Christiansen (28)	XXY and XYY karyotypes rare but associated with criminality

proband had tended to underestimate prevalences in relatives (179), so that family studies in which all available living relatives are studied have become standard. However, between studies widely different frequencies have been reported for illness in relatives of patients, and these appear to be attributable to differences in criteria definition (51).

Mode of Inheritance and the Problem of Heterogeneity

The distribution of illness in relatives of patients with schizophrenia or affective disorders does not approximate the proportions expected in single-locus Mendelian inheritance. Several possible explanations (not mutually exclusive) may account for this. There may be no single-locus inheritance operating—the inheritance may be multifactorial, with many indistinguishable loci having alleles that contribute additively to genetic vulnerability, and with numerous environmental factors acting similarly. Also, there may be reduced penetrance of alleles for illness because of variable age of onset or other factors. These explanations, and others with more complex modes of inheritance, assume that the disease entity is genetically homogeneous, although they can allow for random occurrence of illness in persons not genetically predisposed (phenocopies). But a more complex universe might exist, including situations in which the same clinical manifestations are produced by genetically and biologically separate and independent entities (heterogeneity).

Genetic heterogeneity within affective illness or schizophrenia has been argued on the basis of clinical heterogeneity and different prevalences of illness in relatives (23,88,91,128), but such differences can be compatible with a single underlying type of liability for the different clinical forms, as demonstrated by the mathematical models of Reich et al. (137). This applies particularly to the unipolar and bipolar types of affective disorder for which mathematical models demonstrate that genetic independence of the two forms of disorder is not required by much of the family study data, even though the relatives of each type of patient have different frequencies of illness, since there is overlap of types of illness found in relatives of each type of proband (52,55).

How, then, can heterogeneity be demonstrated? Clinical evidence by itself can demonstrate heterogeneity when there is only nonsignificant overlap of illness types in relatives of probands with the proposed separate illnesses. Heterogeneity would also be suggested when no genetic transmission model fits pedigree data for an illness but clinical or biological characteristics of the patients allow a subgroup to be identified within which the transmission fits one model well and other models can be rejected. The most convincing demonstration of heterogeneity is when a specific gene product or gene locus can be associated with one but not another form of illness. For example, where linkage is present in some but not other pedigrees of the same illness and there are no reasons to doubt that the pedigrees were collected under comparable circumstances, heterogeneity is clearly present. Or if an abnormal gene product is found in some but not other illness pedigrees and it segregates with the illness in the pedigrees where it is found, heterogeneity would also be demonstrated.

Vulnerability Markers

A biological finding can serve as such a marker of genetic vulnerability even where the precise gene product cannot be specified or where its mode of genetic transmission is not known, so long as it is heritable and segregates with the illness in pedigrees where it is found (140). This kind of vulnerability marker can be distinguished from a linkage marker. Linkage of an illness to a single locus marker demonstrates that the illness is transmitted at a single locus and allows identification of persons with nonpenetrant genotype for illness. This would have great value for genetic counseling and for biological research. But it says nothing about the pathophysiology of the illness, because any allele occupying that locus will serve as a marker. When a particular allele, or its gene product, is associated with the illness in pedigrees (except for linkage disequilibrium) such that persons who do not have the allele will not develop the illness, the pathophysiology of illness must be related to that gene product. This is a powerful tool for pathophysiological research. Heterogeneity of illness can be identified by vulnerability factors associated with some but not other patients, provided that in those patients who have it, the factor is associated with illness in their relatives.

Trait Versus State Phenomena

The major psychoses have a variable age of onset, and the affective disorders are also generally episodic, so that even in persons who will suffer the illness there will be significant parts of their lives with no clinically detectable signs of illness. A marker could conceivably be manifest only during illness, but it would be beyond the limits of practicality to demonstrate such a marker in a pedigree study, as discussed elsewhere (140). We do not consider such findings in the present chapter.

SCHIZOPHRENIA

Genetic Analyses of Family Study Data

In schizophrenia, there is only a meager amount of family study data with systematically applied diagnostic criteria. The data of Kallmann (81,83), collected about 40 years ago, remain the most extensive family study data available for analysis. The diagnoses were confirmed by intensive review of the case material after Kallmann's death by Slater, Shields, and Gottesman (149) and are generally accepted as valid.

These data have recently been analyzed (43) using the pedigree analysis models of Elston and Stewart (44). Possible heterogeneity of Kallmann's two clinical types of schizophrenia, nuclear and peripheral, was tested by looking for differences in the prevalences of diagnoses in relatives (154). Since there were differences, data from each type of patient were analyzed separately. But the differences found between the two forms could be compatible with a shared underlying liability, so separate analyses may not have been necessary. Pedigree analyses rejected Mendelian (single-locus) transmisstion for each form of illness. M. Baron analyzed one of Kallmann's data sets using a multiple-threshold analysis of prevalences in relatives (137) and considered nuclear and peripheral schizophrenia to be genetically more and less severe forms of the same disorder. He found that single-locus inheritance and multifactorial inheritance were not tenable as the mode of inheritance (137a).

Debray et al. (33) collected data and then analyzed 25 pedigrees of chronic schizophrenics. Systematic diagnostic criteria and extensive examination of relatives do not appear to have been used. Several models appeared to fit the data, including one-, two-, and four-locus models (the last being functionally similar to a multifactorial model). The authors did not feel the data were well enough conditioned to find any of these models definitively acceptable.

Clinical Entities Related to Schizophrenia

The core clinical syndrome of schizophrenia includes a chronic illness without return to premorbid level of social adjustment and delusions or hallucinations or other formal thought disorder without accompanying perplexity or disorientation (2,46,86). In the adoption study of Kety et al. (86), other disorders were found in relatives, including borderline schizophrenia and inadequate personality. Other investigators recognize similar schizophrenia "spectra" (74,82,84,142). But acute schizophrenia [which may be considered equivalent to schizoaffective disorder (48,132)] is generally not found in relatives of schizophrenics. Family studies of patients with schizoaffective disorders find little schizoaffective illness in the relatives but distinctly elevated frequencies of bipolar manic–depressive illness and some elevation of (chronic) schizophrenia, as reviewed elsewhere (4,120,132,163). Identical co-twins of schizoaffective patients tend to have the same disorder (95). Possibly, specific genetic factors lead to schizoaffective illness, in view of the twin concordance, but these factors are most often superimposed on the genetic diathesis for bipolar illness, since this is the most consistently found disorder in relatives of schizoaffective patients (120). Fowler (48) and Pope and Lipinski (132) have reviewed biological and clinical evidence supporting the same conclusions.

Reductions in IQ at age 7 years are found in children of schizophrenics who suffered perinatal distress but not in children of controls suffering similar distress (139,141). Other evidence of excess perinatal morbidity and mortality in children of schizophrenics was noted in the same study. Mental retardation appears increased in adopted-away offspring of schizophrenics (73) and in nonadopted offspring (111). An increased death rate in adult biological relatives of schizophrenics in the Kety-Rosenthal adoption study (86,142) was the result of suicide, homicide, accidental death, and sudden death of unknown causes (184). In view of all these findings, the phenotypic expressions of the schizophrenia diathesis can be considered to be considerably more inclusive than major and minor forms of schizophrenic psychosis.

Linkage Markers in Schizophrenia

A population isolate in northern Sweden was found to contain three large pedigrees segregating for schizophrenia (18). An association of GC 2-1 haplotype or GC 2 allele with schizophrenia was noted, but the Penrose sib-pair linkage test was not significant. More powerful multigenerational pedigree analyses for linkage have not been performed. A later study in the same area of Sweden did not suggest a population association (12).

Human leukocyte antigens have not been consistently associated with schizophrenia, since several studies have each reported different positive associations (41,49,77,107,130). A pedigree series showing schizophrenia linkage to the HLA locus has been reported, but I have found it difficult to confirm the diagnostic procedures used by Turner (164).

Single pedigrees in which there is association or linkage of schizophrenia to homocystinuria (27) or to albinism (7) have been reported. Whether these relationships are present in any significant number of schizophrenics is unknown. If tests for the heterozygous state for either of these two recessive metabolic disorders were available, association and linkage studies could be performed in an adequate number of informative pedigrees using a strategy recently described (93).

Klinefelter's syndrome (XXY chromosomal pattern) is associated with increased incidence of psychoses which can appear schizophrenic, affective, or drug/alcohol related (117,153). Schizophrenic psychosis in persons with the XYY karyotype has also been reported (34).

Pathophysiological Hypotheses in Genetic Studies

Biological hypotheses abound in schizophrenia (11,99,169), but relatively few of these have been applied in genetic studies.

Enzymes of monoamine metabolism have been studied, stimulated by hypotheses of metabolic and receptor alterations in monoamine neurotransmitters in schizophrenia. Murphy and Wyatt (116) reported decreased platelet monoamine oxidase (MAO) activity in chronic and acute schizophrenia, and Wyatt et al. (185) showed concordance of MAO activity in monozygotic twins discordant for schizophrenia. These studies demonstrated that platelet MAO activity is under genetic control and that decreased activity is associated with schizophrenia but did not demonstrate that reduced activity is part of the genetically transmitted vulnerability to illness. What is required is a demonstration of segregation of MAO activity with schizophrenia within pedigrees (140). On this point the evidence is very meager despite the numerous investigations which have often but not always confirmed the differences between patients and controls [which have been reviewed elsewhere (22)].

In view of the heterogeneity that may exist with regard to low platelet MAO and schizophrenia as implied by the inconsistency of the reported studies, it is necessary to study the ill and well relatives of schizophrenics with low platelet MAO activity. Ask et al. (6) studied a Swedish pedigree identifed by Book et al. (18), with eight schizophrenics and 27 well relatives studied. Lower MAO activity was found in ill than in well pedigree members with certain of the substrates used, as expected by the hypothesis, but problems in the reliability of the assay make this a most tentative finding, as noted by the authors. Studies of well relatives of schizophrenics (16,134) are unfortunately noncontributory to this issue. Baron and Levitt (9) found higher frequency of illness in relatives of low-platelet-MAO schizophrenics than in relatives of schizophrenics with higher MAO activity, but MAO activity in the relatives themselves was not studied. Belmaker et al. (14), on the other hand, found no difference in platelet MAO activity between patients with and without a family history. I conclude that this hypothesis, which has played an important role in schizophrenia research for nearly a decade, has not yet been adequately tested.

The genetics of platelet MAO activity and its possible polymorphisms have also been studied. In normal twins and in a large series of pedigrees with affective illness, MAO activity is clearly familial (56,118,181), but single-locus genetic models (either autosomal or X-linked) do not fit well (56). Bridge et al. (20) have identified thermolability in an increasing proportion of individuals as they age, becoming quite common in persons over 50. A rare polymorphism for thermolability in young persons was identified in one pedigree, but this was not found in schizophrenics. Possible variations in enzyme kinetics and substrate specificities have been noted, but they have not led to identifiable polymorphisms (15,101,102,186).

Serum creatine phosphokinase (CPK) has been studied extensively by Meltzer and his colleagues (98,100), and they have found that there are elevated levels of this enzyme during the course of hospitalization in many patients with affective psychoses and in many with schizophrenic psychoses but not in disturbed nonpsychotic psychiatric patients. They also found abnormal muscle biopsies and increased axonal branching of the motor neuron in a high percentage of patients with psychoses of either type. Each of these abnormalities was found to a greater extent among the first-degree relatives of the patients than among controls. The increased psychiatric vulnerability of the relatives who have this neuromuscular abnormality or CPK elevation has not yet been demonstrated. A review of this and other biological factors that may be associated with schizophrenia has recently appeared (99). No factor has been successfully related to genetic transmission of the illness.

Baron and Witz and their colleagues (10,183) found a globulin in the sera of schizophrenics that bound to human brain tissue was not found in sera of controls. Presence of this binding was found in relatives of schizophrenics. Information was presented on relatives of those patients with positive sera (Table V of ref. 10), but tissue-binding globulins do not distinguish well from ill relatives (hypergeometric distribution, $p = 0.3$). I would disagree with the authors who believe that they have satisfied criteria for a genetic marker for schizophrenia, since the evidence for independent assortment is so weak.

AFFECTIVE DISORDERS

The Affective Spectrum

Bipolar and Unipolar Affective Disorders

The classic manic–depressive patient cycles through episodes of both mania and depression, with each type of mood disorder becoming severe enough at times to impair his major life role and require treatment. Symptom-free intervals are the rule. Chronicity of mood disturbance is generally not seen. Leonhard (91) and Leonhard et al. (92) called patients with both mania and depression bipolar (BP) and hypothesized that they are genetically distinct from patients with recurrent depressions, who are now called unipolar (UP). This hypothesis has been repeatedly tested by comparing the prevalence of illness in relatives for each type of patient, as reviewed elsewhere (55,60). Although BP–UP differences in illness in relatives are repeatedly found, there is a considerable overlap in nearly all studies, particularly in the consistent finding of high rates of UP illness in relatives of BP patients. Bipolar relatives of UP patients are less often found (Table 2).

This overlap is compatible with genetic models of BP and UP disorders in which BP is a genetically more severe and UP is a less severe form of the same disorder (55,151,161,162). These models (87,137) predict that BP patients would have more BP relatives and more UP relatives than do UP patients, as is generally found. Twin studies of affective disorder show a tendency to concordance of polarity, but 19% of identical twins who are concordant for affective disorder consist of one BP and one UP twin (129). Biological and pharmacological studies, reviewed elsewhere (60), have shown similar overlap in the few instances in which statistical differences between BP and UP patients have been reported in more than one study. In that review, my colleagues and I concluded that the most consistent UP–BP differences were in clinical pharmacologic response. These were the antidepressant response

TABLE 2. *Lifetime prevalence of affective illness in first-degree relatives*

Reference	Number at risk	Morbid risk (%)	
		BP	UP
Bipolar probands			
Perris (127)	627	10.2	0.5
Angst (3)	161	4.3	13.0
Winokur and Clayton (178)	167	10.2	20.4
Goetzl et al. (63)	212	2.8	13.7
Helzer and Winokur (72)	151	4.6	10.6
Mendlewicz and Rainer (110)	606	17.7	22.4
James and Chapman (78)	239	6.4	13.2
Gershon et al. (57)	341	3.8	6.8
Smeraldi et al. (151)	172	5.8	7.1
Johnson and Leeman (80)	126	15.5	19.8
Petterson (131)	472	3.6	7.2
Taylor and Abrams (161)	601	4.8	4.2
Unipolar probands			
Angst (3)	811	0.3	5.1
Perris (127)	684	0.3	6.4
Gershon et al. (57)	96	2.1	11.5
Smeraldi et al (151)	185	0.6	8
Taylor and Abrams (161)	96	4.1	8.3

to lithium which is more often found in bipolar patients [although not in one recent study (122)] and in the induction of hypomania or mania by tricyclic antidepressants and by L-DOPA. But the degree of BP–UP overlap in these and in the other reviewed findings suggested that if there is an underlying biologic dichotomy, it does not coincide well with the UP–BP clinical dichotomy.

We can consider two general classes of hypothesis that might account for phenotypic overlap of UP and BP illness, as discussed elsewhere (60). A common genetic diathesis for both UP and BP illness in which BP illness occurs in individuals with a greater genetic and environmental load is one possibility. In this case, biologic and pharmacologic differences would be greatest between BP patients and controls, with UP patients falling in between on any given variable. An alternative hypothesis of at least two distinct genotypes, one of which could be phenotypically variable (manifest as either BP or UP illness), allowing for UP–BP heterogeneity in twin pairs and families could also be considered. In this case, there would be no *a priori* expectations on the biologic differences between UP and BP illness; however, we would anticipate at least two kinds of UP persons—one type who is genetically (and biologically) related to BP illness and others who are not. In either case, we could predict biologic and pharmacologic variables that distinguish between subtypes.

Several investigators (47,89,103,176) have suggested that a subgroup of UP patients may be genetically and biologically related to BP patients. The presence of mania in the family history of UP patients may be a distinguishing characteristic between UP patient subgroups. This hypothesis has not been adequately tested. Kupfer et al. (89) have reported on two groups of patients, one of which they state is related to BP patients in terms of positive response to lithium carbonate, personality variables, and family history features (history of mood swings in relatives).

Winokur and colleagues (13,176,177) have attempted to define different forms of unipolar illness by family history, with validation coming from linkage studies in pedigrees and biological and pharmacological characteristics of patients. They first considered all (UP) patients with a first-degree relative with BP illness to be a variant of BP illness. They divided the remaining patients into pure depressive disease (PDD) (first-degree relative with affective illness), depressive spectrum disease (DSD) (first-degree relative with alcoholism or antisocial personality), and sporadic depressive disease (SDD) with no family history of depression, antisocial personality, or alcoholism. About 25% of UP patients are unclassifiable by this method. Attempts to delineate the groups by clinical features have not been strikingly successful (13), although the DSD had more unstable personalities. Schlesser et al. (143) reported that PDD but not DSD or SDD showed failure of dexamethasone suppression of cortisol, but this was not confirmed by Carroll et al. (26). Linkage studies have been very weakly suggestive of DSD linkage to haptoglobin and group-specific component (158,159).

Subdivisions of BP illness

Dunner et al. (38) describe two BP types: BP I (history of mania) and BP II (history of hypomania). This group reported differential responses to L-DOPA and differences in platelet MAO and 17-OH corticosteroid excretion between BP I and BP II groups. The BP II patients had significantly higher suicide rates. The age of onset for BP II illness is between the younger onset of BP I and the later onset of UP illness. Gershon et al. (54) noted that the antidepressant response to L-DOPA previously reported had occurred in BP II ($N = 5$) patients only. In our own recent data (159a), BPI and BP II patients have very similar prevalences of BPI, BPII, and UP disorders in their relatives, suggesting that they are not genetically distinct.

Schizoaffective Disorder

Schizoaffective disorder represents a taxonomic puzzle, but it is more frequently associated with BP illness in relatives than with any other disorder, as discussed above. In our own ongoing family study (159a), 11 families were studied because a proband was thought to be bipolar, but review of information from the patient, family members, and medical records led to a research diagnosis of schizoaffective. This small group of families had the highest prevalences of affective disorders and of schizoaffective disorders in relatives of all the patient groups studied. One can conceive of these patients, who meet RDC criteria (155) for BP illness but also have the persistent psychotic features required for schizoaffective diagnosis, as having a particularly virulent form of affective disorder, both clinically and genetically.

For biologic and genetic modeling studies of affective disorders, it appears reasonable to assume that whatever the genetic heterogeneity of schizoaffective disorders (4,48,120,132), those cases that occur in pedigrees of bipolar patients represent a variant of affective disorder.

Milder Depressions

The findings suggesting genetic transmission in affective disorders (Table 1) were found in studies of hospitalized or severely disabled patients and relatives. Perris (128) required three episodes with hospitalization or medical treatment, whereas the recent NIMH collaborative study using the Research Diagnostic Criteria (155) would accept (at a minimum) one episode of 2 weeks of depressive mood, four symptoms, and mild social or familial functional impairment. When depression is defined as a more severe disorder, with hospitalization or incapacitation required before diagnosis is made, it is a relatively uncommon disorder with 2½ to 6% lifetime prevalence in the population, having a very strong familial concentration that fits genetic models for either single-gene or multifactorial transmission (55). When defined according to the broader conception of the Research Diagnostic Criteria, the population lifetime prevalence is estimated at 21% (172), and genetic transmission was excluded in a recent careful analysis of family study data (135). The familial influence appeared to be a sibling–sibling cultural effect. In view of this, the milder depressions found in increased frequency in families of patients with major affective disorders (52) cannot be well accounted for. They may be variant expressions of an illness-producing genotype, or they could be part of a familial culture that develops around the presence of severe affective disorder.

Similarly, children (ages 6–16) of depressed inpatients were noted to have more depressive symptoms than children of nonhospitalized normal controls in recent studies (96,174). But children of inpatients with medical diagnoses showed similar depressive symptoms in the later work of D. H. McKnew, Jr. et al. (*unpublished data*, Welner et al., McKnew et al.), which implies no genetic basis to the symptoms.

Other Syndromes

Alcoholism is increased in families of patients with affective disorders, as reviewed elsewhere (60,120). In bipolar families, this appears to be coexistence of two disorders, since alcoholism is increased only in relatives of bipolar patients who themselves have alcoholism (113). For unipolar illness, Winokur has proposed that there are two genetically distinct but clinically identical forms, based on the presence of alcoholism in relatives, as described above.

Other syndromes that may be associated with affective disorder are reviewed elsewhere (60,119). These include emotionally unstable character disorder, cyclothymic personality, anorexia, sociopathy, attention deficit disorder of childhood (hyperkinetic syndrome), and agoraphobia. Of greatest clinical significance is the increase in suicide in biological relatives of adoptees with affective disorder (85).

Mode of Transmission

Genetic models of affective disorders can be tested on family study data, pedigree data, and data on linkage markers. The heterogeneity proposed by Leonhard (91) can be tested as the hypothesis that UP and BP disorders are genetically independent. This hypothesis was applied to prevalences in relatives in three data sets (3,52,127), testing multiple-threshold models of autosomal single-locus and multifactorial transmission (55) (Table 3).

The models have as a null hypothesis that UP and BP represent different thresholds of the same underlying liability to illness. Both models fit the data of Gershon and Angst but not that of Perris. In the Perris data, only relatives with three or more episodes were considered unipolar. There are no unipolar relatives of bipolar patients, but there were about as many relatives of bipolar patients with "other depressions or suicide" in Perris' data as there were unipolar relatives of bipolar patients in the data of Gershon and of Angst. Thus, the models of shared underlying diathesis for UP and BP cannot be excluded. The data, however, do not fit these models strikingly well.

X-chromosome single-locus transmission in bipolar illness has been proposed by Winokur and his colleagues (179) on the basis of a relative lack of male-to-male transmission in family study data (55,60) and later on the basis of linkage studies discussed below. As applied in the linkage studies, this hypothesis is that, in the families of bipolar patients, both bipolar and unipolar illness are transmitted at a

TABLE 3. Threshold models of affective disorder[a]

Reference	Studies on UP and BP probands and their relatives		Studies on BP probands and their relatives: X-chromosome single locus transmission		
	Single major locus (autosomal); multiple thresholds	Multifactorial; multiple thresholds	One threshold for BP only	One threshold for both BP and UP	Two thresholds; BP and UP have separate thresholds
Gershon et al. (57)	Fits	Fits	Fits	No fit	No fit
Winokur et al. (179)	Insufficient data	Insufficient data	Fits	Fits	Fits
Mendlewicz and Rainer (110)	Insufficient data	Insufficient data	Insufficient data	No fit	Insufficient data
Angst (3)	Fits	Fits	Insufficient data	Insufficient data	Insufficient data
Perris (127)	No fit	No fit	Insufficient data	Insufficient data	Insufficient data

[a]Adapted from Gershon et al. (52) and van Eerdewegh et al. (165,166).

single X-chromosome locus. The data of Winokur et al. (179) fit this hypothesis, but other data sets did not (Table 3).

Pedigree and segregation analysis are more powerful analytic techniques than analyses of prevalences in relatives. Models are now applicable to disorders with variable penetrance, requiring maximum likelihood estimation of parameters such as penetrance of each genotype and gene frequency, comparing likelihoods of pedigrees under constraints of genetic transmission with appropriate null hypotheses (44,45). Using these techniques, Bucher (21) ruled out single autosomal locus or X-chromosome transmission in the data of Winokur et al. (179) and in other previously published studies. Goldin and Gershon (64) and Go et al. (62) were unable to fit single autosomal locus models to newly collected pedigrees series from bipolar patients at their respective research centers. Multifactorial models have not yet been applied to affective illness pedigree data to my knowledge.

It is not clear why the single-locus models do not fit in the pedigree analyses but did fit in the analysis of prevalences in relatives. The greater power of the pedigree methods may be correctly rejecting single-locus transmission, whereas multifactorial transmission would fit a pedigree analysis. The parameterization of the new models and their power to identify the correct transmission mode in simulated data are active research issues at this time, so that simply rejecting single-locus transmission of a genetically homogeneous bipolar disorder may be premature. Single-locus effects on a multifactorial background may exist (138). But the possiblity of biologic and genetic heterogeneity remains an alternative interpretation of the data.

Reich and Andreasen (135), analyzing 250 unipolar patients and their relatives in a segregation/path analysis model, found virtually no cross-sex transmissibility. That is, there was a relative absence of mother–son and father–daughter transmission. This is incompatible with any mode of genetic transmission. The strongest transmission seemed to be a sibling-to-sibling cultural effect. The diagnostic criteria for unipolar illness in this study include much milder illness than earlier family studies, so perhaps it is the milder depressions that may be nongenetically transmitted.

The failure to find a close fit of affective illness to a model of genetic transmission has led to attempts to identify genetically homogeneous subgroups using linkage methods or study of transmission of biological deficits in pedigrees.

Linkage Studies

Linkage or association of an illness with a chromosomal marker would establish the role of a single locus in the transmission of affective disorders and would be useful in identification of persons at high risk of illness for counseling and research purposes.

The human leukocyte antigen (HLA) system has been repeatedly studied for association to bipolar illness, and numerous controlled studies have found one or another HLA association, as reviewed elsewhere (60,120). No two controlled studies found the same antigen increase or decrease, even with precise ethnic matching of patients and controls (to control for HLA variance within populations). A linkage study in a pedigree series was clearly negative (160). On the other hand, Smeraldi et al. (152) found increased similarity of HLA haplotypes in sibs concordant for affective illness.

The HLA associations are unusual, since in linkage equilibrium no association is expected. The mechanisms of the numerous associations of certain HLA antigens with certain diseases is hypothesized to be by direct effects of specific HLA antigens, involving molecular mimicry of the HLA antigen by pathogens, or by action of closely linked (presumably in disequilibrium) immune response genes (97). These HLA disease associations are to be distinguished from true disease linkage, where any allele occupying a locus can be linked to the disease in informative pedigree, because there has been recombination in the past on the chromosome somewhere between a disease locus and the marker locus. Such recombination appears to be very uncommon in the HLA-associated disorders. In the absence of replicable associations between HLA antigen and affective disorder, it is difficult to accept reports of true linkage in pedigrees of a psychiatric disease to the HLA loci. However, the clinical example of congenital adrenal hyperplasia shows that linkage to the HLA system without association can exist and can be of clinical significance (40,133,173).

X-chromosome dominant transmission has been hypothesized as an exclusive or very common mode of transmission in bipolar and related affective disorders (179). Mendlewicz and his colleagues have been most prolific in their reports of pedigrees showing linkage to X-chromosome markers (104–106,108,109), although the initial reports (136,179) and the pedigree described by Baron (8) also suggest linkage. However, in another series of pedigrees (61,90), close linkage could be ruled out to either of the X-chromosome regions studied (Xg and color blindness). The evidence favoring X-linkage was criticized by Gershon and Bunney (50,53): Xg and color blindness are too far apart on the X-chromosome, now confirmed by the definite assignment of Xg to the short arm (1,112,175), reported by Mendlewicz and Fleiss (105) was inconsistent with the simultaneously collected family study data (110) which contained a high frequency of male-to-male transmission.

The inconsistency on bipolar illness–color blindness linkage between the two United States series, that of Mendlewicz and Fleiss (105) and that of Gershon et al. (61), remains puzzling, because each series was internally so consistent. A later series reported from Belgium by Mendlewicz et al. (108) was heterogeneous, containing one pedigree that appeared linked and two that clearly did not. I do not believe that the described inconsistencies in the reported results can be satisfactorily explained by genetic heterogeneity within bipolar illness (50). Pedigree series currently being collected by other investigators may demonstrate whether the X-linkage finding is replicable.

Other possible chromosomal linkages in affective disorders have been described with much more tentative data; these have been reviewed elsewhere (65,120).

Pathophysiological Genetic Studies in Affective Disorders

There are hypotheses of abnormal amounts of each of the monoaminergic and cholinergic neurotransmitters in affective disorders as well as hypotheses of altered sensitivities of their receptor sites (60). Few of these have been formulated in terms of a continuing abnormal event that constitutes an ongoing genetic vulnerability, and the number of genetic studies actually carried out to test such hypotheses is meager. In view of this, I include in this discussion some findings in patients with affective illness that persist into the well state and thus seem to be promising candidates for pedigree studies.

Central nervous system (CNS) metabolites of serotonin and dopamine are measurable in cerebrospinal fluid (CSF) without interference from the peripheral sources of these metabolites. 5-Hydroxyindoleacetic acid (5-HIAA), the principal metabolite of serotonin in CSF, was bimodally distributed in two series of depressed patients (5,168). A substantial proportion of depressed patients studied have reduced 5-HIAA in CSF, although not all studies show significant patient–control differences, as reviewed elsewhere (60). Sedvall et al. (145) have demonstrated a significant correlation in CSF 5-HIAA between twins, suggesting genetic control. Van Praag and de Haan (167) reported that in the majority of patients with reduced 5-HIAA accumulation, it persists after recovery. When they compared family history of patients with persistent reduced 5-HIAA to patients with normal 5-HIAA, they found higher depressive admissions in relatives of the reduced-5-HIAA group. These data are compatible with a genetically distinct subgroup of affective patients defined by reduced central serotonin turnover. But the evidence is incomplete, since there are no data on segregation of 5-HIAA and illness in relatives. Also, the family study data are presented without diagnostic definitions or estimates of the number of relatives at risk.

Enzymes of monoamine metabolism measured in peripheral blood have been of persistent interest in schizophrenia and affective illness. Activities of catechol-O-methyltransferase (COMT) and dopamine-β-hydroxylase (DBH) are controlled by single-locus polymorphisms (56,170,171). Although numerous studies have tried to relate these activities to different forms of affective illness, the results have been inconsistent (56,60). With the identification of specific genotypes with ranges of enzyme activity, it became possible to study segregation of alleles for high or low activity in pedigrees of patients with affective illness. No association is present, so these enzyme activities are unrelated to vulnerability to affective illness (56).

Two forms of MAO have been found: MAO-A is predominant in fibroblasts (67,68); MAO-B is found in platelets (29) and in primate brain (114). Platelet MAO has been reported decreased in BP patients by some but not all investigators (42,56,60,115,157). Pandey et al. (124) reported that low MAO is associated with illness in BP and alcoholic relatives of BP probands. Gershon et al. (56), however, reported data that indicate that among relatives of low-platelet-MAO patients, the BP and UP relatives do not tend to have low platelet MAO activity. It is not clear why these data are discrepant from those of Pandey et al. Possibly, inclusion of the alcoholic relatives with bipolar relatives in the Pandey study skewed the data, since alcoholism is associated with decreased platelet MAO for unknown reasons (156). Fibroblast (type A) MAO activity is not altered in affective illness (19) or in schizophrenia (67).

The relationship of these enzyme activities to central monoamine function is obscure, and the failure of the enzymes to provide genetic markers for affective disorders does not test any specific hypothesis on neurotransmitter function in psychiatric illness.

The adrenergic–cholinergic balance hypothesis of Janowsky et al. (79) proposes that increased adrenergic or decreased cholinergic activity is associated with mania and decreased adrenergic or increased cholinergic activity is associated with depression. We have recently looked for state-independent markers of susceptibility to affective disorders according to their formulation. Sitaram et al. (147,148) found that remitted, euthymic, drug-free bipolar affective illness patients had greater sensitivity than controls to stimulation by muscarinic cholinergic agonists, as measured by rapid eye movement (REM) sleep induction by arecoline and by pupillary constriction after pilocarpine. No change in sensitivity to adrenergic stimulation was formed, as measured by behavioral, neuroendocrine, and other responses to injection of d-amphetamine. Genetic studies of these findings have not yet been completed. Gershon, Nurnberger, and their colleagues studied adrenergic stimulation by intravenous injection of d-amphetamine in normal volunteer twins and in BP patients who were euthymic and drug free (58,59,121). Behavioral excitation and neuroendocrine responses were highly correlated in the monozygotic twin volunteers and independent of plasma amphetamine concentrations, suggesting heritable variation in the neurotransmitter systems on which the drug acts. No difference between patients and controls in amphetamine response was present. There was an inverse correlation between the amphetamine excitation response and the REM response to arecoline. It can be hypothesized on the basis of these findings that chronic vulnerability to affective illness consists of cholinergic supersensitivity but not adrenergic supersensitivity.

Lithium transport in erythrocytes is a cell membrane characteristic that is stable in an individual over time (25,39). It may be measured *in vivo* or *in vitro* (thus, individuals not treated with lithium may have an assessment of their ratio by incubating a sample of red cells in a lithium-containing solution), and there is high correlation between these measurements (123). Dorus and co-workers (35,36) have demonstrated genetic control of the ratio.

Dorus et al. (37) reported nonindependent assortment of high lithium ratio and affective disorder within pedigrees. As stated elsewhere (120), there is a good deal of overlap in the three group distributions among the relatives (major affective disorder, minor affective disorder, never ill). Individual pedigree data, data on lithium ratios in the probands, and data on each separate psychiatric entity among the relatives would be helpful in assessing this finding. Since many ill relatives do not have the finding, the difference in Li ratio is at most a contributing factor to vulnerability, not a necessary factor as defined by Rieder and Gershon (140).

Comings (30) used a two-dimensional protein electrophoretic technique to identify a polymorphic protein, Pc 1 Duarte (Pc for perchloric acid extract), with an increased frequency in brain specimens taken at autopsy from individuals with affective disorder and/or alcoholism. In controls ($N = 152$), 31.6% had the abnormal protein (2.6% homozygotes, 28.9% heterozygotes), whereas the frequency for depressed and alcoholic individuals ($N = 42$) was 61.9% (11.9% homozygotes, 50.0% heterozygotes). In 11 bipolar patients, the frequency of the Duarte protein was 72.7% (9.1% homozygotes, 63.6% heterozygotes).

No increased frequency of Duarte protein was found in samples from schizophrenic patients. An association was found, however, with multiple sclerosis ($N = 40$) (52.5% frequency, all heterozygotes) and in a small sample of brains from persons who died with subacute sclerosing panencephalitis and amyotrophic lateral sclerosis. All of these illnesses are suspected of having a viral etiology.

This is the first finding of an abnormal brain protein in persons with psychiatric disorder, and it is highly intriguing. It invites the speculation that affective disorder is associated with increased susceptibility to viral CNS infection. A replication of this finding, possibly by developing a DNA probe for the nucleotide sequence coding for the protein, would be most significant.

REFERENCES

1. Adam, A., Ziprkowski, L., Feinstein, A. (1969): Linkage relations of X-borne ichthyosis to the Xg blood groups and to other markers of X in Israelis. *Ann. Hum. Genet.*, 32:323–332.
2. American Psychiatric Association (1980): *Diagnostic and Statistical Manual (DSM-III)*. American Psychiatric Association, Washington.
3. Angst, J. (1966): Zur Ätiologie und Nosologie endogoner depressiver Psychosen. In: *Monographien aus dem Gesamtgebiete der Neurologie und Psychiatrie, No. 112*. pp. 1–118. Springer-Verlag, Berlin.
4. Angst, J., Felder, W., and Lohmeyer, B. (1979): Schizo-affective disorders: Results of a genetic investigation, I. *J. Affect. Disorders*, 1:137–153.
5. Asberg, M., Thoren, P., Traskman, L., Bertilsson, L., and Ringberger, V. (1976): Serotonin depression—a biochemical subgroup within the affective disorders? *Science*, 191:478–480.
6. Ask, A. L., Book, J. A., Heyden, T. (1979): Platelet monoamine oxidase in a pedigree with schizophrenia: An interlaboratory project. *Clin. Genet.*, 15:289–299.
7. Baron, M. (1976): Albinism and schizophreniform psychosis: A pedigree study. *Am. J. Psychiatry*, 133:1070–1073.
8. Baron, M. (1977): Linkage between an X-chromosome marker (deutan color blindness) and bipolar affective illness. *Arch. Gen. Psychiatry*, 24:721–727.
9. Baron, M., and Levitt, M. (1980): Platelet monoamine oxidase activity: Relation to genetic load of schizophrenia. *Psychiatr. Res.*, 3:69–74.
10. Baron, M., Stern, M., Anavi, R., (1977): Tissue-binding factor in schizophrenic sera: A clinical and genetic study. *Biol. Psychiatry*, 12:199–218.
11. Baxter, C., and Melnechuk, T. (1980): *Perspectives in Schizophrenia Research*. Raven Press, New York.
12. Beckman, L., Beckman, G., and Perris, C. (1980): Gc serum groups and schizophrenia. *Clin. Genet.*, 17:149–152.
13. Behar, D., Winokur, G., Van Valkenburg, C. (1980): Familial subtypes of depression: A clinical view. *J. Clin. Psychiatry*, 41:52–56.
14. Belmaker, R. H., Golan, A., Perez, L., and Ebstein, R. (1977): Platelet MAO in schizophrenics with and without a family history of schizophrenia. *Br. J. Psychiatry*, 131:551–552.
15. Belmaker, R. H., Reches, A., and Ebstein, R. P. (1977): Platelet MAO in schizophrenia. *Lancet*, 2:821.

16. Berretini, W. H., Benfield, T. C., Schmidt, A. O. (1980): Platelet monoamine oxidase in families of chronic schizophrenics. *Schizophrenia Bull.*, 6:235–237.
17. Bohman, M. (1978): Some genetic aspects of alcoholism and criminality. *Arch. Gen. Psychiatry*, 35:269–278.
18. Book, J. A., Wetterberg, L., and Modrzewska, K. (1978): Schizophrenia in a North-Swedish geographical isolate 1900–1977. Epidemiology genetics and biochemistry. *Clin. Genet.*, 14:373–394.
19. Breakefield, X. O., Giller, E. L., Jr., Nurnberger, J. I., Jr. (1980): Monoamine oxidase type A in fibroblasts from patients with bipolar depressive illness. *Psychiatry Res.*, 2:315–322.
20. Bridge, T. P., Wise, C. D., Potkin, S. G., Phelps, B. H., and Wyatt, R. J. (1982): Platelet monoamine oxidase: Studies of activity of thermolability in a general population. In: *Genetic Research Strategies in Psychobiology and Psychiatry*, edited by E. S. Gershon, S. Matthysse, X. O. Breakefield, and R. D. Ciaranello. pp. 95–104. Boxwood Press, Pacific Grove, California.
21. Bucher, K. D. (1977): *The Genetics of Manic Depressive Illness: A Pedigree and Linkage Study.* Ph.D. Thesis, University of North Carolina, Chapel Hill.
22. Buchsbaum, M. S., and Rieder, R. O. (1979): Biologic heterogeneity and psychiatric research. *Arch. Gen. Psychiatry*, 36:1163–1172.
23. Cancro, R. (1979): Genetic evidence for the existence of subgroups of the schizophrenic syndrome. *Schizophrenia Bull.*, 5:453–459.
24. Carey, G., and Gottesman, I. (1980): Genetics of anxiety disorders. Presented at the 69th Annual Mtg. of Amer. Psychopath. Assoc.
25. Carroll, B. J., and Feinberg, M. P. (1977): Intracellular lithium. *Neuropharmacology*, 16:527.
26. Carroll, B. J., Greden, J. F., Feinberg, M. (1980): Neuroendocrine dysfunction in genetic subtypes of primary unipolar depression. *Psychiatry Res.*, 2:251–258.
27. Carson, N. A. J., Cusworth, D. C., Dent, C. E. (1963): Homocystinuria: A new inborn error of metabolism associated with mental deficiency. *Arch. Dis. Child.*, 38:425–436.
28. Christiansen, K. D. (1970): Crime in a Danish twin population. *Acta Genet. Med. Gemellol. (Roma)*, 19:323–326.
29. Collins, G. S., and Sandler, M. (1971): Human blood platelet monoamine oxidase. *Biochem. Psychol.*, 20:389–396.
30. Comings, D. C. (1979): Pc 1 Duarte, a common polymorphism of a human brain protein, and its relationship to depressive disease and multiple sclerosis. *Nature*, 277:28–32.
31. Crowe, R. R. (1974): An adoption study of antisocial personality. *Arch. Gen. Psychiatry*, 31:785–791.
32. Crowe, R. R., Pauls, D. L., Slymen, D. J., and Noyes, R. (1980): A family study of anxiety neurosis: Morbidity risk in families of patients with and without mitral valve prolapse. *Arch. Gen. Psychiatry*, 37:77–79.
33. Debray, Q., Caillard, V., and Stewart, J. (1978): Schizophrenia: A study of genetic models and some of their implications. *Neuropsychobiology*, 4:257–269.
34. Dorus, E., Dorus, W., and Telfer, M. A. (1977): Paranoid schizophrenia in a 47, XYY male. *Am. J. Psychiatry*, 134:687–689.
35. Dorus, E., Pandey, G. N., and Davis, J. M. (1975): Genetic determinant of lithium ion distribution. An *in vitro* and *in vivo* monozygotic–dizygotic twin study. *Arch. Gen. Psychiatry*, 32:1097–1102.
36. Dorus, E., Pandey, G. N., Frazer, A., and Mendels, J. (1974): Genetic determinant of lithium ion distribution: I. An *in vitro* monozygotic–dizygotic twin study. *Arch. Gen. Psychiatry*, 31:463–465.
37. Dorus, E., Pandey, G. N., Shaughnessey, R., Gaviria, M., Val, E., Eriksen, S., and Davis, J. M. (1979): Lithium transport across the red cell membrane: A cell membrane abnormality in manic-depressive illness. *Science*, 205:932–934.
38. Dunner, D. L., Gershon, E. S., and Goodwin, F. K. (1976): Heritable factors in the severity of affective illness. *Biol. Psychiatry*, 11:31–42.
39. Dunner, D. L., Meltzer, H. L., and Fieve, R. R. (1978): Clinical correlates of the lithium pump. *Am. J. Psychiatry*, 135:1062–1064.
40. Dupont, B., Oberfield, S. E., Smithwick, E. M., Lee, T. D., and Levine, L. S. (1977): Close genetic linkage between HLA and congenital adrenal hyperplasia (21-hydroxylase deficiency). *Lancet*, 2:1309.
41. Eberhard, G., Franzen, G., and Low, B. (1975): Schizophrenia susceptibility and HL-A antigen. *Neuropsychobiology*, 1(4):211–217.

42. Edwards, D. J., Spiker, D. G., Kupfer, D. J., Foster, F. G., Neil, J. F., and Abrams, L. (1978): Platelet monoamine oxidase in affective disorders. *Arch. Gen. Psychiatry*, 35:1443–1446.
43. Elston, R. C., Namboodiri, K., Spence, M., and Rainer, J. (1978): A genetic study of schizophrenia pedigrees. *Neuropsychobiology*, 4:193–206.
44. Elston, R. C., and Stewart, J. (1971): A general model for the genetic analysis of pedigree data. *Hum. Hered.*, 21:523–542.
45. Elston, R. C., and Yelverton, K. C. (1975): General models for segregation analysis. *Am. J. Hum. Genet.*, 27:31–45.
46. Feighner, J. P., Robins, E., Guze, S. B. (1972): Diagnostic criteria for use in psychiatric research. *Arch. Gen. Psychiatry*, 26:57–63.
47. Fieve, R. R. (1973): Overview of therapeutic and prophylactic trials with lithium in psychiatric patients. In: *Lithium: Its Role in Psychiatric Research and Treatment*, edited by S. Gershon and B. Shopsin, pp. 317–350. Plenum Press, New York.
48. Fowler, R. C. (1978): Remitting schizophrenia as a variant of affective disorder. *Schizophrenia Bull.*, 4:68–77.
49. Gattaz, W. F., and Beckmann, H. (1980): HLA antigens and schizophrenia. *Lancet*, 1:98–99.
50. Gershon, E. S. (1980): Nonreplication of linkage to X-chromosome markers in bipolar illness. *Arch. Gen. Psychiatry*, 37:1200.
51. Gershon, E. S. (1981): Genetic vulnerability in psychiatric epidemiology. In: *Risk Factor Research in the Major Mental Disorders*, edited by D. A. Regier and G. Allen. pp. 145–162. U.S. Government Printing Office, Washington.
52. Gershon, E. S., Baron, M., and Leckman, J. F. (1975): Genetic models of the transmission of affective disorders. *J. Psychiatr. Res.*, 12:301–317.
53. Gershon, E. S., and Bunney, W. E., Jr. (1977): The question of X-linkage in bipolar manic-depressive illness. *J. Psychiatr. Res.*, 13:99–117.
54. Gershon, E. S., Bunney, W. E., Jr., Goodwin, F. K., Murphy, D. L., Dunner, D. L., and Henry, G. M. (1971): Catecholamines in affective illness: Studies with L-DOPA and alpha-methyl-*para*-tyrosine. In: *Brain Chemistry and Mental Disease*, edited by B. T. Ho and W. M. McIsaac, pp. 135–163. Plenum Press, New York.
55. Gershon, E. S., Bunney, W. E., Jr., Leckman, J. F., Van Eerdewegh, M., and Debauche, B. A. (1976): The inheritance of affective disorders: A review of data and hypotheses. *Behav. Genet.*, 6(3):277–261.
56. Gershon, E. S., Goldin, L. R., Lake, C. R., Murphy, D. L., and Guroff, J. J. (1980): Genetics of plasma dopamine-β-hydroxylase (DBH), erythrocyte catechol-*O*-methyltransferase (COMT), and platelet monoamine oxidase (MAO) in pedigrees of patients with affective disorders. In: *Enzymes and Neurotransmitters in Mental Disease*, edited by E. Usdin, P. Sourkes, and M. B. H. Youdim, pp. 281–299. John Wiley & Sons, New York.
57. Gershon, E. S., Mark, A., Cohen, M., Belizon, N., Baron, M., and Knobe, K. E. (1975): Transmitted factors in the morbid risk of affective disorders: A controlled study. *J. Psychiatr. Res.*, 12:283–299.
58. Gershon, E. S., Nurnberger, J. I., Jr., and Jimerson, D. C. (1980): Sex, plasma prolactin and plasma 3-methoxy-4-hydroxyphenylglycol (MHPG) predict heritable *d*-amphetamine excitation in man. In: *Advances in the Biosciences: Recent Advances in Neuro-Psycho-Pharmacology*, edited by B. Angrist et al., pp. 83–91. Pergamon Press, New York.
59. Gershon, E. S., Nurnberger, J. I., Jr., Sitaram, N., and Gillin, C. (1979): Pharmacogenetics and the pharmacologic challenge strategy in clinical research. Studies of *d*-amphetamine and arecoline. In: *Neuro-Psychopharmacology*, edited by B. Saletu, pp. 75–83. Pergamon Press, New York.
60. Gershon, E. S., Targum, S. D., Kessler, L. R., Mazure, C. M., and Bunney, W. E., Jr. (1977): Genetic studies and biologic strategies in the affective disorders. In: *Progress in Medical Genetics, Vol. II*, edited by A. G. Steinberg, A. G. Bearn, A. G. Motulsky, and B. Childs, pp. 101–164. W. B. Saunders, Philadelphia.
61. Gershon, E. S., Targum, S. D. Matthysse, S., and Bunney, W. E., Jr. (1979): Color blindness not closely linked to bipolar illness. Report of a new pedigree series. *Arch. Gen. Psychiatry*, 36:1423–1430.
62. Go, R. C., Elston, R. C., Fieve, R. R. et al. (1980): Indications of genetic heterogeneity in 68 bipolar I proband families. *Am. J. Hum. Genet.*, 32:143A.
63. Goetzl, V., Green, R., Whybrow, P., and Jackson, L. (1974): X-Linkage revisited: A further family study of manic depressive illness. *Arch. Gen. Psychiatry*, 31:665–672.

64. Goldin, L. R., and Gershon, E. S. (1979): Segregation and linkage analysis in families with major affective disorder. *Am. J. Hum. Genet.*, 31:135A.
65. Goodwin, D. S., Schulsinger, L. F., Hermansen, L. (1973): Alcohol problems in adoptees raised apart from alcoholic biological parents. *Arch. Gen. Psychiatry*, 28:238–243.
66. Goodwin, D. W., Schulsinger, L. F., Moller, N. (1974): Drinking problems in adopted and nonadopted sons of alcoholics. *Arch. Gen. Psychiatry*, 31:164–169.
67. Groshong, R., Baldessarini, R. J., Gibson, A. (1978): Activities of types A and B MAO and catechol-O-methyltransferase in blood cells and skin fibroblasts of normal and chronic schizophrenic subjects. *Arch. Gen. Psychiatry*, 35:1198–1205.
68. Groshong, R., Gibson, D. A., and Baldessarini, R. J. (1977): Monoamine oxidase activity in cultured human skin fibroblasts. *Clin. Chim. Acta*, 80:113–120.
69. Guze, S. B. (1967): Psychiatric illness in families of convicted criminals: A study of 519 first-degree relatives. *Dis. Nerv. Syst.*, 28:651–659.
70. Helzer, J. E., Clayton, P. J., Pambakian, R. (1977): Reliability of psychiatry diagnosis: II. The tests/retests reliability of diagnostic classification. *Arch. Gen. Psychiatry*, 34:136–144.
71. Helzer, J. E., Robins, L. N., Taibleson, M. (1977a): Reliability of psychiatric diagnosis: I. A methodological review. *Arch. Gen. Psychiatry*, 34:129–135.
72. Helzer, J. E., and Winokur, G. (1974): A family interview study of male manic depressives. *Arch. Gen. Psychiatry*, 31:73–77.
73. Heston, L. L. (1966): Psychiatric disorders in foster home reared children of schizophrenic mothers. *Br. J. Psychiatry*, 112:819–825.
74. Heston, L. L. (1970): The genetics of schizophrenic and schizoid disease. *Science*, 167:249–256.
75. Hook, E. B. (1973): Behavioral implications of the human XYY genotype. *Science*, 179:139–150.
76. Hutchings, B., and Mednick, S. A. (1975): Registered criminality in the adoptive and biological parents of registered male criminal adoptees. In: *Genetic Research in Psychiatry*, edited by R. R. Fieve, D. Rosenthal, and H. Brill. pp. 105–116. Johns Hopkins University Press, Baltimore.
77. Ivanyi, D., Zemek, P., and Ivanyi, P. (1976): HLA antigens in schizophrenia. *Tissue Antigens*, 8(3):217–220.
78. James, N. M., and Chapman, C. J. (1975): A genetic study of bipolar affective disorder. *Br. J. Psychiatry*, 126:449–456.
79. Janowsky, D. S., Davis, J. M., El-Yousef, M. K. (1972): A cholinergic–adrenergic hypothesis of mania and depression. *Lancet*, 2:632–635.
80. Johnson, G. F. S., and Leeman, M. M. (1977): Analysis of familial factors in bipolar affective illness. *Arch. Gen. Psychiatry*, 34:1074–1083.
81. Kallmann, F. J. (1946): The genetic theory of schizophrenia: An analysis of 691 twin index families. *Am. J. Psychiatry*, 103:309–322.
82. Kallmann, F. J. (1950): The genetics of psychoses: An analysis of 1,232 twin index families. In: *Congrès Internationale de Psychiatrie*. Hermann et Cie., Paris.
83. Kallmann, F. J. (1959): Genetic aspects of psychosis. In: *The Biology of Mental Health and Disease. The 27th Annual Conference of the Milband Memorial Fund*. Paul B. Hoeber, New York.
84. Kay, D. W. K., Roth, M., Atkinson, M. W. (1975): Genetic hypotheses and environmental factors in the light of psychiatric morbidity in the families of schizophrenics. *Br. J. Psychiatry*, 127:109–118.
85. Kety, S. S. (1979): Disorders of the human brain. *Sci. Am.*, 241(3):202–218.
86. Kety, S. S. Rosenthal, D., Wender, P. H. (1968): The types and prevalence of mental illness in the biological and adoptive familes of adopted schizophrenics. *J. Psychiatr. Res.*, 6:345–362.
87. Kidd, K. K., and Cavalli-Sforza, L. L. (1973): An analysis of the genetics of schizophrenia. *Soc. Biol.*, 20:254–265.
88. Kinney, D. K., and Matthysse, S. (1978): Genetic transmission of schizophrenia. *Annu. Rev. Med.*, 29:459–473.
89. Kupfer, D., Pickar, D., Himmelhoch, J., and Detre, T. P. (1975): Are there two types of unipolar depression? *Arch. Gen. Psychiatry*, 32:866–871.
90. Leckman, J. F., Gershon, E. S., McGinniss, M., Targum, S. D., and Dibble, D. D. (1979): New data do not suggest linkage between the Xg blood group and bipolar illness. *Arch. Gen. Psychiatry*, 36:1435–1441.
91. Leonhard, K. (1957): *Aufteilung der Endogenen Psychosen*. 1st ed., Akademie-Verlag, Berlin.

92. Leonhard, K., Korff, I., and Schulz, H. (1962): Die Temperamente in den Familien der monopolaren und bipolaren phasischen Psychosen. *Psychiatr. Neurol.*, 143:416–434.
93. Matthysse, S. (1980): Genetic detection of cerebral dysfunction. *N. Engl. J. Med.*, 302:516–517.
94. Mazure, C., and Gershon, E. S. (1979): Blindness and reliability in lifetime psychiatric diagnosis. *Arch. Gen. Psychiatry*, 36:521–525.
95. McCabe, M. S. (1975): Reactive psychoses. *Acta Psychiatr. Scand. [Suppl.]*, 259:1–133.
96. McKnew, D. H., Jr., Cytryn, L., Efron, A. M., Gershon, E. S., and Bunney, W. E., Jr., (1979): Offspring of patients with affective disorders. *Br. J. Psychiatry*, 134:148–152.
97. McMichael, A., and McDevitt, H. (1977): The association between the HLA system and disease. *Prog. Med. Genet. (New Ser.)*, 2:39–100.
98. Meltzer, H. Y. (1975): Neuromuscular abnormalities in the major mental illnesses: I. Serum enzyme studies. In: *Biology of the Major Psychoses: A Comparative Analysis*, edited by D. X. Freedman, pp. 165–188. Raven Press, New York.
99. Meltzer, H. Y. (1979): Biology of schizophrenia subtypes: A review and proposal for method of study. *Schizophrenia Bull.*, 5:460–479.
100. Meltzer, H. Y., Cho, H. W., Carroll, B. J., and Russo, P. (1976): Serum dopamine-β-hydroxylase activity in the affective psychoses and schizophrenia. *Arch. Gen. Psychiatry*, 33:585–591.
101. Meltzer, H. Y., Pscheidt, G., Goode, D. (1977): Platelet and plasma monoamine oxidase in schizophrenia. *Sci. Proc of Amer. Psych. Assoc.*, p. 207.
102. Meltzer, H. Y., and Stabl, S. M. (1974): Platelet monoamine oxidase activity and substrate preferences in schizophrenic patients. *Res. Commun. Clin. Pathol. Pharmacol.*, 7:419–431.
103. Mendels, J. (1976): Lithium in the treatment of depression. *Am. J. Psychiatry*, 133:373–378.
104. Mendlewicz, J. (1977): Genetic studies in schizoaffective illness. In: *Impact of Biology on Modern Psychiatry*, edited by E. S. Gershon, R. H. Belmaker, S. S. Kety, and M. Rosenbaum, pp. 229–240. Plenum Press, New York.
105. Mendlewicz, J., and Fleiss, J. L. (1974): Linkage studies with X-chromosome markers in bipolar (manic–depressive) and unipolar (depressive) illnesses. *Biol. Psychiatry*, 9:261–294.
106. Mendlewicz, J., and Linkowski, P. (1978): Linkage between glucose-6-phosphate dehydrogenase deficiency and manic–depressive illness. Second World Congress of Biological Psych., Barcelona, Spain.
107. Mendlewicz, J., and Linkowski, P. (1980): HLA antigens and schizophrenia. *Lancet*, 1:765.
108. Mendlewicz, J., Linkowski, P., Guroff, J. J., and Van Praag, H. M. (1979): Color blindness linkage to manic–depressive illness. *Arch. Gen. Psychiatry*, 36:1442–1449.
109. Mendlewicz, J., Linkowski, P., and Wilmotte, J. (1980): Linkage between glucose-6-phosphate dehydrogenase deficiency and manic–depressive psychosis. *Br. J. Psychiatry*, 137:337–342.
110. Mendlewicz, J., and Rainer, J. D. (1974): Morbidity risk and genetic transmission in manic depressive illness. *Am. J. Hum. Genet.*, 26:692–701.
111. Modrezewska, K. (1980): The offspring of schizophrenic patients in a North Swedish isolate. *Clin. Genet.*, 17:191–201.
112. Mohandas, T., Shapiro, L. J., Sparkes, R. S. (1979): Regional assignment of the steroid sulfatase-X-linked ichthyosis locus. *Proc. Natl. Acad. Sci. U.S.A.*, 76:5779–5783.
113. Morrison, J. R. (1975): The family histories of manic depressive patients with and without alcoholism. *J. Nerv. Ment. Dis.*, 160:227–229.
114. Murphy, D. L., Redmon, D. E., Jr., Garrick, N., and Baulu, J. (1979): Brain region differences and some characteristics of monoamine oxidase type A and B activities in the vervet monkey. *Neurochem. Res.*, 4:53–62.
115. Murphy, D. L., and Weiss, R. (1972): Reduced monoamine oxidase activity in blood platelets from bipolar depressed patients. *Am. J. Psychiatry*, 128:1351–1357.
116. Murphy, D. L., and Wyatt, R. J. (1972): Reduced monoamine oxidase activity in blood platelets from schizophrenic patients. *Nature*, 238:225–226.
117. Nielsen, J. (1969): Klinefelter's syndrome and the XYY syndrome. *Acta Psychiatr. Scand. [Suppl.]*, 209:45.
118. Nies, A., Robinson, D. S., Harris, L. S., and Lamborn, K. R. (1974): Comparison of monoamine oxidase substrate activities in twins, schizophrenics, depressives and controls. *Adv. Biochem. Psychopharmacol.*, 12:59–70.
119. Nurnberger, J. I., Jr., and Gershon, E. S. (1980): Genetics of affective disorders. In: *Depression and Antidepressants: Implications for Cause and Treatment*, edited by E. Friedman. Raven Press, New York.

120. Nurnberger, J. I., Jr., and Gershon, E. S. (1982): Genetics of affective disorders. In: *Handbook of Affective Disorders*, edited by E. Paykel. pp. 126–145. Churchill Livingston, England.
121. Nurnberger, J. I., Jr., Gershon, E. S., Jimerson, D. C., Buchsbaum, M., Gold, P., Brown, G., and Ebert, M. (1981): Pharmacogenetics of d-amphetamine response in man. In: *Genetic Strategies in Psychobiology and Psychiatry*, edited by E. S. Gershon, S. Matthysse, R. D. Ciaranello, and X. O. Breakefield. pp. 257-268. Boxwood Press, Pacific Grove, California.
122. Nurnberger, J. I., Jr., Gershon, E. S., Murphy, D. L., Buchsbaum, M. S., Goodwin, F. K., Post, R. M., Lake, C. R., Guroff, J. J., and McGinniss, M. H. (1979): Biological and clinical predictors of lithium response in depression. In: *International Lithium Conferences: Controversies and Unresolved Issues*, edited by S. Gershon, N. S. Kline, and M. Schou, pp. 241–251. Excerpta Medica, Amsterdam.
123. Pandey, G. N., Dorus, E., Davis, J. M., and Tosteson, D. C. (1979): Lithium transport in human red blood cells. *Arch. Gen. Psychiatry*, 36:902–908.
124. Pandey, G. N., Dorus, E., Shaughnessy, R. (1980): Reduced platelet MAO vulnerability to psychiatric disorders. *Psychiatry Res.*, 2:315–322.
125. Pauls, D. L., Bucher, K. D., Crowe, R. R., and Noyes, R. (1980): A genetic study of panic disorder pedigrees. *Am. J. Hum. Genet.*, 32:639–644.
126. Pauls, D. L., Crowe, R. R., and Noyes, R. (1979): Distribution of ancestral secondary cases in anxiety neurosis (panic disorder). *J. Affect. Dis.*, 1:287–290.
127. Perris, C. (1966): A study of bipolar (manic–depressive) and unipolar recurrent depressive psychoses. *Acta Psychiatr. Scand. [Suppl.]*, 203:15–44.
128. Perris, C. (1966): Genetic transmission of depressive psychoses. *Acta Psychiatr. Scand [Suppl.]*, 203:15–44.
129. Perris, C. (1974): The genetics of affective disorders. In: *Biological Psychiatry*, edited by J. Mendels, pp. 385–415. John Wiley & Sons, New York.
130. Perris, C., Roman, G., and Wahlby, L. (1979): HLA antigens in patients with schizophrenic syndromes. *Neuropsychobiology*, 5:290–293.
131. Petterson, U. (1977): Manic–depressive illness: A clinical, social, and genetic study. *Acta Psychiatr. Scand. [Suppl.]*, 269.
132. Pope, H. G., and Lipinski, F. (1978): Diagnosis in schizophrenia and manic depressive illness. *Arch. Gen. Psychiatry*, 35:811–828.
133. Price, D. A., Klouda, P. T., and Harris, R. (1978): HLA and congenital adrenal hyperplasia linkage confirmed. *Lancet*, 1:930–931.
134. Propping, P., and Friedl, W. (1979): Platelet monoamine oxidase activity in first-degree relatives of schizophrenic patients. *Psychopharmacology*, 65:265–272.
135. Reich, T., and Andreason, N. C. (1979): A preliminary analysis of the segregation distributions of the primary affective disorders. Presented at the Ann. Mtg. of Amer. Coll. of Neuropsychopharm., San Juan, PR.
136. Reich, T., Clayton, P. J., and Winokur, G. (1969): Family history studies. V. The genetics of mania. *Am. J. Psychiatry*, 125:1358–1369.
137. Reich, T., James, J. W., and Morris, C. A. (1972): The use of multiple thresholds in determining the mode of transmission of semi-continuous traits. *Ann. Hum. Genet.*, 36:163.
137a. Baron, M. (1982): Genetic models of schizophrenia. *Acta Psychiat. Scand.*, 65:263–275.
138. Reich, T., Rice, J., and Cloninger, C. R. (1981): The detection of a major locus in the presence of multifactorial variation. In: *Genetic Research Strategies for Psychobiology and Psychiatry*, edited by E. S. Gershon, S. Matthysse, X. O. Breakefield, and R. D. Ciaranello pp. 353–367. Boxwood Press, Pacific Grove, California.
139. Rieder, R. O., Broman, S. H., and Rosenthal, D. (1977): Offspring of schizophrenics. *Arch. Gen. Psychiatry*, 34:789–801.
140. Rieder, R. O., and Gershon, E. S. (1978): Genetic strategies in biological psychiatry. *Arch. Gen. Psychiatry*, 35:866–873.
141. Rieder, R. O., Rosenthal, D., Wender, P (1975): The offspring of schizophrenics: Fetal and neonatal deaths. *Arch. Gen. Psychiatry*, 32:200–211.
142. Rosenthal, D., Wender, P. H., Kety, S. S. (1968): Schizophrenics' offspring reared in adoptive homes. *J. Psychiatr. Res.*, 6(1):377–392.
143. Schlesser, M. A., Winokur, G., and Sherman, B. M. (1979): Genetic subtypes of unipolar primary depressive illness distinguished by hypothalmic–pituitary adrenaline activity. *Lancet*, 1:739–741.
144. Schukit, M. (1972): Family history and one-half sibling research in alcoholism. *Ann. N.Y. Acad. Sci.*, 197:121–125.

145. Sedvall, G., Nyback, H., Oxenstierna, G. (1983): Relationships between aberrant monoamine metabolite concentration in cerebrospinal fluid and family disposition for psychiatric disorders in healthy and schizophrenic subjects. *Prog. Neuro-Psychopharmacol. [Suppl] (in press).*
146. Seixas, F. A., Omenn, G. S., Burk, E. D., and Eggleston, S. (1972): Nature and nurture in alcoholism. *Ann. N.Y. Acad. Sci.*, 197:1–229.
147. Sitaram, N., Moore, A. M., Vanskiver, C., et al. (1981): Hypersensitive cholinergic functioning in primary affective illness. In: *Cholinergic Mechanisms, Vol. 25.*, edited by G. Pepeu and H. Ladinsky, pp. 947–962. Plenum Press, New York.
148. Sitaram, N., Nurnberger, J. I., Jr., Gershon, E. S., and Gillin, E. S. (1980): Faster cholinergic REM sleep induction in euthymic patients with primary affective illness. *Science*, 208:200–202.
149. Slater, E. (1968): A review of earlier evidence of genetic factors in schizophrenia. *J. Psychiatr. Res.*, 6(1):15–26.
150. Slater, E., and Cowie, V. (1971): *Genetics of Mental Disorders.* Oxford University Press, London.
151. Smeraldi, E., Negri, F., and Melica, A. M.(1977): A genetic study of affective disorders. *Acta Psychiatr. Scand.*, 56:382–398.
152. Smeraldi, E., Negri, F., Melica, A. M., and Scorza-Smeraldi, R. (1978): HLA system and affective disorders: A sibship genetic study. *Tissue Antigens*, 12:270–274.
153. Sorensen, K., and Nielsen, J. (1977): Twenty psychotic males with Klinefelter's syndrome. *Acta Psychiatr. Scand.*, 56:249–255.
154. Spence, A., Elston, R. C., Namboodiri, K. K., and Rainer, J. D. (1976): A genetic study of schizophrenia in pedigrees. *Neuropsychobiology*, 2:328–340.
155. Spitzer, R. L., Endicott, J., and Robins, E. (1978): Research diagnostic criteria: Rationale and reliability. *Arch. Gen. Psychiatry*, 35:773–782.
156. Sullivan, J. L., Cavenar, J. O., Jr., Maltbie, A. A. (1979): Familial biochemical and clinical correlates of alcoholics with low platelet monoamine oxidase activity. *Biol. Psychiatry*, 14:385–394.
157. Takahashi, S. (1977): Monoamine oxidase activity in blood platelets from manic and depressed patients. *Folia Psychiatr. Neurol. Jpn.*, 31:37–48.
158. Tanna, V. L., Go, R. C. P., Winokur, G., and Elston, R. C. (1977): Possible linkage between group-specific component (GC protein) and pure depressive disease. *Acta Psychiatr. Scand.*, 55:111–115.
159. Tanna, V. L., Winokur, G., Elston, R. C., and Go, R. C. P. (1976): A linkage study of depression spectrum disease. The use of the sib pair method. *Neuropsychobiology*, 2:52–62.
159a.Gershon, E. S., Hamovit, J., Guroff, J. J. et al. (1983): A family study of Schizoaffective, Bipolar I, Bipolar II, Unipolar, and normal control probands. *Arch. Gen. Psychiat. (in press).*
160. Targum, S. D., Gershon, E. S., Van Eerdewegh, M., and Rogentine, N. (1979): Human leukocyte antigen (HLA) system not closely linked to or associated with bipolar manic–depressive illness. *Biol. Psychiatry*, 14:615–636.
161. Taylor, M. A., and Abrams, R. (1980): Reassessing the bipolar unipolar dichotomy. *J. Affect. Dis.*, 2:195–217.
162. Taylor, M. A., Abrams, R., and Hayman, M. A. (1980): The classification of affective disorders—a reassessment of the bipolar unipolar dichotomy. *J. Affect. Dis.*, 2:95–109.
163. Tsuang, M. T. (1979): Schizoaffective disorder. *Arch. Gen. Psychiatry*, 36:633–634.
164. Turner, W. J. (1979): Genetic markers for schizotaxia. *Biol. Psychiatry*, 14:177–206.
165. Van Eerdewegh, M., Gershon, E. S., and Van Eerdewegh, P. (1976): X-Chromosome threshold models of bipolar manic–depressive illness. *Sci. Proc. Am. Psychiatr. Assoc.*, 129:124–125.
166. Van Eerdewegh, M. M., Gershon, E. S., and Van Eerdewegh, P. (1980): X-chromosome threshold models of bipolar manic–depressive illness. *J. Psychiatr. Res.*, 15:215–238.
167. Van Praag, H. M., and de Haan, S. (1979): Central serotonin metabolism and frequency of depression. *Psychiatry Res.*, 1:219–224.
168. Van Praag, H., and Korff, J. (1971): Endogenous depressions with and without disturbances in the 5-hydroxytryptamine metabolism: A biochemical classification? *Psychopharmacologia*, 19:148.
169. Weinberger, D. R., Torrey, E. F., Neophytides, A. N. (1979): Lateral cerebral ventricular enlargement in chronic schizophrenia. *Arch. Gen. Psychiatry*, 36:735–739.
170. Weinshilboum, R. M. (1979): Serum dopamine-β-hydroxylase. *Pharmacol. Rev.*, 30:133–166.
171. Weinshilboum, R. M., and Raymond, F. A. (1977): Inheritance of low erythrocyte catechol-O-methyltransferase activity in man. *Am. J. Hum. Genet.*, 29:125–135.
172. Weissman, M. M., and Myers, J. K. (1978): Affective disorders in U.S. urban community. *Arch. Gen. Psychiatry*, 35:1304–1312.
173. Weitkamp, L. R., Bryson, M., and Bacon, G. E. (1978): HLA and congenital adrenal hyperplasia linkage confirmed. *Lancet*, 1:931–932.

174. Welner, Z., Welner, A., McCrary, D. (1977): Psychopathology in children of inpatients with depression: A controlled study. *J. Nerv. Ment. Dis.*, 164:408–413.
175. Went, L. N., de Groot, W. P., Sanger, R. (1969): X-Linked ichthyosis: Linkage relationship with the Xg blood groups and other studies in a large Dutch kindred. *Ann. Hum. Genet.*, 32:333–345.
176. Winokur, G. (1973): The types of affective disorders. *J. Nerv. Ment. Dis.*, 156:82–96.
177. Winokur, G., Behar, D., Vanvalkenburg, C. (1978): Is a familial definition of depression both feasible and valid? *J. Nerv. Ment. Dis.*, 166:764–768.
178. Winokur, G., and Clayton, P. (1967): Family history studies. I. Two types of affective disorders separated according to genetic and clinical factors. In: *Recent Advances in Biological Psychiatry, Vol. 9*, edited by J. Wortis, pp. 35–50. Plenum Press, New York.
179. Winokur, G., Clayton, P. J., and Reich, T. (1969): *Manic Depressive Illness.* C. V. Mosby, St. Louis.
180. Winokur, G., Reich, T., Rimmer, J., and Pitts, F. N., Jr. (1970): Alcoholism. III. Diagnosis and familial psychiatric illness in 259 alcoholism probands. *Arch. Gen. Psychiatry*, 23:104–111.
181. Winter, H., Herschel, M., Propping, P., Friedl, W., and Vogel, F. (1978): A twin study on three enzymes (DBH, COMT, MAO) of catecholamine metabolism: Correlations with MMPI. *Psychopharmacology*, 57:63–69.
182. Witkin, H. A., Mednick, S. A., Schulsinger, et al., (1976): Criminality in XYY and XXY men. *Science*, 193:547–555.
183. Witz, I. P., Anavi, R., and Weisenbeck, H. (1977): A tissue-binding factor in the serum of schizophrenic patients. In: *The Impact of Biology on Modern Psychiatry*, edited by E. S. Gershon, R. H. Belmaker, S. S. Kety, and M. Rosenbaum, pp. 1–276. Plenum Press, New York.
184. Worden, G., Childs, B., Matthysse, S. (1976): Frontiers of Psychiatric Genetics. *Neurosci. Res. Prog. Bull.*, 14:37–44.
185. Wyatt, R. J., Murphy, D. L., Belmaker, R. (1973): Reduced monoamine oxidase activity in platelets: A possible genetic marker for vulnerability to schizophrenia. *Science*, 179:916–918.
186. Zeller, E. A., Boshes, B., Davis, J. M. (1975): Molecular aberration in platelet monoamine oxidase in schizophrenia. *Lancet*, 1:1385.

Genetic Heterogeneity in Alcoholism and Sociopathy

C. Robert Cloninger and Theodore Reich

Department of Psychiatry, Washington University School of Medicine, and Jewish Hospital of St. Louis, St. Louis, Missouri 63110

The extensive clinical heterogeneity in common neuropsychiatric disorders is a stumbling block to classical genetic analysis. Classical methods have been fruitful with rare discrete phenotypes such as phenylketonuria but have led to little etiological insight with traits that are common and clinically heterogeneous as are the major psychiatric syndromes.

Such common traits are difficult to categorize discretely (e.g., alcoholic or not) and, even when criteria are defined and validated, much variation within categories is not taken into account. Data relevant to the diagnosis of such traits comes at several different levels of observation: (a) clinical variables such as number, type, severity, sequence, and age of onset of symptoms; (b) neurophysiological variables such as skin conductance recovery (41,58), resting EEG, EEG responses to alcohol (68), and cortical evoked potentials (41,58); (c) biochemical data such as associated variation in metabolism of alcohol (50) and neuroendocrine regulation (52); and (d) variation in chromosomal karyotypes (88) and genetic markers (19,21) that are associated with clinical disorder. In other words, common neuropsychiatric disorders are developmentally complex, multidimensional phenotypes. Unfortunately, we seldom know what level of description is optimal for a particular task related to diagnosis, prognosis, or etiology.

The developmental complexity of a phenotype refers to the length of the chain of development from the primary gene products (enzyme and structural protein subunits) to the observed phenotype. The more complex the phenotype, the more steps in the chain of events and the greater number of loci and environmental factors that are important to phenotypic variation.

Single genes code for the individual polypeptide chains that are the subunits of enzymes and structural proteins, but Wright (90), Thoday (81), Fraser (35), and others have concluded that most complex phenotypes depend largely on a few loci modified by extensive multifactorial (polygenic and environment) variability. Examples from crosses of inbred animal strains have been worked out under experimental control in detail not possible in man. F. Clark Fraser's discussion of cortisone-induced cleft palate defects is an instructive recent example (35). A large number of primary defects leading to cleft palate have been observed, but in any particular susceptible strain only a few factors are frequent, and these few usually differ between strains. Only a proportion of susceptible animals develop cleft palate, depending both on the extent of their predisposition by several independent genetic factors and on the dose of cortisone to which they are exposed during development.

Unfortunately, the number of loci that are involved in the events leading to a particular complex phenotype is probably highly variable. Accordingly, causal models that are applied to the study of these disorders must remain flexible in regard to the number of relevant loci and the magnitude of their individual contributions. Without such flexibility, there is a grave danger that overly specific models will cloud an investigator's perception and lead to premature, erroneous conclusions rather than help him ask testable questions.

It seems most likely that psychiatric disorders such as alcoholism and sociopathy have a complex developmental pathway from genotype to phenotype.

Research about the inheritance of such complex multidimensional phenotypes is now at a watershed. Most work in the past has used simple models of monogenic inheritance of a broad but discrete classification (e.g., Is alcoholism caused by an X-linked recessive gene?). These simplistic models and questions have led to limited insight, leading some to abandon their quest to understand the genotype–phenotype pathway. Others promise to circumvent the difficulty by studying biochemical phenomena closer to the primary gene products or by studying the DNA itself.

These attempts to circumvent the problem of the heterogeneity are still impeded by the inadequacy of the one gene–one enzyme principle for understanding multidimensional phenotypes. In the absence of homogeneous clinical syndromes and/or a detailed understanding of the mechanisms underlying the genotype–phenotype pathway, association studies of a random set of genetic markers or empirically identified biochemical/physiological disturbances are unlikely to be fruitful. The number of primary gene products in man is so large (50,000 to 200,000) that the probability of identifying a true causal factor by empirical association studies of a multidimensional phenotype is very low. The sheer number of primary gene products and the complexity of their possible interactions with one another make it more likely that observed associations will be spurious and unreplicable. Therefore, a more systematic strategy for evaluating the inheritance of multidimensional phenotype is needed, as we have discussed elsewhere (24).

The first stage in a biologically more realistic approach to the heterogeneity problem is to recognize and describe the multiple components that make up a common multidimensional trait. Multivariate methods such as factor analysis or cluster analysis are useful to identify independent factors (21) or patient subgroups (56).

In the second stage, either the inheritance of individual components of a more complex phenotype (23) or more homogeneous subgroups of a heterogeneous set of disorders (56) may be studied. The inheritance of individual component factors is more likely to be simpler than the developmentally more complex ultimate phenotype.

In the third stage, the interaction of individual risk factors in producing the multidimensional phenotype is studied. The inheritance of individual risk factors may be simpler and thereby crucial to understanding the pathophysiological mechanisms underlying the genotype–phenotype pathway. Nevertheless, it is the consequence of the interactions of individual risk factors that is usually most important clinically.

In some cases, distinct syndromes or factors may simply represent different stages in the natural history of the same disorder (early versus late cases), different degrees of the same disorder (mild versus severe cases), or variants modified by nonfamilial environmental factors. In other cases, different components or risk factors may be associated with independent etiological causes. If two syndromes are different manifestations of the same etiological process (i.e., spectrum condition), they are expected to be correlated both within individuals and within families (69). This methodological approach is described in more detail elsewhere (24,69) and is illustrated here for the cases of alcoholism and sociopathy.

Current research on the inheritance of alcoholism and sociopathy are largely limited to the first and second stages. In contrast, the genetic epidemiology of risk factors in cardiovascular disease is well into the third stage (with work continuing at all stages in an iterative process). In this chapter, we review our current knowledge about the inheritance of alcoholism and sociopathy, pointing out gaps in our knowledge. Methodological aspects of current research are detailed elsewhere (20–23,71,72).

GENETIC MODELS OF ALCOHOLISM AND SOCIOPATHY AS UNITARY TRAITS

Systematic investigations of alcoholics' first-degree relatives who have been personally interviewed have been reported by Åmark (1), Bleuler (5) Winokur et al. (84,85), Reich et al. (70), and Cloninger (17). Family interview studies of sociopathy, otherwise called antisocial personality (ASP), have been published by Cloninger (18,19). Other studies in which history about relatives was obtained only from the proband have been reviewed elsewhere (16). The only published data in which both probands and relatives are described according to their primary psychiatric diagnosis is that of Winokur and his associates. These data are summarized in Table 1.

Even when all probands are alcoholics, the disorders segregate according to the primary diagnosis of the proband. The frequency of relatives with the same diagnosis as the proband is significantly increased compared to the relatives of other probands.

Furthermore, the interrelationships among alcoholism and antisocial personality are complicated by the etiological heterogeneity of each of these disorders. The familial relationship of primary alcoholism and primary antisocial personality is shown in Table 2. The data are from Winokur et al. (85) and Cloninger et al. (18) and include only men. There are significant correlations between alcoholism in probands and relatives ($r = 0.59 \pm 0.04, p < 0.05$) and between ASP in probands and relatives ($r = 0.43 \pm 0.07, p < 0.05$). In addition, there is a slight excess

TABLE 1. *Psychiatric disorders in first-degree relatives of alcoholic probands by primary diagnosis[a]*

Primary diagnosis of alcoholic proband	Primary diagnosis of first-degree relatives					
	Alcoholism		Depression		ASP	
	f/n	%	f/n	%	f/n	%
Alcoholism	184/807	23[b]	78/535	15	19/516	4
Depression	19/130	15	24/89	27[b]	3/70	4
ASP	18/106	17	14/72	19	15/68	22[b]

[a]Data from Winokur et al. (85). Diagnoses are based on all available data about subjects over 17 and age adjusted by Stromgren method. Prevalence of ASP is given for male relatives only; other prevalences include both sexes.
[b]Relatives with same primary diagnosis as the proband are significantly increased compared to the others (contingency χ^2, $p < 0.05$).

TABLE 2. *Primary alcoholism (1° Alc) and antisocial personality (ASP) in the general population and the first-degree relatives of alcoholic or antisocial probands (men only)*

Diagnosis of proband	Prevalence in general population		Proportion of affected first-degree relative					
			1° Alc			ASP		
	N	%	N	%	$r \pm SE$	N	%	$r \pm SE$
1° Alc	751	7.6	270	35.2	0.59 ± 0.04[a]	516	3.7	0.03 ± 0.06
ASP	329	3.3	120	15.0	0.18 ± 0.07[a]	120	16.7	0.43 ± 0.07[a]

[a]Correlation between proband and illness in relatives significant ($p < 0.05$).

of primary alcoholics among the relatives of antisocial probands ($r = 0.18 \pm 0.07$, $p < 0.05$) but no such excess of antisocial personality in the relatives of primary alcoholics ($r = 0.03 \pm 0.06$, NS). This suggests either that there is a partial overlap between the liability to transmit alcoholism and ASP in some families or that alcoholism is a milder manifestation of the same underlying process that causes ASP in some families or that the excess of alcoholism in the families of antisocial probands is caused by an intervening variable not controlled in the analysis (e.g., rates of heavy drinking in families of antisocial probands).

In view of the familial segregation of alcoholism and ASP, these traits will be further considered separately. The segregation of alcoholism from affective disorder (21) and from drug dependence (42) is considered elsewhere.

FAMILIAL TRANSMISSION OF ALCOHOLISM

When all recurrent alcohol abusers are grouped together, much of the observed data are compatible with a simple polygenic process with moderate heritability. This is illustrated by a summary of the concordance for recurrent alcohol abuse in different classes of relatives who have been studied in Sweden (Table 3). The

TABLE 3. *Family studies of recurrent alcohol abuse in Swedish men*

Source	Relationship	Probandwise concordance		Tetrachoric correlation
		N	%	
Kaij (48)	MZ twins	27	70	0.92 ± 0.13
Kaij (48)	DZ twins	60	33	0.60 ± 0.09
Åmark (1)	Singleton sib	349	31	0.52 ± 0.03
	No alcoholic parent	252	17	0.44 ± 0.04
	One alcoholic parent	97	33	0.67 ± 0.06
Bohman (6)	Adopted-away sons	50	20	0.55 ± 0.23
Kaij and Dock (49)	Grandsons	270	12	0.14 ± 0.04
Census, 1968	General population		7	—

tetrachoric correlation between relatives was computed by assuming that recurrent alcohol abuse is a threshold character resulting from an underlying multifactorial process whose liability is multivariate normal and correlated between relatives (16,69). Under the hypothesis that familial transmission is polygenic and not cultural, twice the difference between the MZ twin correlation and the DZ twin correlation estimates the heritability of liability to recurrent alcohol abuse: $h^2 = 2(0.92 - 0.60) = 0.64$. Then, first-degree relatives who share half their genes (DZ twins, singleton sibs, and adopted sons) are expected to be correlated as $\frac{1}{2}h^2 = 0.32$ plus any difference caused by home environment or selective placement. Second-degree relatives (e.g., grandsons) should be correlated $\frac{1}{4}h^2 = 0.16$. The figures observed are reasonably close to those expected and suggest only modest influences of home enviroment. Also, having an affected parent increases the risk of sibs to probands (1), so there is evidence of vertical inheritance.

However, this simple polygenic hypothesis was called into question by Shields (79) and Cloninger and associates (21) because it was noted that there was little difference between the concordance of first-degree relatives and second-degree relatives when age and sex were taken into account in studies by Kaij and Dock (49), Bleuler (5), and Schuckit and associates (77). Since one-half of the genes of first-degree relatives are identical by descent compared to one-quarter in second-degree relatives, no simple genetic mechanism can explain the observed pattern. The possibility of interaction between genetic and environmental factors in some cases was suggested by this observation and has recently been confirmed in Swedish adoption data shown later.

MULTIDIMENSIONAL MODELS OF ALCOHOLISM

Since well-designed studies of adoptees (6,9,10,38), half-siblings (77), and twins (48,66) do implicate genetic factors in the development of some forms or components of alcohol abuse, the central questions that must be asked are: What aspects of alcoholism are transmitted predominantly by genetic factors? What aspects of alcoholism are predominantly environmentally determined? How do relevant environmental and genetic factors interact in the development of alcoholism?

Several multivariate studies of symptom patterns in alcoholics indicate that there are many clinical factors that are independent of one another and, most importantly, independent of family drinking histories. Independent factors that have been replicated in different samples are summarized in Table 4 and discussed in the following subsection.

Multivariate Approaches to Phenotypic Heterogeneity

Ever since the work of Jellinek (47), attempts have been made to subdivide alcoholism into more homogeneous subgroups. Some of Jellinek's work implied that alcoholism was a homogeneous disorder, although alcoholics were thought to differ in terms of the phase in which a particular individual was observed. Park and Whitehead (65) tested Jellinek's ideas on American and Finnish samples. Although they found a significant correlation between the order of development of symptoms postulated by Jellinek and the order obtained in their data, they found four factors in these symptoms of alcoholism in both Finnish and American samples of alcoholics. Very high similarities in order of appearance of symptoms existed within factors, although the two samples showed important differences in the temporal sequences between factors. The finding of four orthogonal factors within these samples may indicate that alcoholism is either an etiologically heterogeneous disorder or one with a heterogeneous clinical course modified by familial and nonfamilial environmental influences.

A number of other studies have used multivariate approaches to symptom patterns in alcoholics. Horn and Wanberg (44) reported that 13 factors were found in a study of drinking histories of a large sample of alcoholics. Three independent factors relating to type of beverage were found. Similarly, three orthogonal factors relating to type of drinking pattern (sustained binge, controlled steady, and periodic drinking) were extracted. Physical complications were orthogonal to type of beverage and drinking pattern. These results also indicate considerable heterogeneity within the

TABLE 4. *Multivariate subclassification of heterogeneity in alcoholism: A summary of independent factors*

Type of beverage
Type of drinking pattern
Sustained binge
Controlled steady
Periodic drinking
Medical complications
Early life history
Current functioning
Sociopathic
Depressed
Others based on age, marital status and sex
Social supports and stressors
Parental drinking histories

disorder. A similar analysis by the same authors (45) found seven independent life history factors and eight independent current functioning factors. One life history factor was suggestive of sociopathic personality, and one life history and two current functioning factors resembled those found in affective disorder. Further, a parental drinking factor was found that was orthogonal to the other factors. This implies that variation in the other factors is under independent control.

Wanberg and Horn (45) report that separate analyses in male and female alcoholics showed that 8 of 10 drinking history factors were replicated in each sex. Although in a different area of research, this supports the findings of Reich et al. (73) and Cloninger (16) and associates who concluded that male and female alcoholics have the same underlying liability to alcoholism and that nonfamilial factors are responsible for the differences in incidence.

Horn and Wanberg (45) suggested that their findings supported the presence of independent typologies of abusive drinking. If true, they stated, each pattern might have unique etiological components. Wanberg and Knapp (82) published a study essentially replicating the work of Wanberg and Horn. They concluded that a unitary trait model fails to explain adequately the phenomena of alcoholism.

Evenson and associates (30) reported a multivariate analysis of drinking history, life history, and psychiatric symptomatology for a large sample of alcoholics. The factor structure they obtained differed from the earlier reports of Wanberg and Horn. They found 15 independent factors in the large list of symptoms studied (not confined to drinking history alone). Again, family history of alcoholism was orthogonal to the other variables. Evenson and associates also performed a Q factor analysis on a smaller sample of alcoholics. This analysis yielded three groups of subjects: younger, married employed males; older, unemployed males living alone or with friends; and women. A stepwise discriminant function showed 90% or more of the subjects could be correctly classified. The groups differed significantly on marital and family status, psychological and social reasons for drinking, control over drinking, dependency on alcohol, and pattern of drinking.

In addition to life history and drinking history analyses, multivariate approaches have been used on personality test data. Factor analysis and cluster analysis/classification approaches have been used with MMPI data (36,53,62), with the 16PF (51), the Differential Personality Inventory (67), and the Inventory of Habits and Attitudes (40). Partington and Johnson (67) noted that one type was highly antisocial with many psychological symptoms, whereas another type (Type III) showed many psychological symptoms without signs of sociopathy. Morgar and associates (62) found nine types in both sexes including depressive and sociopathic types. Goldstein and Linden (36) found four types including a sociopathic and affective disorder type. Miller (61) has recently reviewed available alcoholism scales and objective assessment methods. He notes that prior scale development is of doubtful and limited value because of the heterogeneous nature of alcoholism.

These descriptive studies as a whole highlight the phenotypic heterogeneity among alcoholics and indicate that alcoholism is multidimensional. However, heterogeneity among individuals does not demonstrate etiological heterogeneity: the demonstration of etiological heterogeneity requires family studies showing that the clinical differences among individuals are inherited. Unfortunately, there has been little use of multidimensional description in family studies.

Multivariate Approaches to Etiological Heterogeneity

The first multivariate approach to evaluate which aspects of alcoholism are genetically transmitted was carried out by Partanen and associates (66). They reported the drinking histories of 172 monozygous and 557 dizygous twins in Finland. The subjects were a general population and were not selected for alcohol abuse. However, the prevalence of alcohol abuse was high in Finland.

From data provided in their appendices, the prevalence and concordance for the major categorical drinking variables may be calculated (Table 5). This shows that heavy use of alcohol, defined in terms of amount consumed on the last two drinking occasions, showed significantly higher correlations in MZ twins than in DZ twins. In addition, twins were often concordant for arrests for drunkenness and problem drinking, but there was no differences according to degree of genetic relationship.

In addition, they carried out the multivariate analyses summarized in Table 6. A factor analysis was performed on 13 drinking history variables, and five factors were extracted. Factor 1 (density and frequency) had high loadings on time from last drink to second-last drink, regularity of drinking, subject's estimate of frequency of drinking, and co-twin's estimate of subject's frequency of drinking. Factor 2 (amount) had high loadings on amount of alcohol consumed at last two drinking occasions, duration of drinking occasions, degree of intoxication at last two episodes, and subject's own estimate of quantity consumed. The third factor, thought to represent loss of control and dependency on alcohol, loaded most highly on ability to control drinking, ability to stop drinking, amount of alcohol consumed on last two occasions, and estimated degree of intoxication on these occasions. A check on the independence of the extracted factors showed that factors 2 and 3 were correlated ($r = +0.36$).

TABLE 5. *Qualitative drinking variables in Finnish twin study*[a]

Drinking variable	Prevalence	Probandwise concordance	Tetrachoric correlation
Heavy use of alcohol			
MZ twins ($N = 198$)	29.5%	68/117 = 58%	0.61 ± 0.10
DZ twins ($N = 641$)	30.8%	160/395 = 41%	0.08 ± 0.06
Drunkenness arrests[b]			
MZ twins ($N = 135$)	18.1%	22/49 = 45%	0.55 ± 0.13
DZ twins ($N = 409$)	19.4%	69/159 = 43%	0.49 ± 0.07
Problem drinking[c]			
MZ twins ($N = 198$)	23.2%	40/92 = 43%	0.44 ± 0.11
DZ twins ($N = 641$)	30.2%	186/387 = 48%	0.41 ± 0.06

[a]Data from Partanen et al. (66); subjects were all men aged 28 to 37.
[b]There were too few subjects to subdivide according to recurrent arrests.
[c]Includes any user with social complications or addictive symptoms.

TABLE 6. *Inheritance of drinking behavior in 172 MZ and 577 DZ twins in Finland[a]*

Factor	Heritability
Density (frequency)	0.39
Amount	0.36[b]
Lack of control	0.14[b]
Latent root	
First[c]	0.66
Second[d]	0.38

[a]Data from Partanen et al. (66).
[b]Correlation between these factors is 0.36.
[c]Amount and lack of control; loads 0.74 and 0.62, respectively.
[d]Density or frequency.

The heritabilities of four drinking-related quantitative variables were computed. Factor 1 (density) had a heritability of 0.39, whereas heritability for factor 2 (amount) was 0.36. Lack of control and social complications showed little evidence of heritability except in younger subjects.

Canonical correlation analysis was used to find the subset of drinking variables with the greatest heritability. The first latent root showed a heritability of 0.66. Factor 2 (amount of drinking) and factor 5 (lack of control) had loading on the root of 0.74 and 0.62, respectively. The second root was also significant with a heritability of 0.38. This root was defined by factor 1 (density or frequency of drinking).

Partanen and associates (66) interpreted these findings to show that frequency and amount of drinking show considerable hereditary variations, whereas loss of control and social complications are almost entirely the result of environmental variations—an interpretation that is entirely consistent with the heritabilities reported for the individual factor scores. However, the findings of the canonical correlation analysis tend to cast some doubt on it. The first heritable root had high loadings on both amount of drinking and loss of control over drinking. An alternative interpretation is that loss of control is also determined by genetic factors but that before this influence is manifest, exposure to continued heavy drinking is required. If the two sets of hereditary factors operate independently, the appearance of environmental determination of loss of control results in a general sample that has not been selected for heavy drinking.

These alternative interpretations are discussed in more detail elsewhere (21). It should be noted that the latent root derived as a weighted sum of heavy consumption and loss of control is similar to the disease concept of alcoholism and has a heritability (66%) similar to that estimated in Sweden for alcohol abuse (64%). Furthermore, differences in drinking styles in America, Finland, and other Scandinavian countries limit the comparisons that can be made about individual symptoms of alcohol abuse. The rationlike system of alcohol distribution that prevailed in Finland at the time encouraged periods of abstinence alternating with excessive drinking to intoxication when alcoholic beverages were available. As a consequence, individual symptoms such as "daily drinking" or "loss of control" cannot be directly compared between Finnish and American or Danish alcoholics (87). As noted previously, Park and Whitehead (65) have found that social and family problems occur earlier in American than Finnish alcoholics, whereas economic problems occur earlier in Finns. However, core symptoms that can be compared cross-culturally, such as need for more alcohol, occupy the same chronological sequence in the development of alcoholism (87) and are heritable in different countries.

Other multivariate analyses have been carried out to evaluate these issues in adoption studies, and further work on multiple factors in pedigrees is underway by us currently.

ADOPTION STUDIES OF ALCOHOLISM

Adoption studies of alcohol abuse have recently been reviewed by Cloninger, Bohman, and Sigvardsson (15). The Copenhagen adoption study of alcoholism was carried out by Goodwin and associates on a sample of adoptees assembled by Kety and his associates (37). The data about 55 adopted sons and 30 nonadopted sons of hospitalized alcoholics are summarized in Table 7. Control adoptees were selected whose parents had no diagnosed alcoholism, but heavy drinking and lesser forms of alcohol abuse could not be excluded, since the parents were not interviewed. Goodwin and his associates found that there was a fourfold increase in alcoholism in the biological sons whether or not they were adopted out. The correlation between first-degree relatives provides a heritability estimate of about 90% for alcoholism. However, lesser forms of problem drinking are not increased, and heavy drinking is actually significantly decreased in the nonadopted sons. This suggests that the control group may include many children of nonalcoholic heavy drinkers, as suspected by the authors. Goodwin and his associates concluded that genetic factors, but not home environment, were important in the development of severe alcoholism.

Cloninger and associates recently studied the inheritance of alcohol abuse in 862 adopted Swedish men from the Stockholm adoption study (16). The population is a general sample of adoptees and their parents from a birth cohort assembled by

TABLE 7. *Copenhagen adoption study of sons of hospitalized alcoholics*[a]

Drinking variable	Risk in biological sons			
	Adopted %	(N = 55) $r \pm SE$	Nonadopted %	(N = 30) $r \pm SE$
Ever heavy drinker	49	-0.08 ± 0.12	23	-0.44 ± 0.16[b]
Ever problem drinker	27	0.15 ± 0.14	20	0.02 ± 0.17
Ever alcoholic	18	0.45 ± 0.17[b]	17	0.41 ± 0.20[b]

[a] Data from Goodwin et al. (37,38).
[b] Tetrachoric correlation differs from 0 ($p < 0.05$) relative to 78 control adoptees whose parents have no diagnosed alcoholism; most (55%) controls are heavy drinkers, and 5% are alcoholic. Prevalence of alcoholic proband fathers is 9% of 5,483.

Bohman (6). They identified two types of susceptibility to alcohol abuse that have distinct genetic and environmental causes by using discriminant analysis to classify the congenital and postnatal backgrounds of the adoptees (Table 8). One type affects both men and women but is expressed only in particular postnatal environments. The other type of susceptibility is highly heritable from father to son, but mothers of alcoholic sons are seldom alcoholic themselves.

In families with the milieu-limited type of congenital susceptibility, both the biological mothers and fathers are characterized by the adult onset of alcohol abuse and no criminality requiring prolonged incarceration. Alcohol abuse is mild in most such cases and usually does not require treatment. However, more severe disability requiring hospital care may occur in susceptible sons who are exposed to particular postnatal environments, especially sons from the sociocultural background associated with unskilled occupational status.

Both congenital diathesis and postnatal provocation were found to be necessary for an individual to express the milieu-limited type of susceptibility. If there is either a congenital diathesis or a provocation postnatal milieu, but not both, then the risk of alcohol abuse is lower than in the general population. However, if both occur in the same individual, the risk is increased twofold, and the severity of disability is determined by the degree of postnatal provocation.

In families affected with the other type of susceptibility, alcohol abuse is increased ninefold in the adopted sons regardless of their postnatal environment. The heritability in men is about 90%. This male-limited pattern is similar to that reported by Goodwin and his associates.

The demonstration of genetic heterogeneity and gene–environment interaction in alcoholism has practical and theoretical significance: the data reconcile the results

TABLE 8. *A profile of prominent features distinguishing two types of alcoholism in 862 adopted Swedish men*

Distinguishing feature	Type 1 (milieu-limited form)	Type 2 (male-limited form)
Prevalence in adopted men	13%	4%
Biological father's characteristics	Mild alcohol abuse, minimal criminality, no treatment	Severe alcohol abuse, severe criminality, extensive treatment
Biological mother's characteristics	Mild alcohol abuse, minimal criminality	Normal
Postnatal environment	Determines both frequency and severity of alcoholism in susceptible sons	No effect on frequency (may influence severity)
Severity of alcoholism	Usually isolated or mild problems but may be severe	Usually recurrent or moderate problems but may be severe
Relative risk in congenitally predisposed sons[b]	2 with postnatal provocation 1 without postnatal provocation	9 regardless of postnatal milieu

[a]Data from Cloninger et al. (15).
[b]This relative risk is the ratio of the risk of alcoholism in congenitally predisposed sons to that in others. Thus, a relative risk of 1 indicates no difference.

of earlier family adoption studies and demonstrate the critical importance of sociocultural influences in most alcoholics. This suggests that major changes in social attitudes about drinking styles can dramatically change the prevalence of alcohol abuse regardless of genetic predisposition.

Genetic Markers and Subclinical Heterogeneity

There have been a large number of genetic-marker studies to detect association and linkage with alcoholism, but the results have been inconsistent, as might be anticipated with an etiologically heterogeneous phenotype. Conflicting reports of association and linkage of alcoholism with colorblindness, various blood groups, and other markers are reviewed elsewhere (43,86). One observation that has been replicated independently is the association of nonsecretor status with alcoholism, particularly among individuals of blood type A (11,12,80).

Certain specific complications of alcoholism, namely, cirrhosis and Wernicke–Korsakoff syndrome, have recently been reported to be facultative disorders related to simple genetic mechanisms. Only about 5% of alcoholics who are thiamine deficient develop Wernicke–Korsakoff syndrome (57), and Europeans are more susceptible than Asians on comparable diets. Using the techniques of somatic cell genetics, Blass and Gibson (4) found that transketolase in fibroblasts of patients with Wernicke–Korsakoff syndrome bound thiamine pyrophosphate less avidly than normal. Activity of transketolase in the presence of excess thiamine is normal but falls when thiamine levels are low. The abnormality persists through serial passages in cell culture in the presence of excess thiamine and no alcohol. This appears to be an inborn error of metabolism that is clinically important only when dietary thiamine is adequate. It is likely then that Wernicke–Korsakoff syndrome is a facultative recessive disorder, but more family data is needed.

Alcoholics in general are not associated with any known genetic polymorphisms, but alcoholics with cirrhosis have been found to have an excess of the histocompatibility locus antigen (HLA) B8 and a deficiency of HLA A28 compared to other alcoholics and to controls (3). The relationship of the HLA findings to older clinical associations of cirrhosis and hypotrichosis (fine body hair) in alcoholics has not been clarified (64).

Propping et al. (68) have shown that individuals with different types of resting electroencephalogram (EEG) respond to alcohol in different ways. Alcohol has a synchronizing effect on the EEG (increase in alpha activity and decreased variance of frequency) that is most pronounced in the borderline alpha EEG variants who have minimal alpha activity at rest. He has hypothesized that such EEG responses are genetically determined and may predispose to the development of alcohol abuse. However, it is uncertain to what extent such variation is determined by additive genetic effects rather than genetic interactions (dominance and epistasis). Also, the frequency of different EEG types in alcoholics and their relatives has not been studied. Further work with EEG responses to alcohol and with cortical evoked responses in alcoholics is needed.

Other evidence of biochemical variation associated with alcoholism has recently been reviewed elsewhere (63). It is difficult to interpret the inconsistent reports because the clinical heterogeneity we have discussed earlier has not been taken into account in any published work on biological risk factors in alcoholism. Consequently, more refined clinical subclassifications, validated by family studies, are needed to facilitate such etiological research.

ASSESSMENT OF CLINICAL HETEROGENEITY IN SOCIOPATHY

Three general approaches to the classification of antisocial and criminal behavior have each received considerable attention. The earliest and simplest procedure, usually employed by criminologists and sociologists, is to scale such deviant behavior from records or self-reports according to the number and severity of antisocial acts or offenses (13,33). A second procedure, usually employed by psychiatrists, subdivides such behavior into discrete categories, utilizing chronological and mental status characteristics in addition to the history of the number and severity of antisocial acts (2,39,74). A third approach has its advantages and limitations. Although the approaches are not mutually exclusive, little integrative work has been done.

Evidence of the stability and predictive power of unidimensional scales of antisocial behavior has been reviewed recently by Farrington (33). These methods appear reliable and have predictive power in longitudinal studies (83,89). However, psychiatric and psychological studies of delinquents and criminals reveal a marked clinical heterogeneity which is relevant to prognosis (39,54,55,74), physiology (41), and family history (54). This means that unidimensional approaches ignore much important information. Also, ratio scales such as the Sellin–Wolfgang index (89) are conceptually troublesome: what does it mean to equate two rapes with one homicide or with 12 arrests for public drunkeness on a unidimensional scale of severity? More importantly, family and adoption studies of criminals and delinquents give inconsistent results if the clinical heterogeneity of probands is not taken into account.

Categorical approaches to the classification of criminals and delinquents have subdivided subjects into discrete categories (18,19,39,84). The two most prevalent disorders by this procedure are antisocial personality and alcoholism. Antisocial personality, otherwise called sociopathy or psychopathy, is a disorder beginning early in life (before age 15) and characterized by recurrent antisocial and delinquent behavior, such as school troubles (truancy, suspension, expulsion), repeated trouble for fighting or use of weapons, running away from home, prolonged wanderlust or vagrancy, and marked sexual promiscuity or prostitution in addition to criminal behavior. Explicit operational criteria have been defined with moderate test–retest reliability and stability over time (2,34,39). Family studies have demonstrated the usefulness of the criteria (18,19).

Comparison of psychiatric diagnosis and criminal life curves reveals that criminals who are diagnosed as antisocial personality are those with serious offenses, repetitive offenses, or early onset of criminality, all three of which are highly correlated (74,89). However, the diagnostic criteria appear rather arbitrary in selecting a specific exclusion/inclusion point for diagnosis, and quantitative personality scales have been validated that distinguished psychopaths and nonpsychopaths, noncriminals, and first offenders and recidivists (75). In her review, Daisy Schallings concluded that the constructs underlying the scales that most consistently have discriminative and predictive ability include disturbances in inhibitory behavior and self-control (impulsiveness), in empathy and interpersonal relations (low role taking, tough-mindedness, lack of attachments), and increased stimulation seeking or monotony avoidance. These scales provide reliable ranking of subjects, but it is not known whether the same etiological determinants operate at all levels of the scales, i.e., whether the constructs reflect the same underlying processes at all levels of inventory score range. Therefore, evaluation using both categorical and dimensional measures are indicated.

The personality scales that have most consistently differentiated criminals and controls are the Socialization Scale (So) from Gough's California Psychological Inventory, the psychopathic deviate (Pd) scale from the MMPI, and Eysenck's Personality Questionnaire (EPQ). The So scale was first published as the Delinquency Scale in 1952; when included in the CPI, the keying was reversed, and it was renamed socialization. Megaree (60) summarized the extensive data about So as follows:

> In short, an impressive array of data have accumulated demonstrating the concurrent, predictive and construct validity of the CPI socialization scale in the United States and elsewhere. There seems to be little doubt that the So scale is one of the best-validated and most powerful personality scales available; and as with any personality test score, reliable discrimination will even be improved by using So in conjunction with other data, particularly case history material. As one does research on personality assessment devices, one becomes aware of inherent limits on the degree to which any single scale score can correlate with overt behavior. However, the data indicates that few scales approach these limits as closely as does So.

The socialization scale was originally constructed to discriminate delinquents from nondelinquents and is based on Gough's role-taking theory of sociopathy. Gough theorized that sociopathy results from an egocentric inability to perceive the effects of one's behavior on others. Subsequent research indicated that the scale was able to discriminate across the full range of socialization, i.e., within delinquent populations and within nondelinquent populations as well as between the two. It is said to order individuals along a continuum from asocial to social behavior and predicts the likelihood that they will "transgress the mores established by their particular cultures." This claim is based on extensive data involving more than 10,000 subjects in more than 25 samples ranging from subjects nominated "best citizen" through different occupational samples, problem cases, misdemeanants, and convicted felons (point-biserial correlation for the dichotomy more versus less socialized was 0.73).

Cluster analysis indicates that the socialization scale is made up of 4 clusters: (a) lack of cohesiveness in family rearing; (b) feelings of alienation and inferiority; (c) social insensitivity and lack of empathy; and (d) history of overt antisocial behavior or impulses. Thus, the scale appears to have useful empirical characteristics and to measure important aspects of clinical behavior, home environment, and interpersonal/social difficulties. Scores on the different clusters can be used to subdivide subjects for testing of familial transmission of these different aspects of antisocial behavior. In addition, available data indicates that alcoholics score particularly low on the So scale (60), but detailed information is not yet available.

In contrast, other scales used to assess criminals have been based in part on the neurophysiological basis of personality traits such as extroversion, neuroticism, and "tough-mindedness" or psychoticism as measured by EPQ. In 1970, Eysenck (31) presented data indicating that criminals are high in extroversion and neuroticism; that is, they are "neurotic extroverts." In contrast, alcoholics were usually found to be neurotic introverts; that is, they are low in extroversion. Eysenck indicates that extroverts have low "conditionability" because of characteristics of their reticular systems that are related to low cortical arousal. Since then, it has been demonstrated that extroversion includes two factors, impulsivity and sociability (75). Impulsivity and measures of stimulation seeking or monotony avoidance are associated with criminality and antisocial behavior, but sociability is often low in criminals (75,76).

Schallings has derived an inventory termed Extraversion–Impulsivity–Monotony Avoidance (EIM) from item content of EPI, MNT Solidity, Guilford Rhathymia, the Barret Impulsiveness Scale, and some of the Cattell scales. In addition, she has related these to ratings of anxiety. Based on work by Buss, anxiety was subdivided into clusters of psychic anxiety and somatic anxiety (76). Criminals and subjects with low So scores tended to be high in somatic anxiety but not psychic anxiety, as measured by this inventory (75,76). Eysenck has described an additional scale termed tough-mindedness or psychoticism (31) which is high in criminals and is correlated with the So scale (75).

The MMPI Pd scale and others (mania, schizophrenia) distinguished criminals from others, but Crowe (25) found that relatives of antisocial probands who were not themselves antisocial appeared normal on MMPI.

In summary, a wide range of aspects of antisocial behavior and alcoholism may be assessed using such tests as Gough's So, Eysenck's EPQ, and Schalling's EIM. These measures reflect behavioral, sociological, familial, and physiological disturbances. Criminals and alcoholics as groups may be discriminated by measures related to extroversion–introversion. Eaves and Eysenck (27–29) have analyzed extensive twin data about extroversion, neuroticism, and psychoticism using quantitative genetic techniques. If it could be established that such inventories are correlated with the liability to transmit socialization (or its reverse, antisocial behavior) throughout its full range as suggested by analysis of criterion groups and by twin studies of the scales themselves, this would be a major advance. Further work is needed involving simultaneous family studies of both the semicontinuous criterion groups (e.g., mild and severe antisocial personality) and the continuous personality inventory scores. Otherwise, it is uncertain whether the same etiological determinants operate at all levels of the scales. Evaluations using both categorical and dimensional measures in multiple classes of relatives are indicated.

SUBCLINICAL CORRELATES OF SOCIOPATHY

Given the aforementioned clinical heterogeneity, it is difficult to evaluate reports that assess subclinical variation in antisocial subjects. Mednick and Volavka (59) have recently reviewed the available literature. They conclude that repetitively antisocial individuals have slow skin conductance recovery and some slowing of EEG frequencies (excess theta activity) compared to noncriminals. Interpretation of these reports is difficult, however, because of problems in case description, age and attentional effects, and methodological techniques, as discussed by Hare and Schallings (41).

FAMILY AND TWIN STUDIES OF SOCIOPATHY AND CRIMINALITY

Family data about antisocial personality and criminality have been presented in detail elsewhere by Cloninger and Associates (16,18,19). These traits are strongly familial: the correlations for ASP between relatives are 0.29 ± 0.10 in parent–offspring pairs and 0.49 ± 0.09 in singleton sibs. Early twin studies of criminals were based on highly selected samples, but two recent studies are based on general populations of twins (13,26). The correlations were 0.70 ± 0.05 for MZ twins and 0.41 ± 0.12 for DZ twins in the Danish sample (16). The probandwise concordance for male DZ twins was 26% in both the Danish and Norwegian samples, but the concordance was slightly lower for male MZ twins in the smaller Norwegian sample (41% versus 51%). As a result, the difference by zygosity is not significant in the smaller Norwegian sample but is highly significant in the Danish sample.

Dalgard and Kringlen (26) also suggested that their Norwegian twin data indicated that the greater similarity of MZ twins was caused by greater similarity in environmental experiences. To test this hypothesis, they stratified twins according to their degree of psychological closeness to one another, i.e., strong or weak intrapair interdependence. As expected, a slightly greater proportion of MZ twins were psychologically close than were DZ twins (84% of 31 versus 74% of 54). Among pairs who were psychologically close, there was no appreciable difference in criminal concordance by zygosity, so they argued that environmental influences caused criminality. However, this environmental hypothesis would predict increased criminal concordance among close MZ pairs and decreased concordance among distant MZ pairs; in actual fact, the opposite was observed. Therefore, a preferable interpretation is that criminality is associated with lack of psychological attachment or closeness to others, as suggested by Gough's role-taking theory of socialization. Thus, Dalgard and Kringlen have stratified their sample according to a personality trait correlated with criminality, not according to an objective measure of home environment. Nevertheless, the finding that sibling correlations are greater than parent–offspring correlations by others (18) suggests that both genetic factors and home environment are important. More definitive evidence may be obtained from separation experiments (22,23).

ADOPTION STUDIES OF ANTISOCIAL TRAITS

Adoption studies have been carried out using a variety of measures of antisocial behavior with mixed results. Crowe in Iowa studied the offspring of adopted-away children of female felons and found an increase in ASP but not alcoholism or any other disorders (25). The proportions he observed in adoptees were comparable to those observed in nonadopted children of criminal women (18). This confirms the observation that adopted-away children of alcoholics have no increased risk for antisocial personality (37), thereby indicating a remarkable degree of genetic independence between adult alcoholism and antisocial personality.

Cadoret has found an excess of adolescent antisocial symptoms and unexplained somatic symptoms in adoptees with antisocial biological parents, especially in susceptible boys reared in adoptive homes with psychiatrically ill adoptive family members (7,8). This supports the finding of Cloninger and associates that somatization disorders or Briquet's syndrome is a mild manifestation of the same process that causes antisocial personality (19). This association between sociopathy and somatization disorder has been reviewed in detail elsewhere (14).

Hutchings and Mednick (46) found that criminality in both biological and adoptive parents increased the risk of criminality in Danish adoptees. They reported a trend for an interaction between the effect of biological and adoptive parents. However, the interaction was not significant in their preliminary sample, and it will be interesting to learn whether the trend becomes significant in the large sample currently under study. This is a critical point, because Schulsinger was able to show transmission of psychopathy from biological, but not adoptive, parents to Danish adoptees (14,78).

More important, the studies of Bohman and associates show the difficulty in interpreting studies about heterogeneous samples of criminals. Differences in type and severity of crime and association with alcohol abuse can dramatically alter apparent patterns of inheritance (14). Bohman has shown that much criminality is a consequence of alcohol abuse (6). Nevertheless, the adoption studies by Crowe and by Goodwin suggest that adult ASP and alcoholism are inherited independently.

The developmental sequence and age of onset of symptoms are critical aspects of the heterogeneity observed in the families of alcoholic and antisocial probands. Cadoret has found an excess of retrospectively reported adolescent conduct disorders in the adopted-away relatives of antisocial biological parents (7,8). Similarly, Goodwin and others have observed an excess of retrospectively reported adolescent conduct disorder and hyperactivity in the adopted-away relatives of alcoholic biological parents. However, prospective follow-up of the adopted children of criminal and alcoholic parents by Bohman does not confirm these retrospective reports about adolescents, at least in most cases (7,8). Also, twin studies reveal little difference in concordance for juvenile delinquency according to zygosity (16). Better methods are required for distinguishing the subgroups of adolescents with early onset of antisocial behavior whose conduct disorder persists into adulthood from the majority of juvenile delinquents who show no adult criminality.

Overall, much more research is needed both to define precise measures of heritable factors that increase the risk for antisocial behavior and to define more homogeneous subgroups of individuals with different types of antisocial behavior.

CONCLUSION AND SUMMARY

Both alcoholism and sociopathy are traits that are clinically heterogeneous and developmentally complex. Each of these common phenotypes is composed of variations in several simpler factors which may be independent of one another.

Multivariate statistical methods are needed to define clinical syndromes and patient subgroups for genetic and biosocial research with these common multidimensional phenotypes. Current methods for assessment and analysis of heterogeneity are reviewed.

Recent evidence for genetic heterogeneity and gene–environment interaction in alcoholism is presented. Two types of alcoholics are described that have distinct genetic and environmental causes and differ in their association with criminality, severity of alcohol abuse, and frequency of expression in women. In addition, family and adoption data indicate a surprising degree of independence between genetic factors that predispose to alcoholism and those that predispose to sociopathy or antisocial personality.

ACKNOWLEDGMENTS

These studies were supported in part by USPHS grants AA-03539, MH-31302, MH-25430, and MH-00048 to Dr. Cloninger.

REFERENCES

1. Åmark, C. (1951): A study in alcoholism. *Acta Psychiatr. Neurol. Scand. [Suppl.]*, 70:
2. American Psychiatric Association (1980): *Diagnostic and Statistical Manual of Mental Disorders, Third Edition*. American Psychiatric Association, Washington.
3. Bailey, R. J., Krasner, N., Eddleston, A., and Williams, R. (1976): Histocompatibility antigens, autoantibodies and immunoglobulins in alcoholic liver disease. *Br. Med. J.*, 2:727–729.
4. Blass, J. P., and Gibson, G. E. (1977): Abnormality of a thiamine-requiring enzyme in patients with Wernicke–Korsakoff syndrome. *N. Engl. J. Med.*, 297:1367–1370.
5. Bleuler, M. (1955): Familial and personal background of chronic alcoholics. In: *Etiology of Chronic Alcoholism*, edited by O. Diethelm, pp. 110–166. Charles C. Thomas, Springfield, Illinois.
6. Bohman, M. (1978): Some genetic aspects of alcoholism and criminality: A population of adoptees. *Arch. Gen. Psychiatry*, 35:269–276.
7. Cadoret, R. J. (1978): Psychopathology in adopted-away offspring of biological parents with antisocial behavior. *Arch. Gen. Psychiatry*, 35:176–184.
8. Cadoret, R. J., and Cain, C. (1980): Sex difference in predictors of antisocial behavior in adoptees. *Arch. Gen. Psychiatry*, 37:1171–1175.
9. Cadoret, R. J., Cain, C. A., and Grove, W. M. (1980): Development of alcoholism in adoptees raised apart from biologic relatives. *Arch. Gen. Psychiatry*, 37:561–563.

10. Cadoret, R. J., and Gath, A. (1978): Inheritance of alcoholism in adoptees. *Br. J. Psychiatry*, 132:252–258.
11. Camps, F. E., and Dodd, B. E. (1967): Increase in the incidence of non-secretors of ABH blood groups substances among alcoholic patients. *Br. Med. J.*, 1:30–31.
12. Camps, F. E., Dodd, B. E., and Lincoln, P. J. (1969): Frequencies of secretors and non-secretors of ABH group substances among 1000 alcoholic patients. *Br. Med. J.*, 4:457–459.
13. Christiansen, K. O. (1974): Seriousness of criminality and concordance among Danish twins. In: *Crime, Criminology and Public Policy*, edited by R. Hood, pp. 63–77. Heinemann, London.
14. Cloninger, C. R. (1979): The link between hysteria and sociopathy. In: *Psychiatric Diagnosis: Exploration of Biological Predictors*, edited by H. S. Akiskal and W. L. Webb, pp. 189–218. Spectrum, New York.
15. Cloninger, C. R., Bohman, M., and Sigvardsson, S. (1981): Inheritance of alcohol abuse: Cross-fostering analysis of adopted men. *Arch. Gen. Psychiatry*, 38:861–868.
16. Cloninger, C. R., Christiansen, K. O., Reich, T., and Gottesman, I. I. (1978): Implications of sex differences in the prevalence of antisocial personality, alcoholism, and criminality for models of familial transmission. *Arch. Gen. Psychiatry*, 35:841–851.
17. Cloninger, C. R., Lewis, C. E., Rice, J., and Reich, T. (1981): Strategies for resolution of biological and cultural inheritance. In: *Genetic Research Strategies in Psychobiology and Psychiatry*, edited by E. S. Gershon, S. Matthysse, X. O. Breakfield, and R. D. Ciaranella, pp. 319–332. Boxwood Press, Pacific Grove, California.
18. Cloninger, C. R., Reich, T., and Guze, S. B. (1975): The multifactorial model of disease transmission. II. Sex differences in the familial transmission of sociopathy (antisocial personality). *Br. J. Psychiatry*, 127:11–12.
19. Cloninger, C. R., Reich, T., and Guze, S. B. (1975): The multifactorial model of disease transmission. III. Familial relationship between sociopathy and hysteria (Briquet's syndrome). *Br. J. Psychiatry*, 127:23–32.
20. Cloninger, C. R., Reich, T., Suarez, B., Rice, J., and Gottesman, I. I. (1982): The principles of psychiatric genetics. In: *Handbook of Psychiatry*, edited by M. Shepard. Cambridge University Press, Cambridge *(in press)*.
21. Cloninger, C. R., Reich, T., and Wetzel, R. (1979): Alcoholism and affective disorders: Familial associations and genetic models. In: *Alcoholism and Affective Disorders*, edited by D. Goodwin and C. Erickson, pp. 57–86. Spectrum, New York.
22. Cloninger, C. R., Rice, J., and Reich, T. (1979): Multifactorial inheritance with cultural transmission and assortative mating. II. A general model of combined polygenic and cultural inheritance. *Am. J. Hum. Genet.*, 31:176–198.
23. Cloninger, C. R., Rice, J., and Reich, T. (1979): Multifactorial inheritance with cultural transmission and assortative mating. III. Family structure and the analysis of separation experiments. *Am. J. Hum. Genet.*, 31:366–388.
24. Cloninger, C. R., Rice, J., Reich, T., and McGuffin, P. (1982): Genetic analysis of seizure disorders and multidimensional threshold characters. In: *Genetics and Epilepsy*, edited by V. E. Anderson, W. A. Hauser, and C. Sing. pp. 291–309. Raven Press, New York.
25. Crowe, R. D. (1974): An adoption study of antisocial personality. *Arch. Gen. Psychiatry*, 31:785–791.
26. Dalgard, D. S., and Kringlen, E. (1976): A Norwegian twin study of criminality. *Br. J. Criminol.*, 16:213–232.
27. Eaves, L. J., and Eysenck, H. J. (1975): The nature of extraversion: A genetical analysis. *J. Pers. Soc. Psychol.*, 32:102–112.
28. Eaves, L. J., and Eysenck, H. J. (1976): Genetic and environmental components of inconsistency and unrepeatability in twins' responses to a neuroticism questionnaire. *Behav. Genet.*, 6:145–160.
29. Eaves, L. J., and Eysenck, H. J. (1977): A genotype-environmental model for psychoticism. *Adv. Behav. Res. Ther.* 1:5–26.
30. Evenson, R. C., Altman, H., Sletten, I. W., and Knowles, R. R. (1973): Factors in the description and grouping of alcoholics. *Am. J. Psychiatry*, 130:49–54.
31. Eysenck, H. J. (1977): *Crime and Personality*, Third Edition. Routledge and Regan Paul, London.
32. Eysenck, H. J., and Eysenck, S. B. G. (1974): *Eysenck Personality Questionnaire*. Educational and Industrial Testing Service, San Diego.
33. Farrington, D. P. (1979): Longitudinal research on crime and delinquency. In: *Crime and Justice: An Annual Review of Research, Vol. 1*, edited by N. Morris and M. Tonry, pp. 289–348. University of Chicago Press, Chicago.

34. Feighner, J., Robins, E., Guze, S., Woodruff, R., Winokur, G., and Munoz, R. (1972): Diagnostic criteria for use in psychiatric research. *Arch. Gen. Psychiatry*, 26:57–63.
35. Fraser, F. C. (1980): Evolution of a palatable multifactorial threshold model. *Am. J. Hum. Genet.*, 32:796–831.
36. Goldstein, S. G., and Linden, J. D. (1967): Multivariate classification of alcoholics by means of the MMPI. *J. Abnorm. Psychol.*, 74:661–669.
37. Goodwin, D. W., Schulsinger, F., Hermansen, L., Guze, S. B., and Winokur, G. (1973): Alcohol problems in adoptees raised apart from alcoholic biological parents. *Arch. Gen. Psychiatry*, 28:238–243.
38. Goodwin, D. W., Schulsinger, F., Moller, N., Hermansen, L., Winokur, G., and Guze, S. B. (1974): Drinking problems in adopted and non-adopted sons of alcoholics. *Arch. Gen. Psychiatry*, 31:164–169.
39. Guze, S. B. (1976): *Criminality and Psychiatry Disorders*. Oxford University Press, New York.
40. Haertzen, C. A. (1969): Manual for alcoholism scales of the inventory of habits and attitudes. *Psychol. Rep.*, 25:947–973.
41. Hare, R. D., and Schallings, D. (1978): *Psychopathic Behavior: Approaches to Research*. John Wiley & Sons, New York.
42. Hill, S., Cloninger, C. R., and Ayre, F. (1977): Independent familial transmission of alcoholism and opiate abuse. *Alcoholism Clin. Exp. Res.*, 1(4):335–342.
43. Hill, S. Y., Goodwin, D. W., Cadoret, R., Osterland, C. K., and Doner, S. (1975): Association and linkage between alcoholism and eleven serological markers. *J. Stud. Alcohol*, 36:981–992.
44. Horn, J. L., and Wanberg, K. W. (1969): Symptom patterns related to excessive use of alcohol. *Q. J. Stud. Alcohol*, 30:35–58.
45. Horn, J. L., and Wanberg, K. W. (1970): Dimensions of current and background patterns found among excessive drinkers. *Q. J. Stud. Alcohol.*, 31:633–658.
46. Hutchings, B., and Mednick, S. A. (1977): Criminality in adoptees and their adoptive and biological parents. A Pilot Study. In: *Biosocial Bases in Criminal Behavior*, edited by S. A. Mednick and K. O. Christiansen, pp. 127–141. Gardner Press, New York.
47. Jellinek, E. M. (1952): Phases of alcohol addiction. *Q. J. Stud. Alcohol*, 13:573–584.
48. Kaij, L. (1960): *Alcoholism in Twins: Studies on the Etiology and Sequels of Abuse of Alcohol*. Almquist and Wiksell, Stockholm.
49. Kaij, L., and Dock, J. (1975): Grandsons of alcoholics. *Arch. Gen. Psychiatry*, 32:1379–1381.
50. Kopun, M., and Propping, P. (1977): The kinetics of ethanol absorption and elimination in twins and supplementary repetitive experiments in singleton subjects. *Eur. J. Clin. Pharmacol.*, 11:337–344.
51. Lawlis, G. F., and Rubin, S. E. (1971): Study of personality patterns in alcoholics. *Q. J. Stud. Alcohol*, 32:318–327.
52. Loosen, P. T., and Prange, A. J., Jr. (1980): Thyrotropin releasing hormone (TRH): A useful tool for psychoneuroendocrine investigation. *Psychoneuroendocrinology*, 5:63–80.
53. MacAndrew, C. (1967): Self-reports of male alcoholics. *Q. J. Stud. Alcohol*, 29:43–51.
54. Martin, R. L., Cloninger, C. R., and Guze, S. B. (1978): Female criminality and the prediction of recidivism: A prospective six year follow-up. *Arch. Gen. Psychiatry*, 35:207–214.
55. Martin, R. L., Cloninger, C. R., and Guze, S. B. (1979): The evaluation of diagnostic concordance in follow-up studies: II. A blind prospective follow-up of female criminals. *J. Psychiatr. Res.*, 15:107–125.
56. Matthysse, S., and Kidd, K. K. (1981): Pattern recognition in genetic analysis. In: *Genetic Research Strategies in Psychobiology and Psychiatry*, edited by E. S. Gershon, S. Matthysse, X. O. Breakefield, and R. D. Ciaranello, pp. 333–340. Boxwood Press, Pacific Grove, California.
57. McKusick, V. (1978): *Mendelian Inheritance in Man*, Fifth Edition. Johns Hopkins University Press, New York.
58. Mednick, S. A., and Christiansen, K. O. (1977): *Biosocial Bases of Criminal Behavior*. Gardner Press, New York.
59. Mednick, S. A., and Volavka, J. (1980): Biology and crime. In: *Crime and Justice: Annual Review of Research, Vol. 2*, edited by N. Morris and M. Tonry, pp. 86–158. University of Chicago Press, Chicago.
60. Megargee, E. I. (1972): *The California Psychological Inventory Handbook*. Jossey-Bass, San Francisco.
61. Miller, W. R. (1976): Alcoholism scales and objective assessment methods: A review. *Psychol. Bull.*, 83:649–674.

62. Mogar, R. E., Wilson, W. M., and Helm, S. T. (1970): Personality subtypes of male and female alcoholic patients. *Int. J. Addict.*, 5:99–133.
63. Murray, R. M., and Gurling, H. M. D. (1980): Genetic contributions to normal and abnormal drinking. In: *Psychopharmacology of Alcohol*, edited by M. Sandler, pp. 90–105. Raven Press, New York.
64. Omenn, G. S., and Motulsky, A. G. (1972): A biochemical and genetic approach to alcoholism. *Ann. N.Y. Acad. Sci.*, 197:16–23.
65. Park, P., and Whitehead, P. (1973): Developmental sequence and dimensions of alcoholism. *Q. J. Stud. Alcohol*, 34:887–904.
66. Partanen, J., Brunn, K., and Markkanen, T. (1966): *Inheritance of Drinking Behavior*. The Finnish Foundation for Alcohol Studies, Helsinki.
67. Partington, J. T., and Johnson, F. G. (1969): Personality types among alcoholics. *Q. J. Stud. Alcohol*, 30:21–34.
68. Propping, P., Kruger, J., and Janah, A. (1980): Effect of alcohol on genetically determined variants of the normal electroencephalogram. *Psychiatry Res.*, 2:85–90.
69. Reich, T., Cloninger, C. R., and Guze, S. B. (1975): The multifactorial model of disease transmission. I. Description of the model and its use in psychiatry. *Br. J. Psychiatry*, 127:1–10.
70. Reich, T., Rice, J., Cloninger, C. R., and Lewis, C. (1980): Path analysis of the segregation distribution for alcoholism. In: *Social Consequences of Psychiatric Illness*, edited by L. Robins, P. Clayton, and J. Wing, pp. 91–119. Brunner/Mazel, New York.
71. Reich, T., Rice, J., Cloninger, C. R., Wette, R., and James, J. (1979): The use of multiple thresholds and segregation analysis in analyzing the phenotypic heterogeneity of multifactorial traits. *Ann. Hum. Genet.*, 42:371–390.
72. Reich, T., Suarez, B., Rice, J., and Cloninger, C. R. (1980): Current direction in genetic epidemiology. In: *Current Developments in Anthropological Genetics, Vol. I*, pp. 229–326. Plenum Press, New York.
73. Reich, T., Winokur, G., and Mullaney, J. (1975): The transmission of alcoholism. In: *Genetic Research in Psychiatry*, edited by R. R. Fieve, D. Rosenthal, and H. Brill, pp. 259–272. Johns Hopkins University Press, Baltimore.
74. Robins, L. N. (1966): *Deviant Children Grown Up*. Williams & Wilkins, Baltimore.
75. Schallings, D. (1978): Psychopathy-related personality variables and the psychophysiology of socialization. In: *Psychopathic Behavior*, edited by R. D. Hare and D. Schallings, pp. 85–106. John Wiley & Sons, New York.
76. Schallings, D., Cronholm, B., Asberg, M., and Espmark, S. (1973): Ratings of psychic and somatic anxiety indicants: Interrater reliability and relations to personality variables. *Acta Psychiatr. Scand.*, 49:353–368.
77. Schuckit, M. A., Goodwin, D. A., and Winokur, G. (1972): A study of alcoholism in half-siblings. *Am. J. Psychiatry*, 129:1132–1136.
78. Schulsinger, F. (1972): Psychopathy: Heredity and environment. *Int. J. Ment. Health*, 1:190–206.
79. Shields, J. (1978): Genetics and alcoholism. In: *Alcoholism: New Knowledge and New Responses*, edited by G. Edwards and M. Grant, pp. 117–136. Croon Helm, London.
80. Swinson, R. P., and Madden, J. S. (1973): ABO blood groups and ABH substance secretion in alcoholics. *Q. J. Stud. Alcohol*, 34:64–70.
81. Thoday, J. M. (1967): New insights into continuous variation. In: *Proceedings of the Third International Congress of Human Genetics*, edited by J. F. Crow and J. V. Neel, pp. 339–350. Johns Hopkins University Press, Baltimore.
82. Wanberg, K. W., and Knapp, J. (1970): A multidimensional model for research and treatment of alcoholism. *Int. J. Addict.*, 5:69–98.
83. West, D. J., and Farrington, D. P. (1973): *Who Becomes Delinquent?* Heinemann, London.
84. Winokur, G., Reich, T., Rimmer, J., and Pitts, F. N. (1970): Alcoholism. III. Diagnosis and familial psychiatric illness in 259 alcoholic probands. *Arch. Gen. Psychiatry*, 23:104–111.
85. Winokur, G., Rimmer, J., and Reich, T. (1971): Alcoholism. IV. Is there more than one type of alcoholism? *Br. J. Psychiatry*, 118:525–531.
86. Winokur, G., Tanna, V., Elston, R., and Go, R. (1976): Lack of association of genetic traits with alcoholism: C3, Ss, and ABO systems. *J. Stud. Alcohol*, 37:1313–1315.
87. Wirt, R. D., and Briggs, P. F. (1959): Personality and environmental factors in the development of delinquency. *Psychol. Monogr.*, 73:(15 No. 415).

88. Witkin, H. A., Mednick, S. A., Schulsinger, F., Bakkestrom, E., Christiansen, K. O., Goodenough, D. R., Herschorn, K., Lundsteen, C., Owen, D. R., Phillip, J., Rubin, D. B., and Stocking, M. (1976): Criminality in XYY and XXY men. *Science*, 193:547–555.
89. Wolfgang, M. E. (1973): Crime in a birth cohort. *Proc. Am. Phil. Soc.*, 117:404–411.
90. Wright, S. (1968): *Evolution and the Genetics of Populations, Vol. I.* University of Chicago Press, Chicago.

Genetics of Neurological and Psychiatric Disorders, edited by Seymour S. Kety, Lewis P. Rowland, Richard L. Sidman, and Steven W. Matthysse. Raven Press, New York © 1983.

Application of Recombinant DNA Techniques to Neurogenetic Disorders

*David Housman and **James F. Gusella

*Center for Cancer Research, Massachusetts Institute of Technology, Cambridge, Massachusetts 02139; and **Department of Neurology, Massachusetts General Hospital, Boston, Massachusetts 02114*

In this chapter we analyze the application of recombinant DNA methodology to the diagnosis and understanding of neurogenetic disorders. Although we shall use Huntington's Disease (HD) as a model to illustrate the basic principles of the approach that we believe will be most fruitful, a similar approach is applicable to a wide variety of inherited neurological disorders.

The scientific literature about HD reveals two basic facts: (a) it is inherited through an autosomal dominant gene that has a high degree of penetrance (3), and (b) the disease results in a progressive degeneration of particular areas of the brain (1). The strategy that we propose to approach the problem of HD is based initially on the first of these facts. Our first goal will be to determine the precise genetic site at which the HD gene is located. The underlying principles that permit localization of specific genes on the human genetic map have been outlined previously (2). The strategy commonly employed is to use genetic linkage of markers that are segregating in a family to determine on which chromosome and in which segment of that chromosome the gene causing the disease is located. The limitation of this approach is that too few genetic markers are available to saturate the human genetic map. The markers currently available have been useful only in testing for potential linkage of HD in approximately 15% of the human genome (4). The present data indicate that the HD gene is not likely to be located within these areas of the human genome. In order to determine the site of the HD gene, many more markers are clearly required. The most promising strategies for generating these additional markers depend on recombinant DNA technology.

RESTRICTION FRAGMENT LENGTH POLYMORPHISM AS GENETIC MARKERS FOR LINKAGE STUDIES

All markers used in genetic linkage studies, whether they refer to visible characteristics such as eye color, or biochemical characteristics such as the mobility of an enzyme in a gel electrophoretic separation system, depend on the same basic principle: the two alleles must differ in the order of those nucleotides in the DNA that are responsible for the phenotype characteristic of the gene. The advantages of recombinant DNA technology for creating new genetic markers are that the sequence of the nucleotides in the DNA is used directly to create a genetic marker for linkage studies and that the technology can be applied to all segments of the genome with equal ease. One of the difficulties in previous linkage studies is that genetic markers have not been available for all segments of all chromosomes. With the recombinant DNA methodology this difficulty can be resolved.

An important tool in this undertaking is the ability to split DNA molecules at specific sequences of nucleotides. The four nucleoside bases adenosine, guanosine, cytosine, and thymidine are strung together in a linear sequence in the DNA, and the order of the bases encodes the information content of the DNA molecule. Specific enzymes, termed restriction endonucleases, which have been isolated from a variety of bacteria, have the capacity to cut DNA molecules at specific points at which a particular order of bases is found. Thus, for example, the restriction endonuclease EcoRI will cleave DNA from any source at any point at which the base sequence GAATTC is found. The DNA that is cleaved by the restriction endonuclease can then be separated on the basis of its size by gel electrophoresis. For a simple organism like a virus, the DNA cleaved by a restriction enzyme gives a very simple pattern of fragments that serves as a signature for that particular virus. If a variant virus arose that contained an extra piece of DNA, a deleted fragment of DNA, or a change in the recognition sites for one of the cleavage sites of the restriction enzyme, then it would yield a different pattern of DNA fragments after digestion with the restriction endonuclease and gel electrophoresis. For such a simple organism, the small number of DNA fragments created by this procedure can be observed directly by staining the DNA after the gel electrophoresis has been completed. A schematic example is shown in Table 1. The simple procedure of digesting the DNA with the restriction endonuclease and separating the fragments by gel electrophoresis allows us to visualize directly a difference between a variant virus DNA and the normal virus DNA.

For more complex organisms, however, the number of DNA fragments produced by cleavage with a restriction endonuclease is quite large. The human genome digested with the restriction endonuclease EcoRI, for example, yields approximately 500,000 to a million different fragments. Although any one of these fragments could be subject to allelic variation in a fashion analogous to the variations in the virus DNA given in the example, the individual fragment that varies cannot be seen directly by this staining procedure because of the large number of other fragments that mask the variant fragment. A procedure can be used, however, that allows one to observe a single fragment or group of fragments apart from all of the others produced by digestion by the restriction endonuclease. This approach relies on the ability to isolate specific segments of the human genome by recombinant DNA techniques. A number of strategies for isolating such DNA segments are available. For the purposes of this discussion, it is most important to realize that an isolated DNA segment is a useful reagent for determining the allelic variation at a single

TABLE 1. *Map of the DNA molecule of a virus 10,000 base pairs (bp) long*

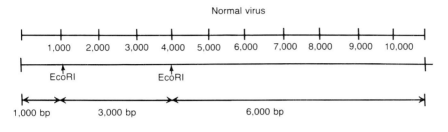

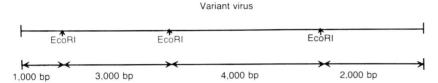

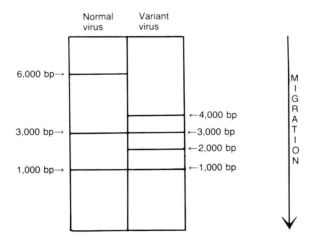

The sequence GAATTC occurs twice in the sequence of the normal virus so the restriction enzyme EcoRI cuts the virus into three pieces 1,000 base pairs, 3,000 base pairs, and 6,000 base pairs in length, respectively. The base sequence of the variant virus is similar to that of the normal virus except that the sequence GAATTC appears three times in the viral DNA sequence. As a result, the variant virus DNA is cut into four fragments by the reduction enzyme EcoRI. These fragments include the 1,000 and 3,000 base-pair fragments found in the normal virus, but instead of the normal 6,000 base-pair fragment, two smaller fragments 4,000 and 2,000 base pairs in length, respectively, are found. The diagram below shows the pattern that is observed when the DNA of the normal and variant virus digested with EcoRI is separated by agarose gel electrophoresis. This separation technique causes DNA fragments to migrate as a function of molecular weight, with smaller fragments migrating most rapidly. DNA fragments can be visualized directly in the gel by several alternative staining procedures.

point on a human chromosome. The isolated DNA segment is able to act as a probe to identify the segment in the genome of interest in the restriction endonuclease digest apart from all others. The principle underlying this process is that the DNA molecule is a double helix in which the nucleotide bases are paired. Adenosine residues will match only with thymidine residues, and guanosine residues will match only with cytosine residues. The two strands of the DNA helix can be physically separated by heating the DNA or subjecting the DNA to alkaline conditions. The DNA strands, however, have a specific affinity for each other that is dependent on the precise order of the nucleotides in the strand. Two DNA strands with exactly complementary sequences have a much higher physical affinity for each other than for any other sequence of DNA (hybridization). This property of matching strands of the DNA molecule can be used to locate a single DNA sequence in the presence of a large number of other DNA sequences. To accomplish this goal, a single DNA sequence that can be isolated by recombinant DNA techniques is made radioactive. It can then be mixed with the separated DNA fragments that compose the entire human genome. Only in the positions in the gel at which the sequences are complementary to this radioactive probe DNA molecule will radioactive DNA be bound. X-ray film can be used to detect the position at which the radioactive probe DNA has bound. In this way the cloned DNA segment from the human genome can be used as a probe to allow us to observe the position at which sequences complementary to it are found in the gel.

If the DNA from two individuals is run in parallel lanes on the gel and it has the same allele with respect to a given restriction endonuclease fragment length polymorphism, then the same pattern will be seen in the two lanes when the appropriate hybridization probe is reacted with the DNA of the gel. If, however, the two individuals exhibit an allelic difference for this particular DNA segment, a different pattern will be observed between them after the probe molecule is reacted with the DNA that has been separated on the gel. One advantage of this procedure is that the basic technology is the same for markers on all chromosomes. In fact, the same gel can be reacted with one hybridization probe, the result observed by placing X-ray film against the gel, and the radioactive probe can be washed off the gel. A second radioactive probe representing a DNA sequence from a different part of the genome can then be applied and the gel can be reused in this way a second, third, or fourth time. The only limiting factor in this process is the isolation and identification of a sufficient number of probe DNA molecules that expose useful polymorphisms segregating in the pedigrees in which we may be interested. The current effort in this area of research is being directed towards collecting enough of these probe DNA molecules so that the entire human genome is essentially covered.

USE OF MARKERS TO DETERMINE THE CHROMOSOME ON WHICH THE HD GENE IS LOCATED

Every human being receives half of his or her chromosomes from each parent. Therefore, each time a sperm or egg is formed, there is a 50% chance that a particular genetic region will be derived from paternal genes and a 50% chance that it will be derived from maternal genes. If for any individual in a pedigree in which the HD gene was segregating we could visualize precisely the contributions that were originally derived from each paternal and maternal grandparent, we would be in a position to see which genetic region was always associated with the presence of HD in that pedigree. Although direct visualization of differences between the homologous chromosomes in an individual is not readily achievable under the microscope, the DNA techniques outlined above allow this distinction to be made. In principle, if no exchange between paternal and maternal chromosomes occurred during the formation of gametes, only 22 pairs of alleles would be required to determine on which chromosome the HD gene was located. However, because crossing over can occur at meiosis, in larger chromosomes the ability to place a marker on one portion of the chromosome cannot determine the parent of origin of another region of the chromosome. The solution to this problem is to place many markers over the length of each chromosome. It has been calculated that approximately 150 to 300 markers would be required to cover adequately the human genetic map to detect linkage for a gene such as the HD gene with the sizes of pedigrees that are available.[1]

A further advantage of the recombinant DNA technology applied in this context is that when the HD gene is tentatively located, many more DNA probes can be obtained that cover the region in which the gene is likely to be found. Two advantages accrue to the saturation of a genetic region with markers: the ability to confirm or deny with great precision the linkage of the HD gene to the region in question, and the ability to use the DNA found in this region as a starting point to assess the function that is abnormal in HD.

USING CHROMOSOMAL DNA SEGMENTS TO UNDERSTAND THE BIOCHEMICAL BASIS FOR HD

After the precise region in which the HD gene is located has been determined, it is technically feasible to isolate all of the chromosomal DNA in this region using recombinant DNA techniques. This might amount to perhaps 1 to 2 million base pairs of DNA. Our second task will be to determine what function coded for by this DNA is aberrant in HD. Although this may seem formidable considering the large amount of information coded for by such a genetic region, it is useful to realize that by reducing the region in which the HD gene is located to approximately a million base pairs we have simplified the problem over 1,000-fold: the human genome contains approximately 3 billion base pairs. Our goal will be to determine which genes from the region to which the HD gene has been localized are functional in the cells in which the primary lesion in HD is present. The technical tool used in this endeavor is similar to the method used for the hybridization reaction described

---

[1] If it were not possible to space the markers evenly along the chromosomes, a much larger number of markers would be needed (K. Lange and M. Boehnke, *in preparation*).

above. The genetic information in the DNA is translated into an active program of gene activity via an intermediate type of nucleic acid, the mRNA. The hybridization reaction, which allows one DNA molecule to identify another, also allows a radioactive DNA molecule to identify a mRNA molecule on the basis of its nucleotide sequence. Using techniques analogous to those described in the previous section, we shall be able to construct a complete map of all mRNA sequences that are coded by the region in which the HD gene is located. This map can be taken a step further to identify the polypeptide chains that are coded in this region. In this way the abstract information on the location of the HD gene can be translated into a functional map of the most likely candidates for the polypeptide chain that is aberrant in HD.

At this point the HD problem again enters the realm of protein biochemistry. The successful application of the genetic approach to the problem should greatly simplify efforts to identify the abnormal gene product and function in this disease. Furthermore, a similar approach could potentially be used for any number of genetic diseases for which no primary biochemical defect has yet been identified.

REFERENCES

1. Bruyn, G. W., Bots, G. T. A. M., and Dom, R. (1979): Huntington's chorea: Current neuropathological status. *Adv. Neurol.*, 23:83–93.
2. McKusick, V. A., and Ruddle, F. H. (1977): The status of the gene map of the human chromosomes. *Science*, 196:390–405.
3. Wallace, D. C. (1979): Distortion of mendelian segregation in Huntington's disease. *Adv. Neurol.*, 23:73–81.
4. Went, L. N., and Volkers, W. S. (1979): Genetic linkage. *Adv. Neurol.*, 23:37–42.

Genetics of Neurological and Psychiatric
Disorders, edited by Seymour S. Kety, Lewis P.
Rowland, Richard L. Sidman, and Steven W.
Matthysse. Raven Press, New York © 1983.

Recombinant DNA and the Analysis of Cytogenetic Disorders Associated with Mental Retardation

P. S. Gerald and *G. A. Bruns

Clinical Genetics Division, Children's Hospital Medical Center, Boston, Massachusetts 02115; and *Department of Pediatrics, Harvard Medical School, Boston, Massachusetts 02115

Chromosome abnormalities collectively are the etiology in perhaps 10% or more of individuals with mental retardation (MR). In the past, trisomy 21 (Down's syndrome) has been numerically predominant, and only a few other cytogenetic disorders have been singled out as contributing significantly to the population of individuals with MR. This position is changing in the light of recent findings, as exemplified by the developments in nonspecific, X-linked MR and in the Prader-Willi syndrome.

CYTOGENETIC TECHNIQUES FOR DISCOVERY OF CHROMOSOMAL ABNORMALITIES

In general, these new developments have resulted from the introduction of new cytogenetic techniques and from a better appreciation of the significance of subtle cytogenetic changes. Because of its numerical importance, this chapter will begin with an account of what currently is called "nonspecific, X-linked MR." During the last 20 years or so, a number of large kindreds exhibiting X-linked MR has been reported (21). The affected individuals in these kindreds have generally had few distinctive phenotypic features, other than the MR. A large family described by Renpenning et al. (10) has frequently been used as the prototype for this category of MR, which led initially to the use of the label "Renpenning's syndrome."

Although X-linked MR as a whole is now realized to be approximately as frequent among males as is Down's syndrome (22), this information initially had little impact on clinical investigators. This was in consequence of the presumptive absence of any distinctive physical features. The situation changed dramatically and abruptly with the recognition of a cytogenetic abnormality of the X chromosome in some individuals with Renpenning's syndrome (or nonspecific, X-linked MR, as this is now more commonly called). The discovery of this abnormality was accomplished by Dr. Herbert Lubs while working, most appropriately, at an NICHD-supported

Mental Retardation Research Center (8). The abnormality described by Lubs consisted of a constriction (narrowing) of the X chromosomes at a specific region [tentatively identified as Xq27 or Xq28 (16)]. The thinned region apparently affects the stability of the chromosome, since breaks in the constricted region occur frequently and serve as justification for the commonly used term "fragile X."

The absence of reports of additional patients with the fragile X during the years immediately following Lubs' publication led many to believe this was a unique observation. The full significance of Lubs' discovery had to await the investigations of Sutherland, who demonstrated that the detectability of the fragile site was highly dependent on the culture medium used (13). It is now evident that the culture medium must be deficient in folate and thymidine in order for the fragile X phenomenon to be expressed (14). When this deficient form of culture medium is employed, fragile sites are found on other chromosomes as well as on the X (15). Each of these fragile sites occurs at a specific location on a given chromosome, and each is heritable independently of the fragile sites on other chromosomes. Only the fragile site on the X chromosome, however, is known to be associated with clinical findings.

Notwithstanding the information now available, the molecular basis for these constricted, fragile regions is quite unknown. Indeed, it is likely that no single explanation will suffice, since some fragile sites are expressed even when nondeficient medium is used, whereas one of them requires the presence of 5-bromodeoxyuridine (BUdR) for its expression (12,18). The constancy in location of each fragile site and their heritability may possibly indicate that the fragile site is the result of an alteration in the deoxyribonucleic acid (DNA) sequence of the region.

Even when medium is used that optimizes the expression of the fragile site on the X chromosome, only a portion (10 to 50%) of metaphases from an affected male exhibits the X chromosome abnormality. In the heterozygous female, an even smaller proportion of metaphases shows the abnormality. Indeed, as many as two-thirds of obligate heterozygotes have fewer than 1% of metaphases with a fragile X (20). Even this low level of expression of the fragile site is attained only with peripheral leukocyte cultures; metaphases from cultured fibroblasts derived from known hemizygotes and heterozygotes frequently do not show any abnormality (4). A recent publication suggests that conditions may be found that will enhance the formation of a fragile site in fibroblastic metaphases (2).

The determination of culture conditions required for the demonstration of the fragile X in cultured leukocytes made it possible to survey families with nonspecific, X-linked MR for the presence of this abnormality. A rather large number of fragile X families has been detected as the result of such surveys, and the phenotype of affected males has now been rather well delineated (6,20). Surprisingly, most affected males have significant macroorchidism, although the enlargement is minimal before puberty (17). In some instances, only one testis is enlarged (9). An occasional adult fragile X male will lack macroorchidism, even though other males in his kindred have typical testicular enlargement (5). The intellectual deficit is usually mild to moderate, although severe retardation can occur (6,20). The head circumference is normal or increased, which distinguishes these patients from many other retarded individuals (20). The external ears are commonly enlarged and the symphysis of the jaw is prominent (20). A few investigators believe there is a characteristic speech pattern (5).

Although women heterozygous for the fragile X were initially believed to be unaffected, this viewpoint now appears to be incorrect. In a survey of special schools for women (with media appropriate for identifying the fragile site), 5 out of 128 women examined had the fragile X (19). Among the females in this group with normal head circumference and without gross physical abnormalities, the incidence of the fragile X was 7%. This incidence is striking, particularly in view of the recognized difficulty in cytogenetically demonstrating the fragile X in obligate heterozygotes. The proportion of metaphases with the fragile X is apparently greater in younger females (<20 years), as well as in those females with intellectual deficit (19). From a study of female relatives of affected proposita, it has been estimated that one-third of heterozygotes have a degree of intellectual deficit (19). The range of intelligence quotient (IQ) found among heterozygotes could be a consequence of the Lyonization of the X chromosome, but there is no objective evidence to support this impression. The fragile site, for example, can be demonstrated with nearly equal frequency in the early and late replicating X's (5,9).

The fragile X is found in only a portion, perhaps half, of families with nonspecific, X-linked MR. The Renpenning family, for example, does not possess the fragile X and appears to represent a separate syndrome (1). It is probable that several other syndromes are included within the category of nonspecific, X-linked MR without the fragile X (21).

The discovery of the fragile X disorder has focused an appropriate level of interest on X-linked MR. It is unfortunate that, despite the ease of demonstrating the fragile X, there has not yet been any increased insight into the pathogenesis of the MR in this disorder.

Our present understanding of the fragile X disorder was an outgrowth of a technologic development that, in turn, resulted from the study of a new class of cytogenetic phenomena, the fragile sites on chromosomes. A different type of technologic advance—high resolution cytogenetics (24)—has resulted in the discovery of a cytogenetic abnormality in Prader-Willi syndrome. This relatively rare syndrome is well documented in medical texts and is characterized by obesity, short stature, hypogonadism, MR, and infantile hypotonia.

Initially it was recognized that abnormalities of chromosome 15 occurred with remarkably increased frequency in patients with the Prader-Willi syndrome (23). Cytogenetic studies demonstrated various translocations involving chromosome 15 in these patients, although the translocation at times appeared to be balanced (without obvious missing or excess chromosomal material). Ledbetter et al. (7) have recently demonstrated, in 4 Prader-Willi patients, a very small interstitial deletion in the proximal region of 15q, in the absence of any gross translocation.

The ability to demonstrate these very small deletions is a tribute to the usefulness of high-resolution cytogenetics. Yunis nearly single-handedly has demonstrated that synchronized leukocyte cultures yield a high frequency of cells in prophase (25,26). The highly extended chromosomes produced by these procedures reveal much more detail than do conventional chromosome preparations. The extended chromosomes permit the detection of small alterations, such as deletions, that otherwise would frequently be missed. It is expected that small deletions may play a significant role in the etiology of a number of diseases for which no cytogenetic basis has been suspected. This has already proven to be true, for example, for some instances of retinoblastoma and Wilms' tumor (27,28).

USE OF RECOMBINANT DNA EXPERIMENTS FOR CYTOGENETIC ANALYSIS

In both the fragile X and deletion-based Prader-Willi syndromes a further understanding of the disorders will require a more detailed understanding of the genetic elements involved. It seems unlikely that much more information can be obtained with present cytogenetic techniques. New developments in the area of recombinant DNA, however, appear likely to provide the needed information.

Published recombinant DNA experiments have tended to concentrate on genetic elements whose products [messenger ribonucleic acid (RNA) and protein] have been relatively easily isolated. This approach is not applicable to such disorders as the fragile X or Prader-Willi syndromes, which are known only as clinical or cytogenetic entities. Fortunately, new techniques such as those described by Housman and Gusella *(this volume)* may permit an attack upon genetic elements whose location in human gene map is known. An example of this approach, as applied to X-linked diseases, will illustrate how this may be accomplished.

The authors, in collaboration with Drs. Gusella and Housman, have begun isolating random DNA segments from the human X chromosome, as part of an investigation of the fragile X syndrome. The procedures are the same as those initially used by Gusella et al. for the isolation of DNA segments from human chromosome 11 (3). In outline form these are: (a) segregation of the X chromosome from the rest of the human chromosome complement in human-rodent hybrid cells; (b) fragmentation of the hybrid cell DNA by endonuclease and insertion of the DNA fragments into bacteriophage; (c) amplification and isolation of the individual recombinant bacteriophage with a human DNA insert.

The segregation of the human X chromosome in human-rodent hybrids is a well-known procedure (11). Hybrid cells, prepared by fusing human cells (such as leukocytes) with cells from an established rodent line, can be readily produced. The resulting hybrid cell will multiply indefinitely in culture and will progressively lose human chromosomes. Retention of the human X can be forced by requiring that the hybrid cell depend on the presence of a specific human X-linked gene for survival. [The enzyme hypoxanthine-guanine phosphoribosyl transferase (HPRT) is generally chosen for this purpose.] Since the loss of the human chromosomes is relatively random, it is possible to identify a hybrid cell line that contains only the human X, or at most one or two other human chromosomes.

The incorporation of the endonuclease-digested hybrid cell DNA into bacteriophage is a standard procedure (3) and requires no further discussion. The means of identifying bacteriophage with a human DNA insert was developed by Gusella et al. (3). Their procedure depends on the distribution throughout the human genome of a particular repetitive DNA sequence. One or more clusters of this repetitive sequence occurs in nearly every DNA segment that is 15 to 20,000 DNA bases long (the length of the DNA segment inserted into phage in this procedure). The presence of this interspersed repetitive DNA confers a degree of affinity between random segments of human DNA. Because of this property, labeled whole human DNA will preferentially hybridize (label) to any phage that contains a human DNA insert. As has been noted, this procedure was successfully used to isolate DNA segments from human chromosome 11. In collaboration with Drs. Gusella and Housman, we have shown that the same procedure can be used to isolate DNA segments from the human X chromosome.

The next step is the determination of the region of the X chromosome from which the DNA segments are derived. Before this can be done, however, it is necessary to isolate from each of these human recombinant segments a subfragment that lacks the interspersed repetitive sequence—that is, the subfragment must contain essentially only "unique sequence" DNA. The "unique sequence" subfragment will then hybridize preferentially with its identical copy from the human genome. The "unique

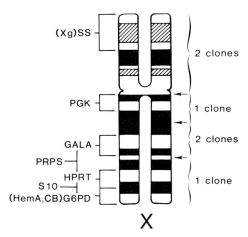

FIG. 1. Schematic diagram of the human X chromosome showing the banding pattern at early metaphase. Many of the genes whose locations are known are indicated: SS (steroid sulfatase); Xg blood group (linked to SS); PGK (phosphoglycerokinase); GALA (alpha-galactosidase A); PRPS (phosphoribosyl pyrophosphate synthetase); HPRT (hypoxanthine-guanine phosphoribosyl transferase); S10 (an X-linked surface antigen); and G6PD (glucose-6-phosphate dehydrogenase). HemA (hemophilia A) and CB (color blindness, two forms) are known to form a linkage group with G6PD. The horizontal lines adjacent to each of these genes designate the subregion of the X to which each has been assigned. PRPS and S10 are sublocalized indirectly from the known gene order of GALA, PRPS, HPRT, S10, and G6PD.

The braces to the right of the chromosome designate the subregion to which the presently available DNA segments from the authors' laboratory have been tentatively assigned.

sequence" piece is prepared by fragmenting (with an endonuclease) each human DNA-containing phage and by isolating a subfragment that lacks affinity for random pieces of human DNA. Each "unique sequence" segment is then labeled and used to probe the DNA from a hybrid cell that contains only a portion of the human X. If this is done with a series of hybrid cells, each of which contains a different portion of the X, it is possible to assign the "unique sequence" segment to a specific region of the X chromosome.

As illustrated in Fig. 1, several DNA fragments have been isolated and assigned by this procedure to specific subregions of the human X. It is only a matter of time before a very large number of segments derived from the human X will be available. Attaining the final goal of isolating the specific segment corresponding to the fragile X defect will nonetheless be accomplished only after an enormous labor, since the X chromosome is an estimated 200 to 300 million DNA base pairs in length. It is possible, though, that the task may prove simpler than these figures suggest. Since the constricted region that accounts for the fragile site on the X chromosome is large enough to be seen through the microscope, it is conceivable that it represents an alteration in the DNA that affects many hundreds, or even thousands, of adjacent DNA bases. If such a massive change were present, it would be correspondingly easier to detect.

Since technologic advances in the field of recombinant DNA have been appearing at such a rapid rate, it is not beyond the realm of possibility that a large portion of the human DNA sequence will be known within the next decade. It is reasonable to expect that this new knowledge will provide many insights into neurologic and mental diseases.

ACKNOWLEDGMENTS

Supported in part by NIH grants HD-04847 and HD-06285, also The Children's Hospital Mental Retardation Center CORE grant HD-06276.

REFERENCES

1. Fox, P., Fox, D., and Gerrard, J. W. (1980): X-linked mental retardation: Renpenning revisited. *Am. J. Med. Genet.*, 7:491–495.
2. Glover, T. W. (1981): FUdR induction of the X chromosome fragile site: Evidence for the mechanism of folic acid and thymidine inhibition. *Am. J. Hum. Genet.*, 33:234–242.
3. Gusella, J. F., Keys, C., Varsanyi-Breiner, A., Kao, F.-T., Jones, C., Puck, T. T., and Housman, D. (1980): Isolation and localization of DNA segments from specific human chromosomes. *Proc. Natl. Acad. Sci. USA*, 77:2,829–2,833.
4. Jacky, P. B., and Dill, F. J. (1980): Expression in fibroblast culture of the satellited X chromosomes associated with familial sex-linked mental retardation. *Hum. Genet.*, 53:267–269.
5. Jacobs, P. A., Glover, T. W., Mayer, M., Fox, P., Gerrard, J. W., Dunn, H. G., and Herbst, D. S. (1980a): X-linked mental retardation: A study of 7 families. *Am. J. Med. Genet.*, 7:471–489.
6. Jacobs, P. A., Glover, T. W., Mayer, M., Fox, P., Gerrard, J. W., Dunn, H. G., and Herbst, D. S. (1980b): X-linked mental retardation. *Am. J. Med. Genet.*, 7:503–505.
7. Ledbetter, D. H., Riccardi, V. M., Airhart, S. D., Strobel, R. J., Keenan, B. S., and Crawford, J. D. (1981): Deletions of chromosome 15 as a cause of the Prader-Willi syndrome. *N. Engl. J. Med.*, 304:325–329.

8. Lubs, H. A. (1969): A marker X chromosome. *Am. J. Hum. Genet.*, 21:231–244.
9. Martin, R. H., Lin, C. C., Mathies, B. J., and Lowry, R. B. (1980): X-linked mental retardation with macroorchidism and marker X chromosomes. *Am. J. Med. Genet.*, 7:433–441.
10. Renpenning, H., Gerrard, J. W., Zaleski, W. A., and Tabota, T. (1962): Familial sex-linked mental retardation. *Can. Med. Assoc. J.*, 87:954–956.
11. Ringertz, N. R., and Savard, R. E. (1976): *Cell Hybrids*. Academic Press, New York.
12. Scheres, J. M., and Hustinx, T. W. (1980): Heritable fragile sites and lymphocyte culture medium containing BrdU. *Am. J. Hum. Genet.*, 32:628–629.
13. Sutherland, G. R. (1977): Fragile sites on human chromosomes: Demonstration of their dependence on the type of tissue culture medium. *Science*, 197:265–266.
14. Sutherland, G. R. (1979*a*): Heritable fragile sites on human chromosomes. I. Factors affecting expression in lymphocyte culture. *Am. J. Hum. Genet.*, 31:125–135.
15. Sutherland, G. R. (1979*b*): Heritable fragile sites on human chromosomes. II. Distribution, phenotypic effects, and cytogenetics. *Am. J. Hum. Genet.*, 31:136–148.
16. Sutherland, G. R. (1979*c*): Heritable fragile sites on human chromosomes. III. Detection of fra(X)(q27) in males with X-linked mental retardation and in their female relatives. *Am. J. Hum. Genet.*, 53:23–27.
17. Sutherland, G. R., and Ashforth, P. L. C. (1979): X-linked mental retardation with macroorchidism and the fragile site at Xq27 or 28. *Hum. Genet.*, 48:117–120.
18. Sutherland, G. R., Baker, E., and Seshadri, R. S. (1980): Heritable fragile sites on human chromosomes. V. A new class of fragile sites requiring BrdU for expression. *Am. J. Hum. Genet.*, 32:542–548.
19. Turner, G., Brookwell, R., Daniel, A., Selikowitz, M., and Zilibowitz, M. (1980): Heterozygous expression of X-linked mental retardation and the marker X: fra(X)(q27). *N. Engl. J. Med.*, 303:662–664.
20. Turner, G., Daniel, A., and Frost, M. (1980): X-linked mental retardation, macroorchidism, and the Xq27 fragile site. *J. Pediatr.*, 96:837–841.
21. Turner, G., and Opitz, J. M. (1980): Editorial Comment: X-linked mental retardation. *Am. J. Med. Genet.*, 7:407–415.
22. Turner, G., and Turner, B. (1974): X-linked mental retardation. *J. Med. Genet.*, 11:109–113.
23. Wisniewski, L. P., Witt, M. E., Ginsberg-Fellner, F., Wilner, J., and Desnick, R. (1980): Prader-Willi syndrome and a bisatellited derivate of chromosome 15. *Clin. Genet.*, 18:42–47.
24. Yunis, J. J. (1976): High-resolution human chromosomes. *Science*, 191:1,268–1,270.
25. Yunis, J. J. (1981): Midprophase human chromosomes. The attainment of 2,000 bands. *Hum. Genet.*, 56:293–298.
26. Yunis, J. J., and Chandler, M. E. (1977): High-resolution chromosome analysis in clinical medicine. *Prog. Clin. Pathol.*, 7:267–288.
27. Yunis, J. J., and Ramsay, N. K. C. (1978): Retinoblastoma and sub-band deletion of chromosome 13. *Am. J. Dis. Child*, 132:161–163.
28. Yunis, J. J., and Ramsay, N. K. C. (1980): Familial occurrence of the aniridia-Wilms' tumor syndrome with deletion 11p13-14.1. *J. Pediatr.*, 96:1,027–1,030.

Genetics of Neurological and Psychiatric Disorders, edited by Seymour S. Kety, Lewis P. Rowland, Richard L. Sidman, and Steven W. Matthysse. Raven Press, New York © 1983.

Genetic Errors and Enzyme Replacement Strategies

Roscoe O. Brady

Developmental and Metabolic Neurology Branch, National Institute of Neurological and Communicative Disorders and Stroke, National Institutes of Health, Bethesda, Maryland 20205

The discovery that diminished activity of specific lipid-catabolizing enzymes is the biochemical basis of the sphingolipid storage diseases has permitted the development of specific enzymatic tests for: (a) the facile diagnosis of patients with these disorders; (b) the identification of heterozygous carriers of these traits; and (c) the diagnosis of any of these conditions prenatally. These procedures are currently in widespread use throughout the world, and their application has provided effective genetic counseling for many families in which these disorders have occurred. Thus, the theoretical basis for the control of the incidence of such conditions is available.

However, public screening for lipid storage diseases and other metabolic disorders, such as mucopolysaccharide storage diseases, is, in fact, now used only for Tay-Sachs disease. This limitation is due to the relatively infrequent occurrence of these disorders in a readily accessible segment of the population. Even in the case of Tay-Sachs disease, which is largely confined to infants of Ashkenazi Jewish ancestry, a considerable number of infants with this disorder will be born despite intensive screening efforts (32). Therefore, a continuing and expanding effort to develop effective therapy is under way in a number of research centers. This chapter will deal with the status of the therapeutic modalities under consideration, and an attempt will be made to indicate the strategies that may be helpful in the future. Although summaries of the metabolic lesions in the lipid storage diseases are readily available (e.g., 6–8), a brief overview of the clinical manifestations and enzymatic defects is provided in Fig. 1.

ENZYME REPLACEMENT BY ORGAN ALLOGRAFTS

Since the metabolic lesion in the lipid, mucopolysaccharide, and other classes of storage diseases is a deficiency of a specific, catabolic, hydrolytic enzyme, the possibility of replacing the missing catalyst was quickly perceived. It was apparent, however, that the isolation of enzymes of satisfactory purity and sufficient quantity

DISEASE	SIGNS AND SYMPTOMS	MAJOR LIPID ACCUMULATION	ENZYME DEFECT
GAUCHER'S DISEASE	SPLEEN AND LIVER ENLARGEMENT, EROSION OF LONG BONES AND PELVIS, MENTAL RETARDATION ONLY IN INFANTILE FORM	GLUCOCEREBROSIDE — CERAMIDE — GLUCOSE	GLUCOCEREBROSIDE β-GLUCOSIDASE
NIEMANN-PICK DISEASE	LIVER AND SPLEEN ENLARGEMENT, MENTAL RETARDATION, ABOUT 30 PERCENT WITH RED SPOT IN RETINA	SPHINGOMYELIN — PHOSPHORYLCHOLINE	SPHINGOMYELINASE
KRABBE'S DISEASE (GLOBOID LEUKODYSTROPHY)	MENTAL RETARDATION, ALMOST TOTAL ABSENCE OF MYELIN, GLOBOID BODIES IN WHITE MATTER OF BRAIN	GALACTOCEREBROSIDE — GALACTOSE	GALACTOCEREBROSIDE- β-GALACTOSIDASE
METACHROMATIC LEUKODYSTROPHY	MENTAL RETARDATION, PSYCHOLOGICAL DISTURBANCES IN ADULT FORM, NERVES STAIN YELLOW-BROWN WITH CRESYL VIOLET DYE	SULFATIDE — GALACTOSE-3-SULFATE	SULFATIDASE
CERAMIDE LACTOSIDE LIPIDOSIS	SLOWLY PROGRESSING BRAIN DAMAGE, LIVER AND SPLEEN ENLARGEMENT	CERAMIDE LACTOSIDE — GLUCOSE — GALACTOSE	CERAMIDELACTOSIDE- β-GALACTOSIDASE

FABRY'S DISEASE	REDDISH-PURPLE SKIN RASH, KIDNEY FAILURE, PAIN IN LOWER EXTREMITIES	CERAMIDE TRIHEXOSIDE 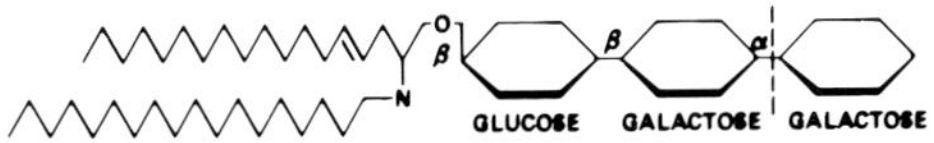 Ceramide—β—GLUCOSE—GALACTOSE—α—GALACTOSE	CERAMIDETRIHEXOSIDE α—GALACTOSIDASE
TAY-SACHS DISEASE	MENTAL RETARDATION, RED SPOT IN RETINA, BLINDNESS, MUSCULAR WEAKNESS	GANGLIOSIDE GM_2 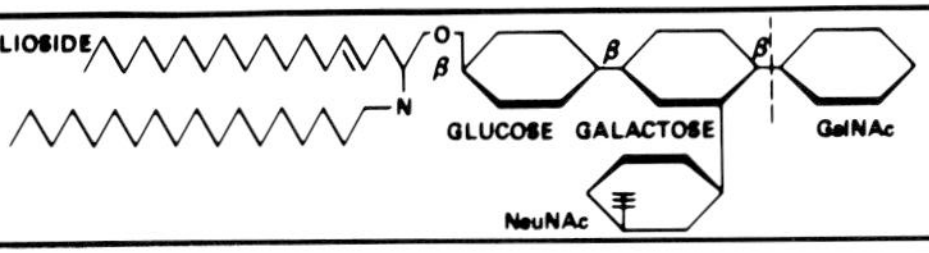 Ceramide—β—GLUCOSE—GALACTOSE—β—GalNAc NeuNAc	HEXOSAMINIDASE A
TAY-SACHS VARIANT	SAME AS TAY-SACHS DISEASE BUT PROGRESSING MORE RAPIDLY	GLOBOSIDE (AND GANGLIOSIDE GM_2) 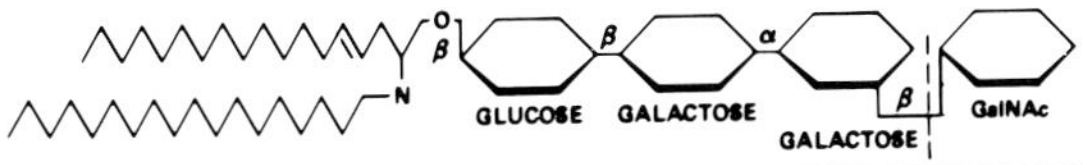 Ceramide—β—GLUCOSE—GALACTOSE—α—GALACTOSE—β—GalNAc	HEXOSAMINIDASE A AND B
GENERALIZED GANGLIOSIDOSIS	MENTAL RETARDATION, LIVER ENLARGEMENT, SKELETAL DEFORMITIES, ABOUT 50 PERCENT WITH RED SPOT IN RETINA	GANGLIOSIDE GM_1 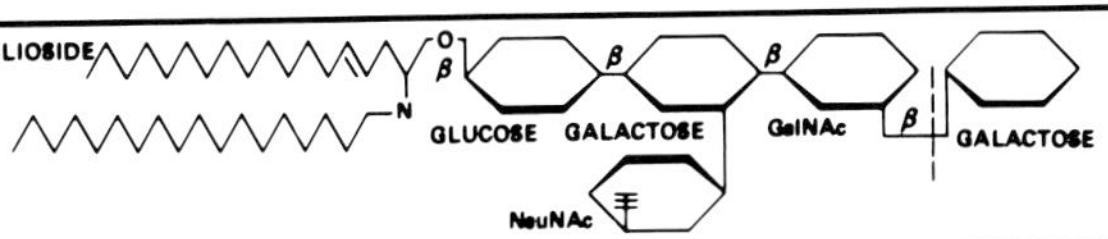 Ceramide—β—GLUCOSE—GALACTOSE—β—GalNAc—β—GALACTOSE NeuNAc	β—GALACTOSIDASE
FUCOSIDOSIS	CEREBRAL DEGENERATION, MUSCLE SPASTICITY, THICK SKIN	H-ISOANTIGEN 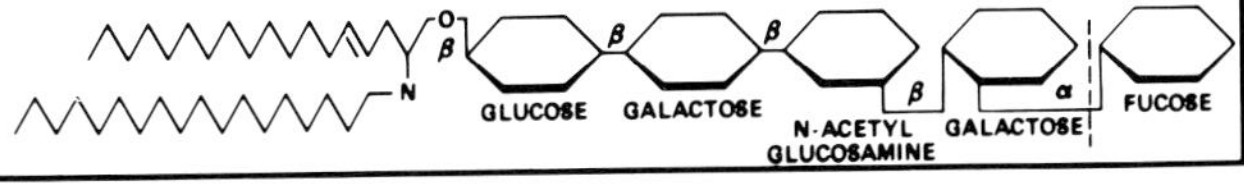 Ceramide—β—GLUCOSE—GALACTOSE—β—N-ACETYL GLUCOSAMINE—β—GALACTOSE—α—FUCOSE	α—FUCOSIDASE

FIG. 1. Principal manifestations, stored lipids, and sites of metabolic lesions in the sphingolipidoses. (From ref. 10 with permission.)

to be clinically effective would be a difficult and time-consuming task, considering the state of knowledge at the time. Accordingly, the alternative of organ grafting was suggested as an interim procedure (5). Several such trials have been attempted in patients with lipid storage diseases. The first was a spleen allograft in a patient with the juvenile (Type III) form of Gaucher's disease (28). No benefit was noted. The patient experienced a severe tissue incompatibility reaction and died a few months after the operation (26). The next attempt was carried out in a patient with the infantile (Type II) form of Gaucher's disease who received a kidney graft (18). Again, no improvement was noted. This procedure was performed again in 1978 by Groth et al. on a patient with Type III Gaucher's disease, again without benefit (27). A more complicated response was reported by a Canadian team that performed an orthotopic liver transplantation on a patient with Niemann-Pick disease (16). There seemed to be some indications of improvement in the first 6 months, but no subsequent benefit was noted and the patient died 2 years after the operation.

Somewhat more promising results have been obtained in patients with Fabry's disease who received kidney grafts. Complete renal shutdown is a frequent terminal manifestation in affected boys. It appears that kidney transplantation can be beneficial, ameliorating the uremic state (19,42). It is generally agreed, however, that the other manifestations of the disorder are not significantly improved by this procedure (15,48). Perhaps the most definitive study in this area was carried out by Van den Bergh et al. (53). Although the grafted kidney did not accumulate ceramidetrihexoside, there was no evidence that the graft had provided supplemental ceramidetrihexosidase, the deficient enzyme, to other organs. A number of lessons about enzyme replacement strategies may be derived from this important investigation. The most critical of these is that each organ seems to synthesize its own specific complement of enzymes in direct proportion to the metabolic load that must be handled by the particular tissue (33).

Summing up, organ allografts have only limited benefit for the treatment of heritable metabolic disorders. This approach has, therefore, been largely discontinued as a therapeutic measure. These developments emphasize the need for careful investigations of enzyme replacement in metabolic disorders.

REPLACEMENT TRIALS WITH PURIFIED ENZYMES

Early Investigations of Metachromatic Leukodystrophy

The first examination of the effect of direct enzyme replacement in a human lipid storage disorder was carried out in a patient with metachromatic leukodystrophy (Fig. 1, line 5). Arylsulfatase A was partially purified from human urine and injected intrathecally (1). The recipient had a marked pyrogenic reaction, and no clinical improvement was observed. A similar investigation was carried out 2 years later by Greene et al. (24) who used beef brain enzyme preparation. Again, fever followed intrathecal injection of this preparation and no benefit was observed. Also, a considerable amount of intravenously injected arylsulfatase was detected in the patient's liver. Both groups of investigators reported that the intrathecally infused enzyme did not penetrate the brain substance. Furthermore, neither group detected antibody directed against the injected enzyme, although the beef brain preparation might have contained nonhuman antigenic determinants. It was concluded that neither intravenous nor intrathecal injection of arylsulfatase A was likely to be beneficial for the treatment of this metabolic disorder.

In contrast to these negative results *in vivo*, two groups of investigators reported that addition of urinary arylsulfatase A to skin fibroblasts cultured from patients with metachromatic leukodystrophy corrected the defective degradation of sulfatide in these cells *in vitro* (43,55). This observation provided considerable incentive for continuing investigations of enzyme replacement, although the difficulty of delivering an exogenous enzyme to the brain was realized.

Enzyme Replacement Trials in Tay-Sachs Disease

The first investigation of enzyme replacement in a human ganglioside storage disorder was carried out in 1971 in a patient with the Sandhoff variant of Tay-Sachs disease (Fig. 1, line 8) (31). Although both the A and B hexosaminidase isozymes are lacking in patients with this condition, only hexosaminidase A is believed necessary for the catabolism of the accumulating ganglioside. This isozyme was isolated from human urine and aliquots of it were infused into the patient on 2 consecutive days. The enzyme was cleared from the circulation with extraordinary rapidity and most of it was localized in the liver. None of the infused enzyme reached the brain. As might be anticipated from this finding and the results of the previous experiments with arylsulfatase A, there was no improvement in the patient's condition. However, the injected enzyme caused a 46% reduction in the quantity of globoside (see Fig. 1, line 8) in the circulation of the recipient. In addition to ganglioside G_{M2}, significant quantities of globoside accumulate in patients with the Sandhoff form of Tay-Sachs disease. This reduction was the first indication that an exogenous enzyme could affect the amount of an accumulating metabolite *in vivo*, and it provided strong encouragement for replacement trials in patients with storage diseases that do not involve the central nervous system.

Other enzyme replacement trials in Tay-Sachs disease have been carried out in Israel. Human placental hexosaminidase A was injected intracisternally and intrathecally into 3 (23) [or 2, according to (54)] patients with this disorder. The enzyme preparation was administered as a mixture of free and polyvinylpyrrolidone-bound hexosaminidase A. The infusions were accompanied by pyrogenic reactions on several occasions. Again, there was no clinical improvement in the recipients. Two potentially useful observations were made in the course of this investigation, however. No humoral immune response was detected in the recipients and there was a dramatic decrease in the concentration of ganglioside G_{M2} in the serum of 1 of the patients shortly after infusion of the enzyme.

Enzyme Replacement in Fabry's Disease

As a consequence of the unsuccessful attempts to deliver exogenous enzymes to the central nervous system by intrathecal or intracisternal infusion, realistic considerations dictated that success in enzyme replacement would be more likely in patients with storage disorders that did not involve the brain. A condition for which enzyme replacement seemed likely to exert a beneficial effect was Fabry's disease, in which most symptoms arise from involvement of peripheral tissues. Accordingly, we set about to purify this enzyme from human placental tissue. We eventually succeeded in isolating some highly purified ceramidetrihexosidase (29), and small quantities of this preparation were infused intravenously into 2 patients with Fabry's disease (13). Injections of this enzyme significantly decreased the elevated quantity of ceramidetrihexoside in the plasma in both recipients. The amount of this lipid that was cleared was proportional to the quantity of enzyme administered. It was deduced that the ceramidetrihexoside had been hydrolyzed within tissue stores such as the liver rather than within the bloodstream.

Several other important observations were made in these trials: (a) Although there was no clear evidence of a change in the clinical status of the recipients, there were no untoward reactions to the injected enzyme. (b) The patients did not seem to have been sensitized to the placental protein because no reaction was elicited when they were challenged with enzyme a year after the infusion. (c) The amount of ceramidetrihexoside in the plasma returned to the preinfusion level by 72 hr after administration of the enzyme. (d) Considerably larger quantities of enzyme will apparently have to be given to produce a beneficial clinical response. Unfortunately, further replacement trials with placental ceramidetrihexosidase have been delayed because of a pyrogenic contaminant in the larger preparations that are considered necessary. This difficulty should eventually be surmounted for further replacement investigations.

Desnick and co-workers (17) carried out a similar study using ceramidetrihexosidase preparations from human spleen tissue and outdated plasma. Their results were essentially the same as the those observed earlier with placental enzyme. Enzyme derived from plasma seemed to be more effective than the splenic preparation in clearing circulating lipid. Again, no evidence of sensitization to the exogenous enzyme preparations was detected in the recipients, who each received six injections over a period of 117 days. The confirmation of the lack of antibody production to the exogenous ceramidetrihexosidase was significant because it has not been possible to demonstrate cross-reacting, catalytically inactive protein in patients with this disorder (46). Although the results obtained in these investigations are encouraging, the usefulness of enzyme replacement in Fabry's disease remains to be established.

Enzyme Replacement in Gaucher's Disease

The most comprehensive trials of enzyme replacement have been in patients with Type I Gaucher's disease. These patients have extensive hepatosplenomegaly, thrombocytopenia, osteoporosis, and, frequently, gastrointestinal symptoms related to intestinal hypomotility and abdominal distention. Glucocerebrosidase was first purified from human placental tissue by conventional enzyme isolation procedures (41). When this enzyme was injected into 2 patients with Gaucher's disease, there was a 26% reduction in the quantity of glucocerebroside in the liver (11). The amount of this lipid in the circulation, which is primarily associated with erythrocytes, gradually returned to normal during the 3 days following infusion of the enzyme. This effect seemed to be the result of redistribution of glucocerebroside from the circulation to the vacated storage sites in the liver and other organs caused by the exogenous enzyme. The reduction of red cell-associated glucocerebroside persisted for a long time (6,40). This reduction of lipid in the circulation was seen only when a significant quantity of stored material was catabolized. For example, glucocerebrosidase infusion in a 3rd Gaucher patient resulted in only an 8% decrease in the amount of glucocerebroside in the liver. This patient had accumulated much more glucocerebroside than the previous recipients, and, in this case, no reduction of circulating glucocerebroside was observed (12).

It was apparent that much larger quantities of enzyme would have to be administered to patients like the 3rd one for biochemical and clinical benefit. Because the enzyme isolation procedure employed at that time could not be scaled up satisfactorily, a new method was devised to provide highly purified material in good yield and in a form suitable for injection into humans (20). Comparatively large quantities of this enzyme have been injected into several patients with Gaucher's disease. We obtained encouraging results in 2 young boys in whom there was a moderate reduction of liver and circulating glucocerebroside after infusion of the enzyme (9).

We therefore formulated a protocol of enzyme infusion on a prospective bimonthly basis. We believe that: (a) the general health of these individuals has improved; (b) there has been an arrest of the progression of organomegaly; and (c) blood platelet levels have improved. Because of these encouraging observations, we have initiated a similar regimen in 4 additional young patients. Less consistent effects on hepatic glucocerebroside levels have been obtained in older Gaucher's patients who received the enzyme on a sporadic basis. Whether these results are because of a difference in the subcellular distribution of the stored material in the older patients or because repeated injections have not been given to these individuals is not known.

It is fair to say that some clinical benefit seems to have been obtained by the younger recipients of exogenous glucocerebrosidase. It is also desirable to improve the effectiveness of the enzyme. We have shown that the placental enzyme can catalyze the hydrolysis of all of the stored glucocerebroside in liver biopsy specimens *in vitro* (39). We therefore conclude that a major obstacle to be surmounted is the delivery of the enzyme to the site of storage. Glucocerebrosidase is a glycoprotein, and the oligosaccharide moiety directs the enzyme primarily to hepatocytes in the liver (21,36). Since the major portion of glucocerebroside is stored in reticuloendothelial cells (such as the Kupffer cells in the liver), we wished to increase the delivery of the enzyme to these cells. We have demonstrated that selective enzymatic cleavage of the oligosaccharide chains (so that the terminal sugar is mannose) causes a fivefold increase in the amount of glucocerebrosidase taken up by Kupffer cells in experimental animals (22). We shall examine the effectiveness of mannose-terminated enzyme in patients when sufficient quantities of this preparation are available.

An alternative strategy under investigation is the possibility of covalently linking additional molecules of mannose to the native enzyme. If the enzyme can be modified in this manner and this preparation is also more efficiently endocytosed by Kupffer cells than the native protein, clinical trials with this preparation will also be undertaken. Although we (14) and others (4,25) have demonstrated that infusion of placental glucocerebrosidase does not evoke antibody production in humans, careful monitoring of patients who receive enzymes with modified carbohydrate structures will be required.

STRATEGIES FOR ENZYME REPLACEMENT IN THE BRAIN

Although most patients afflicted with a sphingolipid storage disorder are those with Type I Gaucher's disease and, to a lesser extent, Fabry's disease, in which the central nervous system is minimally, if at all, involved, there is extensive brain damage in all of the other forms, including Type II and Type III Gaucher's patients. Thus, consideration of therapy for these disorders, and for most of the mucopolysaccharide storage disorders as well, necessitates careful attention to the possibility of providing adequate enzymatic function in the brain and peripheral nervous system. It is abundantly clear that the intravenous, intrathecal, or intracisternal administration of enzymes is totally inadequate for this purpose. We must develop effective alternative procedures to provide for the delivery of sufficient quantities of the required enzyme to the nervous system.

The most likely way to accomplish this appears to be by temporarily altering the blood-brain barrier so that the enzyme can reach the cells in which the offending substance is stored. Intracarotid infusion of hyperosmolar solutions of mannitol or arabinose causes a temporary shrinking of the endothelial cells that form the blood-brain barrier (44). Under these conditions, a physiologically normal complement of a lysosomal enzyme, such as mannosidase, can be delivered to the substance of the brain (3). This infusion must be carried out under rigorously controlled conditions with regard to the concentration of solute and velocity of infusion so that adequate barrier disruption is obtained without untoward neurological sequelae. The study has been extended to primates, and no evidence of brain damage was observed in comprehensive examinations of this organ (47).

Several other important observations have been made along this line. Barranger et al. (2) demonstrated that mannose-terminated glycoproteins, such as horseradish peroxidase, are specifically endocytosed by neuronal cells in the central nervous system after barrier modification. Steer et al. (49) showed that human hexosaminidase A is a mannose-terminated enzyme. In accordance with this finding, Kusiak and co-workers (34,35) demonstrated that radioiodinated hexosaminidase A was specifically bound by a high-affinity receptor on rat brain synaptosomes and synaptic plasma membranes. Furthermore, mannose-terminated glycoproteins, such as horseradish peroxidase, eventually become localized within lysosomes after endocytosis (2,3). This subcellular site is precisely the locus of the accumulating ganglioside in Tay-Sachs disease (51). Thus, if a sufficient quantity of hexosaminidase A could be delivered to the nervous system by barrier modification, we would have a rational basis for enzyme replacement trials in this disorder. Intensive efforts are under way to determine the feasibility of this approach.

A final note of encouragement may be derived from recent studies of Neuwelt et al. (37) who successfully altered the blood-brain barrier in humans without untoward neurological sequelae. This critical demonstration, coupled with the fundamental investigations concerning the uptake of hexosaminidase A by neuronal cells and its packaging within lysosomes, provide a firm basis for enzyme replacement trials in Tay-Sachs disease. The principal constraint at the moment is the comparatively small quantity of hexosaminidase A that enters the brain of animals even after barrier modification. Thus, a realistic enzyme replacement trial in Tay-Sachs disease would require the infusion of a prohibitively large quantity of the purified enzyme to expect a beneficial response. We are engaged in a collaborative effort with Neuwelt and his associates to overcome this impediment. The chances for a successful outcome of these efforts are good, and we anticipate that enzyme replacement trials of this type will be undertaken in the future.

Rationale for Enzyme Replacement in Disorders with Brain Involvement

Data from various investigations have been cited that indicate that enzyme replacement may be beneficial in several deficiency disorders. The strongest positive indications have come from replacement trials in conditions without brain damage, such as Gaucher's and Fabry's diseases. We hope that these encouraging results will be substantiated and that increasingly effective enzyme replacement procedures will be developed. One of the most compelling observations was reported by Rattazzi et al. (45) who infused human placental hexosaminidase A into cats with the Sandhoff form of Tay-Sachs disease. They observed virtually complete clearance of the globoside that had accumulated in the liver of these animals. Although the critical experiment of determining globoside levels in a particular cat before and after enzyme infusion has not been carried out, the quantity of globoside in the liver of nontreated homozygotes was always much more than in normal or treated animals. Even greater emphasis will be placed on enzyme replacement trials in humans if these results can be confirmed.

In addition to the initial difficulty of delivering an exogenous enzyme to the brain, other critical factors must be kept in mind with regard to the enzyme replacement in disorders such as Tay-Sachs disease. These considerations include the frequency of administration of the enzyme and the possibility that, in addition to intracarotid injections, alternate routes of delivery may also be necessary to control the disorder. Large amounts of hexosaminidase A might be necessary only in the neonatal period when the turnover of gangliosides is at its maximum (50). This concept seems strengthened by the fact that the catabolism of the accumulating ganglioside G_{M2} can be initiated either by cleavage of the molecule of N-acetylgalactosamine by hexosaminidase A or by the hydrolysis of N-acetylneuraminic acid by a neuroaminidase known to be in human brain tissue (52). The product of the latter reaction is asialo-G_{M2} (N-acetylgalactosaminyl-galactosyl-glucosyl-ceramide). This compound may be further metabolized through the action of either hexosaminidase A or hexosaminidase B. Although hexosaminidase A is lacking in patients with the conventional form of Tay-Sachs disease, the activity of hexosaminidase B is greatly increased over normal. It is evident that this alternate pathway for G_{M2} catabolism is not sufficient to prevent ganglioside accumulation in Tay-Sachs disease. If enough exogenous hexosaminidase A could be delivered to the brain during the period of maximal ganglioside turnover, though, the escape route through neuraminidase and hexosaminidase B might prevent further accumulation of significant quantities of G_{M2}. (Of course, these considerations are not applicable to the Sandhoff form of Tay-Sachs disease.) However, emerging reports of patients with late onset of signs and symptoms of Tay-Sachs disease (30,38) make this hypothesis less attractive, and exogenous enzyme may have to be supplied throughout the life of the patient.

A second major consideration that must be taken into account is the distribution of exogenous enzymes within the brain after barrier modification. Since only the barrier in the hemisphere ipsilateral to the injected artery is altered by the hyperosmolar solution, bilateral barrier modification would seem to be necessary if this strategy is to be successful. Also, the accumulation of G_{M2} in cerebellum and spinal cord is not likely to be reduced if the barrier modification is restricted to infusions via the carotid artery. We simply do not know enough about the pathophysiology of G_{M2} accumulation in the cerebellum and spinal cord, although it seems likely that the reported ataxias are consequences of neuronal involvement in these regions. It is apparent that serious additional considerations will be required to expect complete amelioration of the consequences of sphingolipid or mucopolysaccharide accumulation within the nervous system.

CONCLUDING REMARKS

It is clear that the ultimate benefit of enzyme replacement in hereditary metabolic disorders remains to be established. Although encouraging results have been obtained in Gaucher's disease and Fabry's disease and there have been spectacular findings in hexosaminidase-deficient cats, substantiation of the apparent clinical improvement observed in the young patients with Gaucher's disease is mandatory. There is sufficient interest in this type of investigation to expect additional trials that will clarify the feasibility of enzyme replacement. We cannot predict how swiftly this will be determined, since a number of the strategies outlined in this chapter for improving the efficiency of exogenous enzymes will require the concentrated efforts of many investigators and the commitment of substantial research resources for a long time.

Ultimately, even more novel approaches to the therapy of human metabolic disorders will be conceived and tested. Among the strategies under consideration are: (a) plasmapheresis to remove excess lipid and other metabolites from the circulation; (b) bone marrow transplantation; and (c) procedures to correct harmful distortions of the genetic code by insertion of new genetic material or selective modification of abnormal deoxyribonucleic acid (DNA) sequences. It is difficult to predict when investigations of the latter type will be undertaken in humans with metabolic disorders involving the nervous system. Meanwhile, competent investigations of enzyme replacement techniques occupy a justifiably prominent place in consideration of the therapy of heritable disorders of the nervous system.

REFERENCES

1. Austin, J. H. (1967): Some recent findings in leukodystrophies and in gargoylism. In: *Inborn Disorders of Sphingolipid Metabolism*, edited by S. M. Aronson and B. W. Volk, pp. 38–64. Academic Press, New York.
2. Barranger, J. A., Rapoport, S. I., and Brady, R. O. (1980): Access of enzymes to brain following osmotic alteration of the blood-brain barrier. In: *Enzyme Therapy in Genetic Diseases, Vol. 2*, edited by R. J. Desnick, pp. 195–205. Alan R. Liss, New York.
3. Barranger, J. A., Rapoport, S. I., Frecricks, W. R., Pentchev, P. G., MacDermot, K. D., Steusing, J. K., and Brady, R. O. (1979): Modification of the blood-brain barrier: Increased concentration and fate of enzymes entering the brain. *Proc. Natl. Acad. Sci. USA*, 76:481–485.
4. Beutler, E., Dale, G. L., and Kuhl, W. (1980): Replacement therapy in Gaucher's disease. In: *Enzyme Therapy in Genetic Diseases*, Vol. 2, edited by R. J. Desnick, pp. 369–381. Alan R. Liss, New York.
5. Brady, R. O. (1966): The sphingolipidoses. *N. Engl. J. Med.*, 275:312–318.
6. Brady, R. O. (1977): Heritable catabolic and anabolic disorders of lipid metabolism. *Metabolism*, 26:329–345.
7. Brady, R. O. (1978a): Sphingolipidoses. *Ann. Rev. Biochem.*, 47:687–713.
8. Brady, R. O. (1978b): Elucidation of clinical lysosome deficiencies. In: *Molecular Basis of Biological Degradative Processes*, edited by R. Berlin, H. Herrmann, I. H. Lepow, and J. Tanzer, pp. 39–44. Academic Press, New York.
9. Brady, R. O., Barranger, J. A., Gal, A. E., Pentchev, P. G., and Furbish, F. S. (1980): Status of enzyme replacement therapy for Gaucher's disease. In: *Enzyme Therapy in Genetic Diseases, Vol. 2*, edited by R. J. Desnick, pp. 361–368. Alan R. Liss, New York.
10. Brady, R. O., Pentchev, P. G., and Gal, A. E. (1975): Investigations in enzyme replacement therapy in lipid storage diseases. *Fed. Proc.*, 34:1,310–1,315.
11. Brady, R. O., Pentchev, P. G., Gal, A. E., Hibbert, S. R., and Dekaban, A. S. (1974): Replacement therapy for inherited enzyme deficiency: Use of purified glucocerebrosidase in Gaucher's disease. *N. Engl. J. Med.*, 291:989–993.
12. Brady, R. O., Pentchev, P. G., Gal, A. E., Hibbert, S. R., Quirk, J. M., Mook, G. E., Kusiak, J. W., Tallman, J. F., and Dekaban, A. S. (1976): Enzyme replacement therapy for the sphingolipidoses. In: *Current Trends in the Sphingolipidoses and Allied Disorders*, edited by B. W. Volk and L. Schneck, pp. 523–532. Plenum Press, New York.
13. Brady, R. O., Tallman, J. F., Johnson, W. G., Gal, A. E., Leahy, W. E., Quirk, J. M., and Dekaban, A. S. (1973): Replacement therapy for inherited enzyme deficiency: Use of purified ceramidetrihexosidase in Fabry's disease. *N. Engl. J. Med.*, 289:9–14.
14. Britton, D. E., Leinikki, P. O., Barranger, J. A., and Brady, R. O. (1978): Gaucher's disease: Lack of antibody response to intravenous glucocerebrosidase. *Life Sci.*, 23:2,517–2,520.
15. Clarke, J. T. R., Guttmann, R. D., Wolfe, L. S., Beaudoin, J. G., and Morehouse, D. D. (1972): Enzyme replacement therapy by renal allotransplantation in Fabry's disease. *N. Engl. J. Med.*, 287:1,215–1,218.

16. Daloze, P., Delvin, E. E., Glorieux, F. H., Corman, J. L., Bettez, P., and Toussi, T. (1977): Replacement therapy for inherited enzyme deficiency: Liver orthotopic transplantation in Niemann-Pick disease Type A. *Am. J. Med. Genet.*, 1:229–239.
17. Desnick, R. J., Dean, K. J., Grabowski, G. A., Bishop, D. F., and Sweeley, C. C. (1979): Enzyme therapy in Fabry's disease: Differential *in vivo* plasma clearance and metabolic effectiveness of plasma and splenic α-galactosidase A isozymes. *Proc. Natl. Acad. Sci. USA*, 76:5,326–5,330.
18. Desnick, S. J., Desnick, R. J., Brady, R. O., Pentchev, P. G., Simmons, R. L., Najarian, J. S., Swaiman, K., Sharp, H. L., and Krivit, W. (1973): Renal transplantation in type 2 Gaucher's disease. In: *Enzyme Therapy in Genetic Diseases*, edited by R. J. Desnick, R. W. Bernlohr, and W. Krivit, pp. 109–119. Williams and Wilkins, Baltimore.
19. Desnick, R. J., Simmons, R. C., Allen, K. Y., Woods, L. E., Anderson, C. F., Najarian, J. S., and Krivit, W. (1972): Correction of enzymatic deficiencies by renal transplantation: Fabry's disease. *Surgery*, 72:203–211.
20. Furbish, F. S., Balir, H. E., Shiloach, J., Pentchev, P. G., and Brady, R. O. (1977): Enzyme replacement therapy in Gaucher's disease: Large-scale purification of glucocerebrosidase suitable for human administration. *Proc. Natl. Acad. Sci. USA*, 74:3,560–3,563.
21. Furbish, F. S., Steer, C. J., Barranger, J. A., Jones, E., A., and Brady, R. O. (1978): The uptake of native and deasialylated glucocerebrosidase by rat hepatocytes and Kupffer cells. *Biochem. Biophys. Res. Commun.*, 81:1,047–1,053.
22. Furbish, F. S., Steer, C. J., Krett, N. L., and Barranger, J. A. (1981): Targeting of a lysosomal enzyme: Uptake and distribution of placental glucocerebrosidase in rat hepatic cells and effects of sequential deglycosylation. *Biochim. Biophys. Acta*, 673:425–434.
23. Godel, V., Blumenthal, M., Goldman, B., Keren, G., and Padeh, G. (1978): Visual functions in Tay-Sachs-diseased patients following enzyme replacement therapy. *Metab. Ophthalmology*, 2:27–31.
24. Greene, H. L., Hug, G., and Schubert, W. K. (1969): Metachromatic leukodystrophy. Treatment with arylsulfatase A. *Arch. Neurol.*, 20:147–153.
25. Gregoriadis, G., Neerunjun, D., Meade, T. W., Goolamali, S. K., Weereratne, H., and Bull, G. (1980): Experiences after long-term treatment of a Type I Gaucher's disease patient with liposome-entrapped glucocerebroside-β-glucosidase. In: *Enzyme Therapy in Genetic Diseases*, Vol. 2, edited by R. J. Desnick, pp. 383–392. Alan R. Liss, New York.
26. Groth, C. G., Bergström, K., Collste, L., Egberg, N., Högman, C., Holm, G., and Möller, E. (1972): Immunologic and plasma protein studies in a splenic homograft recipient. *Clin. Exp. Immunol.*, 10:359–365.
27. Groth, C. G., Collste, H., Dreborg, S., Håkansson, G., Lundgren, G., and Svennerholm, L. (1980): Attempt at enzyme replacement in Gaucher's disease by renal transplantation. *Acta Paediatr. Scand.*, 68:475–479.
28. Groth, C. G., Hagenfeldt, L., Dreborg, S., Löfström, B., Öckerman, P. A., Samuelsson, K., Svennerholm, L., Werner, B., and Westberg, G. (1971): Splenic transplantation in a case of Gaucher's disease. *Lancet*, 1:1,260–1,264.
29. Johnson, W. G., and Brady, R. O. (1972): Ceramidetrihexosidase from human placenta. *Methods Enzymol.*, 28:849–856.
30. Johnson, W. G., Cohen, C. S., Miranda, A. F., Waran, S. P., and Chutorian, A. M. (1980): α-Locus hexosaminidase genetic compound with the phenotype of juvenile gangliosidosis. *Am. J. Hum. Genet.*, 32:508–518.
31. Johnson, W. G., Desnick, R. L., Long, D. M., Sharp, H. L., Krivit, W., Brady, B., and Brady, R. O. (1973): Intravenous injection of purified hexosaminidase A into a patient with Tay-Sachs disease. In: *Enzyme Therapy in Genetic Diseases*, edited by R. L. Desnick, R. W. Bernlohr, and W. Krivit, pp. 120–124. Williams and Wilkins, Baltimore.
32. Kaback, M. M. (1981): Tay-Sachs screening—World-wide update. In: *Lysosomes and Lysosomal Storage Diseases*, edited by J. A. Lowden and J. W. Callahan. Raven Press, New York *(in press)*.
33. Kampine, J. P., Kanfer, J. N., Gal, A. E., Bradley, R. M., and Brady, R. O. (1967): Response of sphingolipid hydrolases in spleen and liver to increased erythrocytorrhexis. *Biochim. Biophys. Acta*, 137:135–139.
34. Kusiak, J. W., Toney, J. H., Quirk, J. M., and Brady, R. O. (1979): Specific binding of ^{125}I-labeled β-hexosaminidase A to rat brain synaptosomes. *Proc. Natl. Acad. Sci. USA*, 76:982–985.
35. Kusiak, J. W., Quirk, J. M., and Brady, R. O. (1980): Specific binding of β-hexosaminidase A to rat brain synaptic plasma membrane. In: *Enzyme Therapy in Genetic Diseases, Vol. 2*, edited by R. J. Desnick, pp. 93–102. Alan R. Liss, New York.

36. Morrone, S., Pentchev, P. G., Thorpe, S., and Baynes, J. (1981): *In vivo* studies in rat of the tissue uptake, cellular distribution, and catabolic turnover of exogenous glucocerebrosidase. *Biochem. J.*, 194:733–742.
37. Neuwelt, E. A., Frenkel, E. P., Diehl, J., Vu, L. H., Rapoport, S., and Hill, S. (1980): Reversible osmotic blood-brain barrier disruption in man: Implications for chemotherapy of malignant brain tumors. *Neurosurgery*, 7:44–52.
38. O'Brien, J. S. (1978): The gangliosidoses. In: *The Metabolic Basis of Inherited Disease*, edited by J. B. Stanbury, J. B. Wyngaardern, and D. S. Fredrickson, pp. 841–865. McGraw-Hill, New York.
39. Pentchev, P. G., Barranger, J. A., Gal, A. E., Furbish, F. S., and Brady, R. O. (1978): Incorporation of exogenous enzymes into lysosomes: A theoretical and practical means for correcting lysosomal blockage. In: *Glycoproteins and Glycolipids in Disease Processes*, edited by E. F. Walborg, Jr., pp. 150–159. Am. Chem. Soc. Symposium Series 80, Washington, D.C.
40. Pentchev, P. G., Brady, R. O., Gal, A. E., and Hibbert, S. R. (1975): Replacement therapy for inherited enzyme deficiency: Sustained clearance of accumulated glucocerebroside in Gaucher's disease following infusion of purified glucocerebrosidase. *J. Mol. Med.*, 1:73–78.
41. Pentchev, P. G., Brady, R. O., Hibbert, S. R., Gal, A. E., and Shapiro, D. (1973): Isolation and characterization of glucocerebrosidase from human placental tissue. *J. Biol. Chem.*, 248:2,526–2,527.
42. Philippart, M., Fanklin, S. S., and Gordon, A. (1972): Reversal of an inborn sphingolipidosis (Fabry's disease) by kidney transplantation. *Ann. Intern. Med.*, 77:195–200.
43. Porter, M. T., Fluharty, A. L., and Kihara, H. (1971): Correction of abnormal cerebroside sulfate metabolism in cultured metachromatic leukodystrophy fibroblasts. *Science*, 172:1,263–1,265.
44. Rapoport, S. I. (1976): *Blood-Brain Barrier in Physiology and Medicine*. Raven Press, New York.
45. Rattazzi, M. C., Appel, A. M., and Baker, J. H. (1980): Enzyme replacement in feline GM_2 gangliosidosis: Reduction of glycolipid storage in visceral organs. Presented at the 31st meeting of the American Society of Human Genetics, September 24–27, New York City.
46. Rietra, P. J. G. M., Molenaar, J. L., Hamers, M. N., Tager, J. M., and Borst, P. (1974): Investigation of the α-galactosidase deficiency in Fabry's disease using antibodies against the purified enzyme. *Eur. J. Biochem.*, 46:89–98.
47. Smith, M. T., Girton, M., Rapoport, S. I., Brady, R. O., and Barranger, J. A. (1980): Pathology of reversible blood-brain barrier opening. *J. Neuropathol. Exp. Neurol.*, 39:389.
48. Spence, M. W., MacKinnon, K. E., Burgess, J. K., d'Entremont, D. M., Belitsky, P., Lannon, S. G., and MacDonald, A. S. (1976): Failure to correct the metabolic defect by renal allotransplantation in Fabry's disease. *Ann. Intern. Med.*, 84:13–16.
49. Steer, C. J., Kusiak, J. W., Brady, R. O., and Jones, E. A. (1979): Selective hepatic uptake of human β-hexosaminidase A by a specific glycoprotein recognition system on sinusoidal cells. *Proc. Natl. Acad. Sci. USA*, 76:2,774–2,778.
50. Suzuki, K. (1965): The pattern of mammalian brain gangliosides. III. Regional and developmental differences. *J. Neurochem.*, 12:969–979.
51. Tallman, J. F., Jr., Brady, R. O., and Suzuki, K. (1971): Enzymic activities associated with membranous cytoplasmic bodies and isolated brain lysosomes. *J. Neurochem.*, 18:1,775–1,777.
52. Tallman, J. F., Johnson, W. G., and Brady, R. O. (1972): The metabolism of Tay-Sachs ganglioside: Catabolic studies with lysosomal enzymes from normal and Tay-Sachs brain tissue. *J. Clin. Invest.*, 51:2,339–2,345.
53. Van den Bergh, F. A. J. T. M., Rietra, P. J. G. M., Kolk-Vegter, A. J., Bosch, E., and Tager, J. M. (1976): Therapeutic implications of renal transplantation in a patient with Fabry's disease. *Acta Med. Scand.*, 200:249–256.
54. Von Specht, B. U., Geiger, B., Arnon, R., Passwell, J., Keren, G., Goldman, B., and Padeh, B. (1979): Enzyme replacement in Tay-Sachs disease. *Neurology*, 29:848–854.
55. Wiesmann, U. N., Rossi, E., E., and Herschkowitz, N. N. (1972): Correction of the defective sulfatide degradation in cultured fibroblasts from patients with metachromatic leukodystrophy. *Acta Paediatr. Scand.*, 61:296–302.

Genetics of Neurological and Psychiatric Disorders, edited by Seymour S. Kety, Lewis P. Rowland, Richard L. Sidman, and Steven W. Matthysse. Raven Press, New York © 1983.

Dominant Ataxias

Roger N. Rosenberg

Laboratory of Cellular Neurobiology, Department of Neurology, University of Texas Health Science Center, Southwestern Medical School, Dallas, Texas 75235

A resurgence of interest and research has taken place in the past 5 years to reclassify and rechart the clinical and neuropathologic features of, and seek a molecular basis for, the dominantly inherited ataxias. These are disorders in which ataxia is the predominant or sole neurologic feature expressed in the heterozygote due to the dominance of one mutant autosomal allele, in which one-half of both daughters and sons of an affected parent develop disease. Credit for the heightened interest of neurologists in these dominantly inherited disorders must be given to the impressive progress being made in recent years in elucidating biochemical bases for several recessively inherited ataxias (7). A recent survey of the literature for defined biochemical defects in syndromes with inherited ataxia indicated that there are at least 18 separate biochemical entities, all inherited as X-linked or autosomal recessive traits (17). The only dominantly inherited ataxic syndrome in which biochemical abnormalities have been found consistently is Joseph disease, a unique spinocerebellar degeneration present only in the Portuguese.

Why the disparity between recessively and dominantly inherited diseases in the ability of investigators to find biochemical clues, a molecular marker of disease, or, ideally, an altered primary gene product? Why have the dominantly inherited diseases not succumbed to biochemical scrutiny, as reviewed by Brady and Rosenberg (2), with the possible exception of hepatic porphyria, in which uroporphrinogen synthetase is reported to be defective? Intensive enzymologic studies, particularly of regulatory enzymes of biochemical cycles, have not been systematically undertaken in the dominant ataxias. Also, the molecular basis of dominantly inherited genetic disorders, including those affecting the nervous system and particularly the cerebellum, may be entirely different from recessive diseases. Progress may come with conventional studies of enzyme activity, but a more creative new approach may be required, as will be discussed later. Research groups in the United States, Canada, United Kingdom, France, Portugal, Norway, and Japan are exploring and devising research strategies, as expressed at international symposia in Los Angeles (1977) and Lisbon (1980). Progress is being made in understanding these disorders, and this chapter will focus on the current status and future directions of research of the dominant ataxias.

CLASSIFICATION

For more than 100 years, classifications of the inherited ataxias have been based on variations of clinical or neuropathologic features (see Table 1). Friedreich (1863), Menzel (1891), Greenfield (1954), Konigsmark and Weiner (1970) (8), Rosenberg (1977), Kark et al. (1978) (7), and Blass (1980) (1), among others, have presented logical categories of disease. The most recent classification by Blass (1980) (1) seems particularly worthy, offering three broad groups of syndromes based on the major clinical-neuropathologic features: (a) hereditary spinal ataxias, in which the primary pathology is caudal to the brainstem; (b) hereditary cerebellar ataxias, including both the cortical cerebellar atrophies (e.g., of Holmes) and the cerebellar brainstem atrophies [e.g., olivopontocerebellar atrophy (OPCA)]; and (c) hereditary ataxic encephalopathy, in which there is evidence of widespread degeneration of the nervous system, including the cerebral hemispheres, but with ataxia as the predominant clinical feature (1). This nosologic approach is simple, broad-based within any of the three groups, and allows for flexibility in assigning biochemical defects in the future within and between groups.

The scholarly review of the OPCAs by Konigsmark and Weiner (8) was a landmark in assigning specific clinical findings and neuropathologic features, mode of

TABLE 1. *Dominant ataxias*

I. OPCAs[a]
   Type I   Menzel type (Marie type)
      Ataxia, intention tremor, dysarthria, choreoathetosis, upper motor neuron signs
   Type III
      Ataxia, tremor, dysarthria, progressive visual loss due to retinal degeneration
   Type IV   Schut-Haymaker type
      Ataxia, dysarthria, dysphagia, dystonia, upper motor neuron, signs, late onset of dementia
   Type V
      Ataxia, dysarthria, progressive external ophthalmoplegia, extrapyramidal signs (tremor, rigidity), dementia
II. Ataxia with peroneal muscular atrophy
III. Ataxia with optic atrophy
    Ataxia with optic-acoustic nerve degeneration
IV. Ataxia with myoclonic epilepsy and neuropathy
V. Joseph disease: Autosomal dominant motor system degeneration of the Portuguese
   Type I
      Onset in 20–30 years; progressive spasticity, rigidity, dystonia, choreoathetosis; spastic-dysarthria, facial and lingual fasciculations, prominent eyes due to lid retraction; death by age 45; homozygous patient: onset at 8 years and died at 15 years
   Type II
      Onset in 20–45 years; progressive ataxia of gait and extremities; progressive spasticity, rigidity, dystonia, choreoathetosis, spastic-dysarthria, facial and lingual fasciculations; death by age 60
   Type III
      Onset in 40–65 years; more slowly progressive ataxia of gait and extremities, cerebeller scanning speech, motor-sensory distal polyneuropathy, hand and leg muscular atrophy

[a]After Konigsmark and Weiner (8).

inheritance, and natural history of disease into five clear types, with types I, III, IV, and V being dominantly inherited. Differences were minimal between types I (Menzel type) and IV (Schut-Haymaker type) and represented marginal variations in two large families. Type III was unique in having severe retinal degeneration, but shared all other features with types I and IV. Type V was accorded a separate designation because patients manifested progressive external ophthalmoplegia, extrapyramidal signs, and dementia in addition to the usual features of progressive ataxia and neuropathologic changes in the medulla, pons, and cerebellum. This report was important because it cited clear differences in familial involvement superimposed on a general matrix of essential, invariate, cerebellar clinical and neuropathologic abnormalities. It was a crucial contribution to the documentation of the variation in penetrance and expressivity of a dominantly inherited disorder within a family and between families.

Future biochemical studies will have to deal with these nosologic categories, because any given dominantly inherited syndrome may result from several genotypes, and the clinical expression of any one genotype may differ within a family or between families. The present classifications are useful, particularly that of Blass (1), but, in the final analysis, the specific molecular defect will reorder events and will be the constant, independent of clinical and neuropathologic variations and current arbitrary categories.

The dominant ataxias are largely comprised of OPCAs, or the hereditary cerebellar ataxias, as listed by Blass (1). Joseph disease (14,18,19,21) is a dominantly inherited ataxia and the second most common form of the hereditary cerebellar ataxias. It is not an OPCA because the inferior olivary nuclei are not involved, but it has emerged as an important dominant ataxia because of the intense neuroepidemiologic, clinical, neuropathologic, and biochemical studies conducted on the Joseph family and other families with this disease. From these studies have emerged several principles applicable to all dominant ataxias; the OPCAs and Joseph disease will therefore serve as subsequent major areas for discussion. Ataxia in families with dominantly inherited Charcot-Marie-Tooth disease (15), dominant ataxia with optic atrophy (3), and myoclonic epilepsy with ataxia and neuropathy (23) are interesting but minor types which should be noted for possible variations.

CLINICAL AND NEUROPATHOLOGIC FEATURES

OPCAs

The olivopontocerebellar syndromes are disorders in which there is progressive impairment of cerebellar functions with structural alterations in neurons of the inferior olivary nuclei of the medulla, as well as of the basis pontis, cerebellar cortex, and deep cerebellar nuclei.

Pathology

The cerebellum, cerebellar peduncles, and basis pontis become grossly atrophic. There is severe alteration of Purkinje cells at the light microscopic level, with reduction in the number of granule cells of the cerebellar cortex, and marked neuronal loss in the deep cerebellar nuclei, including the dentate nucleus. According to Landis et al. (9), an increased number of stacked cisternae, curvilinear densities, pleomorphic membranous tubules, and crystalloid inclusions are found as early lesions in Purkinje cells by electron microscopy. Although these ultrastructural alterations in the cerebellar cortexes obtained from live patients do disclose a variety of abnormalities, these changes nevertheless do not reveal anything of unequivocal pathogenetic significance. The possibility that infectious agents may participate in the pathogenesis of some autosomal dominant olivopontocerebellar degenerations was raised by the finding of paramyxovirus-like tubular structures and crystalloid inclusions in the degenerating cerebellar neurons.

Clinical Manifestations

Progressive ataxia, dysarthria, dysmetria, dysadiodochokinesia, nystagmus, loss of fast saccadic eye movements, and subsequent development of spasticity, optic nerve atrophy, distal sensory involvement, and late intellectual dysfunction are the essential clinical features of the olivopontocerebellar degenerations. There may be variations or new clinical phenomena in different families, expressed as either autosomal recessive or dominant diseases (8). The variances described by early investigators (8) were separated by arbitrary minor differences in clinical features. The olivopontocerebellar degenerations that are inherited as autosomal dominant traits may comprise only a few unique clinical diseases in which penetrance and expression of a single point gene mutation are altered by other modifying allelic and nonallelic genes.

In the 2nd or 3rd decade of life, there is progressive ataxia of limbs and trunk, with dysmetria and dysarthria. This is followed by spasticity associated with clonus, hyperreflexia, and extensor plantar responses. Nystagmus, optic nerve atrophy, and loss of fast saccadic eye movements may occur. Late muscle atrophy and fasciculations may involve muscles of the face, jaw, and tongue because lower motor neurons are affected. Distal sensory loss may be seen. Intellectual deterioration and dementia may affect some patients late in the disease. Ophthalmoplegia, extrapyramidal signs, and optic atrophy with visual loss may also be encountered.

Cerebrospinal fluid (CSF) protein and cell counts are normal. Jackson et al. (6) have presented convincing evidence of linkage between the genes for ataxia and the HLA complex situated on the sixth chromosome. These data have become useful as a marker of disease for purposes of genetic counseling.

In general, the OPCAs are symmetrical disorders, pathologically and clinically, and often have a clear autosomal dominant inheritance or suggestion of familial involvement. Abnormal organic acid composition of the urine (20) and abnormally low aspartate concentrations in the CSF (12) may be markers of the disease and useful in genetic counseling, but the basic molecular defects in these disorders remain to be elucidated.

The OPCAs do not have any special ethnic or racial selectivity, and males and females are affected equally. The frequency of the OPCAs is not exactly known.

The prevalence of all cerebellar ataxias is estimated to be about 5 per 100,000 population; the prevalence of the OPCAs can only be a fraction of this rate. No specific therapy is available, and only early genetic counseling of the at-risk population may be effective in reducing subsequent incidence of disease. Baclofen or L-DOPA therapy may benefit some individuals with prominent pyramidal or extrapyramidal clinical features.

Joseph Disease

Historical Review

In 1972, two papers described dominantly inherited neurologic disease in two Azorean-Portuguese families residing in Massachusetts. The first report (11) referred to "Machado disease" to acknowledge the proband family in which there was a progressive cerebellar disease with distal extremity atropy and sensory loss beginning in the 5th decade. A sural nerve biopsy from 1 patient documented segmental demyelination. The second report (24) described "nigro-spinal-dentatal degeneration with nuclear ophthalmoplegia" in the Thomas family, presenting in the 3rd to 5th decades with gait ataxia, external ophthalmoplegia, facial-lingual fasciculations, elements of spasticity, and extremity rigidity. The neuropathologic features included neuronal loss in the substantia nigra, dentate nuclei, and Clark's column in the spinal cord. Demyelination was present in the posterior columns, lateral corticospinal, and spinothalamic tracts of the spinal cord, most evident in the thoracic portion.

In 1976, Rosenberg et al. (19) described a progressive motor system disease beginning in the 2nd and 3rd decades in another Portuguese family, the Joseph family of California, consisting of 329 persons in 9 generations at that time. The clinical picture differed from the two previous reports in that Joseph family affected persons had progressive spasticity, lurching unsteadiness of gait due to spasticity (without cerebellar signs), spastic dysarthria, loss of fast saccadic eye movements, ophthalmoparesis for upward gaze, and facial-lingual fasciculations, with prominent dystonia of head, face, extremities, and trunk in some individuals (Figs. 1 to 3). The neuropathology consisted of neuronal loss in the striatum, substantia nigra, basis pons, dentate of cerebellum, cerebellar cortex, and anterior horns of the spinal cord. CSF homovanillic acid levels were reduced, correlating with neuronal loss in the zona compacta of the substantia nigra. The Joseph family disorder differed from the Massachusetts cases because: (a) there was striatonigral degeneration; (b) dystonia was severe in some patients, and (c) cerebellar signs were not evident in any family member examined in 1975. The disease was also different from olivopontocerebellar degeneration because the medullary interior olives were unaffected. It was unlike Huntington disease (HD) because intelligence remained normal in all Joseph patients and the cerebral cortex was histologically normal.

Epidemiologic studies were necessary to establish the spectrum of clinical and neuropathologic findings. The logical place to proceed was in the Azores Islands,

FIG. 1. Patient with Joseph disease, type II, with progressive gait ataxia and spasticity of the legs. This 36-year-old man is patient NS (V, 32) in our Sousa family pedigree of the Island of Flores, Azores. He was examined in June, 1977. (See Fig. 4 and ref. 7.)

particularly on the island of Flores, from which both the Joseph and Thomas families originated. Other suggestions of variable clinical and neuropathologic features were made as personal communications to Dr. Rosenberg from Paula Coutinho and Corino Andrade of the Neurology Service, Hospital Santo Antonio, Porto, Portugal. They traveled to Flores in December, 1976, where they found patients with prominent spasticity and extrapyramidal signs, and some with cerebellar disorders or peripheral neuropathy. Coutinho and Andrade (4) emphasized types I, II, and III as variants of the autosomal dominant disorder in affected individuals on Flores. Type I corresponded to the typical Joseph phenotype with only extrapyramidal and pyramidal signs, including dystonia, athetosis, and rigidity with associated spasticity. Type II included patients with cerebellar deficits and spasticity. Type III was reserved for patients with the Machado phenotype—patients with a dominantly inherited form of motor polyneuropathy and true cerebellar deficits (4,14) (Table 1).

Romanul et al. (13) described another Massachusetts family of Portuguese ancestry with a similar clinical spectrum of gait ataxia, parkinsonian rigidity, ophthalmoparesis, muscle fasciculation, areflexia and associated nystagmus, cerebellar tremor, and extensor plantar responses. Postmortem study of 1 patient showed

FIG. 2. Patient with Joseph disease, illustrating ophthalmoparesis. The patient is attempting to gaze to the left and has a paresis of the left lateral rectus muscle. He is a type II patient with gait ataxia and a moderate spastic paresis. He was examined on Flores, Azores, in 1977.

striatonigral degeneration with severe neuronal loss and gliosis in the putamen; in another patient, the basal ganglia were normal. Their observations supported the concept of a clinical and neuropathologic variation. It seemed likely that all the reported families of similar ancestry represented a single genetic entity with variable expression, and that a better understanding of this disease might accrue from a visit to Flores and San Miguel, from which the family reported by Romanul et al. migrated.

Azorean Neuroepidemiology

In 1977, Rosenberg and Nyhan joined Coutinho in the Azores to investigate the presence of this genetic disease, especially on the island of Flores. Rosenberg, Nyhan, and Coutinho encountered typical genetic disease as described in the Joseph and Thomas families on Flores in the Sousa family (14) (Fig. 4).

Two main variants of the disease were encountered. Type I corresponded to the Joseph phenotype, with spasticity, clonus, hyperreflexia, extensor plantar responses, facial and lingual fasciculations, and nystagmus with ophthalmoparesis beginning about age 25. Type II included cerebellar signs, with truncal ataxia and cerebellar

FIG. 3. Patient with Joseph disease, type I, with dystonic posturing of his jaw and left arm. He had evidence of spasticity in his legs and arms. Disease began at age 8 years; he is shown at age 12. He died at age 15. Of note is the fact that his parents were both affected with type II disease, having late-onset cerebellar deficits with some leg upper motor neuron signs.

deficits of extremities, that began in the early 30s and were less rapid in progression than Type I. Typical examples of type I were brothers AS (V, 31), age 38, and NS (V, 32), age 36 (14). Both noted onset of symptoms in their late 20s, starting with impairment of gait caused by a lurching spasticity. Dysarthria and dysphagia also occurred early. AS developed striking prominence of his eyes, whereas NS

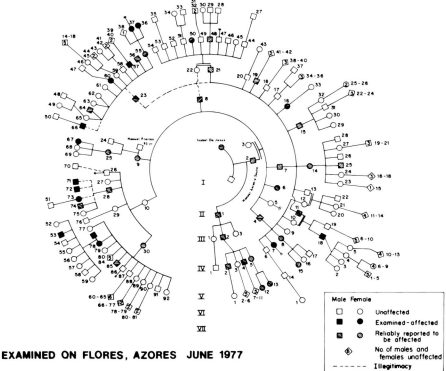

FIG. 4. Sousa family pedigree from the Island of Flores, Azores. Descendants of Manual Antonio Sousa (II, 2) on Flores are shown who indicate dominantly inherited disease of the Joseph disease-type.

did not. Opthalmoparesis for upward gaze and nystagmus on horizontal gaze were noted. There was stiffness of the arms, spasticity of the legs, and bilateral ankle clonus with extensor plantar responses in 1 brother and bilateral flexor plantar responses in the other brother. Both brothers demonstrated dystonia of the face, head and extremities, exaggerated by talking, standing, and especially walking. Dysphagia and spastic dysarthria were present in both patients.

In the same family was EN (V, 38), a 38-year-old woman who was a first cousin of these two brothers; her symptoms began at age 30. She manifested ophthalmoparesis, nystagmus, facial fasciculations and atrophy, spasticity, and increased tone of the legs and hyperreflexia, but with flexor plantar responses. She had truncal ataxia and heel-to-shin and finger-to-nose dysmetria (14).

The husband of EN (V, 38) also developed an impaired gait at about age 55. He was examined at age 59; tandem walking was impaired and there was ataxia of the heel-to-shin maneuver. Their 8-year-old son, Mauricio (VI, 63), had impaired coordination of hands, speech, and walking. When we examined him at age 12, there was striking dystonia of facial movements, extremity functions, and walking. He had marked spasticity of the arms and legs, with sustained clonus at the knee and ankle and equivocal plantar responses. Walking and standing increased the dystonic posturing and elicited superimposed athetotic twisting movements of his hands and face. Drinking water could provoke choking, and his speech was spastic and dysarthric. He was seriously incapacitated, unable to eat or walk unassisted. No cerebellar signs were present (14) (Fig. 3).

Both parents expressed type II disease, with definite cerebellar signs and spasticity, whereas the son expressed a prominent type I form at an early age. It is presumed that he had a double dose of, or was a homozygote for, the mutant autosomal dominant gene, and therefore represented the full expression of the disorder. It might be argued that type I disease, with pure extrapyramidal signs of dystonia and athetosis along with clonus, spasticity, and hyperreflexia, but without cerebellar signs, is also a fairly complete expression of the dominant gene. Type II, having cerebellar signs, a milder clinical course, and later onset (as expressed by the two parents), might represent less complete expression of the gene.

Although we encountered many examples of type I and type II disease in the large kindred on Flores, we did not encounter a single example of type III, the Machado form, with peripheral neuropathy, distal atrophy, and cerebellar disorder.

Patient JTS, a man of 67 years, was examined in Santa Cruz. He had a steppage gait and foot drop with absent tendon reflexes in the legs and severe atrophy below the knees. This corresponds to the Machado form, or type III. He indicated that he had a family history of similar illness, but could not trace his ancestry sufficiently to relate him to the Machado family in the United States or to the Sousa family manifesting type I and type II disease. Type III disease is present on Flores and patient JTS resembles patient MB (II, 7) in the Thomas family; both have polyneuropathy and cerebellar disorder.

Several patients in the Sousa family developed progressive ocular prominence (14). Patient AM (V, 36) was a 19-year-old woman who noted ataxic gait at 15 years and, when examined, had leg spasticity, ankle clonus, flexor plantar responses, and limb ataxia. Her gait was spastic, with lurching from side to side. Her eyes were prominent, and there was ophthalmoparesis and horizontal nystagmus. Ocular prominence also occurred in her mother.

Single Genetic Disease Hypothesis

Synthesizing the data from the Joseph, Thomas, and Sousa families (14), the data of Coutinho and Andrade (4), and the data of Romanul et al. (13), we believe we are witnessing the expression of a single gene that causes (a) early-onset disease with pyramidal and extrapyramidal signs (type I); (b) intermediate age of onset and progression, with cerebellar deficits as well as pyramidal and extrapyramidal findings (type II); and (c) later onset, with peripheral neuropathy and cerebellar disorder (type III) (Table 1 and Fig. 5). Dominantly inherited parkinsonism with distal atrophy in one family (13) could constitute a type IV disease. It took a systematic

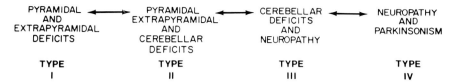

FIG. 5. Spectrum of clinical findings in Joseph disease. It is proposed that a single gene mutation can have clinical and neuropathologic variation, with at least four types now being recognized.

evaluation of the California, Massachusetts, and Flores families, however, to document the variations of clinical expression. The additional neuropathologic findings of Romanul et al. (13), indicating the wider pathologic involvement with striatonigral degeneration in only some patients, also implied varied expression of this mutant gene.

Alternatively, types I and II may represent one distinct genetic disease with variable cerebellar involvement that is present only on Flores and in descendants from the island. The Machado variant, type III, present in San Miguel, may be a separate dominant entity. The family reported by Romanul et al. (13) was originally from the same region of San Miguel and had typical Machado, or type III, disease. In favor of a single basic genetic mutation, however, were the findings in 1 autopsied member of the Romanul-Fowler family (13) of a striatonigral degeneration (patient III, 7); the presence of parkinsonian features in 2 others (III, 3 and III, 7); the fact that 2 Thomas family members, a family originally from Flores, had type III disease; and, finally, the fact that 1 Joseph family member, EJ (III, 19), may have a variant expression (type III) of the family disorder (Fig. 6A and B). The hypothesis of a single genetic disease seems plausible, but lacks the critical support of a biochemical marker that is a constant abnormality regardless of clinical expression.

More recently, two reports have cited this genetic disorder in two families, one in northeastern Portugal (10) and one in New York (5)—a black family traced to North Carolina. They had clinical features similar to those of the Joseph and Thomas families, but neither family had any identifiable linkage to the Azores. Neuropathologic studies are required to prove that these families have the same disease.

It is generally agreed that, despite the three major phenotypes, there is only one genetic disorder. Type I disease refers to onset in the 2nd or 3rd decades with an expression of pyramidal or extrapyramidal features, including spasticity, rigidity, and dystonia. Type II disease includes pyramidal and extrapyramidal findings with cerebellar ataxia of gait and extremities. Type III disease refers to late-onset disease, beginning in the 5th to the 7th decades, and expressing progressive cerebellar-type gait ataxia, hyporeflexia, hypotonia, and distal sensory loss and distal atrophy. These three phenotypes are probably clinical variations of the same gene mutation because (a) all three types have been recorded in each of three separate families; (b) type I disease occurred in a parent, with types I and III in that parent's two children (Fig. 7); and (c) type II disease presented in both parents with onset at 8 years of type I disease in their child, who presumably was homozygous for the dominant gene.

FIG. 6. A: Patient EJ (III, 19) (7,19) is shown at age 75 with Joseph disease, type III. Onset of progressive cerebellar deficits began at age 65.

Biochemical Findings

In 1979, Rosenberg et al. (21) published two-dimensional acrylamide gels from 1 patient and control brain samples. There were several significant changes, with increases of 50,000 MW and 40,000 MW proteins in patient cerebellum and putamen corresponding to gliosis seen histologically (Fig. 8A and B). These findings were later confirmed in samples of cerebellum and putamen from 5 patients (18).

Joseph disease results from neuronal loss with attendant glial reaction and replacement, i.e., gliosis. The disease may occur because the gene mutation results in the loss of a glial trophic factor required for neuronal function. Two changes would follow. First, glial cells would increase in dynamic compensation to provide an increased total amount of glial factor, but this is ultimately insufficient, resulting in secondary neuronal dysfunction and loss. This view is supported by the evidence of increased Jc, Jd, L_1, L_2, and h proteins which are synthesized by glia (Fig. 8A and B). Multiple sclerosis plaques are also foci of gliosis, and only these J and L specific classes of proteins are increased (9). The h class of proteins was determined by comigration studies to be glial filamentous acidic proteins (21).

FIG. 6. B: Atrophy distally at age 75 years. Of note is the fact that he is the grandson of Antone Joseph and is the exception in the Joseph family, having type III disease whereas others in the family express type I disease.

Increased glial proteins might be merely a biochemical reflection of nonspecific gliosis, but these changes may be due to an increased glial protein response per cell and by a total increase in glial numbers, attempting specifically to correct another primary gene mutation. The neuropathology and these biochemical changes are still compatible with the provocative interpretation that Joseph disease and the other dominant ataxias are the result of a genetic mutation that causes the loss of glial trophic factor, leading to neuronal loss and glial reaction. A genetic reciprocal interaction between glial cells and neuroblasts had been documented in cell culture (22). A glial gene product induces a repressed neuronal gene and results in the expression of a specific neuronal protein (22). Joseph disease and other dominant ataxias may be the clinical counterpart of defective glial-neuronal interaction.

Joseph Disease Designation

A preference for the term "Joseph disease" is generally expressed because it is the largest known pedigree, including more than 600 persons in 10 generations. The Joseph family includes all three types of disease, and brain samples of all three types show similar protein patterns on two-dimensional gels. It has been possible to reconstruct the spectrum of clinical disease, propose the unitary hypothesis of a single gene mutation resulting in several clinical phenocopies from analysis of this family, and provide direct biochemical evidence in support of this single genotype from brain samples obtained principally from this family. It may be possible to

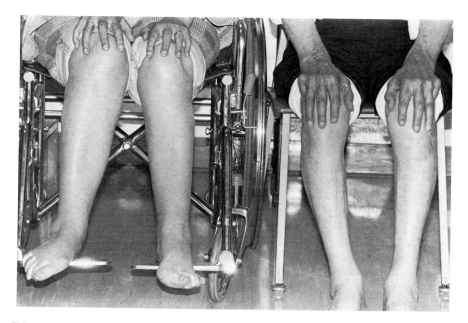

FIG. 7. A brother *(right)* and sister *(left)* having type III and I Joseph disease, respectively. She presented in her late 30s with leg spasticity, and he presented in his 60s with progressive cerebellar gait ataxia. He is developing distal extremity atrophy as well, characteristic of type III disease. By history, their mother had type I disease—evidence supporting the hypothesis of a single gene producing either type I or III phenotype.

identify expression of disease by patterns of skin fibroblast proteins labeled with ^{35}S-methionine obtained from Joseph family members (18). For these compelling reasons, Joseph disease seems to be the proper designation. The term "Azorean disease" was objected to by physicians and scientists from Portugal, the Azore Islands, the United States, and Canada for the reasons cited above and also because some affected families have no known connection with any Azorean family (5,10). Azoreans and the Portuguese medical community find the term objectionable because of the stigma it gives to the Azorean people.

OPCA and Joseph Disease

There is considerable historical evidence from the explorations and subsequent reciprocal migrations of the Dutch, the Portuguese, the Portuguese-Azoreans, and the Portuguese-Brazilians between the 12th and 17th centuries to speculate that autosomal dominant OPCA of the Schut-Swier type (8) (of Dutch origin) and Joseph disease (of the Portuguese) could be the result of the same mutant gene. Clinical and neuropathologic variations of OPCA and Joseph disease, and among families with Joseph disease, could be the result of both heterogeneous nonallelic and allelic genetic modification of the mutant gene expression. Population genetic analyses,

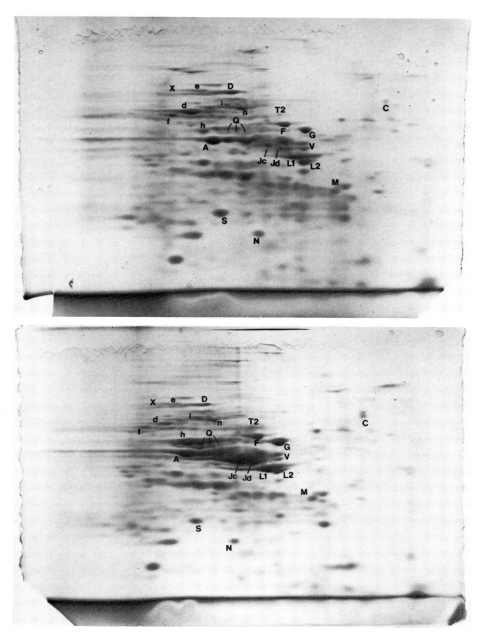

FIG. 8. Two-dimensional gel protein separations of control **(Top)** and Joseph disease **(Bottom)** cerebellar proteins. E. C. was a 73-year-old, type III patient reported previously in detail (21). The patient gel shows a significant increase in the h, J, and L classes of proteins. 1 mg of protein from a brain homogenate was added to each gel.

gene mapping, and precise identification of the altered gene product in patients with both types of disease will be required to resolve this point.

Joseph Disease as a Unique Spinocerebellar Degeneration

Joseph disease is a unique, dominantly inherited spinocerebellar degeneration, rarely encountered in ethnic groups other than the Portuguese. Its clinical phenotype includes a specific constellation of features: prominent eyes due to lid retraction, ophthalmoparesis, loss of fast saccadic eye movements, facial and lingual fasciculations without atrophy, dystonia, spasticity and parkinson-like features of tremor-at-rest, bradykinesia, and rigidity. It is a spinocerebellar degeneration with invariant symmetrical degeneration of the cerebellum and thoracic cord in all patients and variable degeneration of the striatum, substantia nigra, basis pons, oculomotor nuclei, and peripheral nerves. The variation in penetrance and expressivity follows the pattern encountered in other dominantly inherited disorders but was verified only by virtue of neuroepidemiologic, clinical, and biochemical analyses (14,16,18). Patients with elements of spasticity or parkinsonism, despite associated cerebellar deficits, may benefit from treatment with L-DOPA.

MECHANISMS OF DOMINANT ATAXIAS

It is expected that several molecular bases for the pathogenesis of disease will be uncovered for any given single clinical syndrome. Conversely, any one genotype will include variation of clinical features within a family and between families. The biochemical challenge is formidable, but, with an appreciation of the molecular basis of any of the dominantly inherited neurologic diseases, rapid progress should be made in all the ataxias. Several biochemical clues are emerging. Two other dominant conditions, HD and myotonic muscular dystrophy (MMD), are probably diffuse and universal disorders of cell surface membranes. Huntington fibroblasts overgrow in cell culture compared to control cell lines. Biophysical studies of HD and MMD fibroblast, red cell, and lymphocyte membranes indicate alterations in membrane fluidity and dynamics (2). Dominantly inherited neurofibromatosis results in part from defective and elevated serum levels of nerve growth factor and reduced numbers of cell surface receptor sites for lymphocyte epidermal growth factor. These defects may result in improper cell-cell recognition and biochemical cell-cell interaction. Neuronal loss and gliosis, as discussed above, result in HD and Joseph disease; tumor formation occurs in neurofibromatosis, along with such other architectonic abnormalities as neuronal heterotopias and cortical disorganization. Glial-neuronal biochemical interactions, so indispensible for neuronal life, fail, perhaps because of cell surface alterations resulting in compensatory glial proliferation. This idea of impaired glial-neuronal interaction is based on these biochemical clues and should be pursued (18).

Viral factors must be mentioned in view of the finding of paracrystalline and myovirus-like structures in neurons of OPCA patients by Landis et al. (9). Further, a dominantly inherited pattern of Creutzfeldt-Jakob disease (C-J-D) has been reported by Gajdusek and colleagues, with several patients having ataxia (2). No evidence of vertical transmission has been maintained, and it is assumed there is increased susceptibility on a genetic basis for the C-J-D agent. A brain specimen from a Joseph patient was inoculated into primates and other small mammals, and as of 1980, no disease resulted in the host animals. The necessity for careful pathologic examination is emphasized, since C-J-D has now been diagnosed in a family with HD and is suspected in a family with Alzheimer disease. Inoculation of biopsy material from some Alzheimer patients caused spongiform encephalopathy in animals.

Other mechanisms productive of the dominant ataxias due to a single gene mutation include: alteration of a subunit of a cell surface receptor site for a neurotransmitter or neuromodulator; alteration of an antigenic site on the cell surface resulting in an autoimmune state; an altered soluble repressor substance within the cell nucleus irreversibly inhibiting the transcription of a gene; altered regulatory enzymatic activity in a crucial biochemical cyclic pathway; enzyme alteration producing a neurotoxin or an inhibitor; and alteration in the intron region of deoxyribonucleic acid (DNA), a region that normally splits the exon portion of a gene and that may be essential for efficiency or rate of ribonucleic acid (RNA) transcription for protein synthesis or for final protein modification by phosphorylation, glycosylation, or methylation.

CONCLUSION

The dominant ataxias are a group of disorders characterized by a regular, progressive, and symmetrical degeneration of the cerebellum and its pathways. Disease results in half of the sons and daughters of an affected parent, and disease therefore

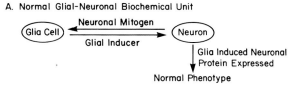

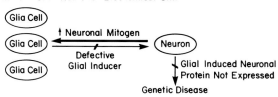

FIG. 9. A model for the biochemical basis of gliosis. It is postulated that glia and neurons form a biochemical unit with reciprocal intercellular biochemical regulation. It is also postulated that glial genes exist that induce neuronal genes for the expression of neuronal proteins; in turn, neurons regulate the glial population by the release of a mitogen at low levels in the normal brain. The mutation results in the loss of glial inducer for a neuronal protein and, thus, in neuronal degeneration. The neuronal compartment, in turn, secretes increased amounts of mitogen to cause an increase in glial number to provide more glial inducer and more glialtrophic support. Gliosis and associated neuronal degeneration, therefore, are the end-stage results of a compensatory biochemical interaction between these two cell types.

occurs in the heterozygote. Variation in age of onset, clinical findings, neuropathologic features, and penetrance of disease is the rule (14,16,18). New mutations must be rare. HLA linkage to the ataxia gene on chromosome 6 in one large family but not others suggests at least two separate genotypes for the OPCAs (6). It is not clear if all the dominantly inherited ataxias are due to a single gene mutation that has entered various ethnic populations and produces different syndromes. Such a supposition is possible but must be supported by a common molecular defect, independent of clinical or neuropathologic variation. Joseph disease is a dominantly inherited ataxia in which it has been demonstrated that a single gene mutation can cause wide variation in phenotype of affected persons within a single family and between families. Increased 50,000- and 40,000-MW classes of proteins have been found in 6 Joseph disease brains compared to appropriate control brain as shown on two-dimensional acrylamide gels (18,21). These increased proteins, in part, are glial proteins expressed most abundantly in cerebellum and basal ganglia and represent a glial response to compensate, it is hypothesized, for defective glial-neuronal interaction (16,18,21). The dominant ataxias may be due to allelic mutations within a single gene, resulting in the clinical variation noted. The altered gene product is responsible for an impairment in cell-cell recognition (glial-neuronal biochemical interaction), leading to the characteristic neuropathology of neuronal loss and gliosis (Fig. 9). The finding of the altered gene product in nonneural tissue would provide a marker for effective genetic counseling and elimination of disease in a family in one generation. The goals are now clearly defined and current research approaches appear promising for new insights into these disorders in the immediate future.

REFERENCES

1. Blass, J. (1981): *Hereditary Ataxias.* In: *Current Neurology, Vol. 3*, pp. 66–91. John Wiley and Sons, New York.
2. Brady, R., and Rosenberg, R. N. (1978): Autosomal dominant neurological disorders. *Ann. Neurol.*, 4:548–552.
3. Budka, H., Seeman, D., and Danielczyk, W. (1979): Hereditary cerebellar atrophy (Holmes type) with optic atrophy: A clinicopathological study of four generations of a family. *Arch. Psychiatr. Nervenkv.*, 6:311–318.
4. Coutinho, P., and Andrade, C. (1978): Autosomal dominant system. Degeneration in Portuguese families of the Azores Islands. *Neurology*, 28:703–709.
5. Healton, E., Brust, J., Kerr, D. (1979): Familial cerebellar ataxia, dystonia, and abnormal eye movements in a non-Portuguese family. *Neurology*, 29:559–560.
6. Jackson, J., Currier, R., Teraski, P., and Morton, N. (1977): Spinocerebellar ataxia and HLA linkage risk prediction in HLA typing. *N. Engl. J. Med.*, 296:1,138–1,141.
7. Kark, P., Rosenberg, R. N., Schut, L. (editors) (1978): *The Inherited Ataxias. Adv. Neurol.*, 21:1–424. Raven Press, New York.
8. Konigsmark, B., and Weiner, L. (1970): The olivopontocerebellar atrophies: A review. *Medicine*, 49:227–241.
9. Landis, D., Rosenberg, R. N., Landis, S., Schut, L., and Nyhan, W. (1974): Olivopontocerebellar degeneration. *Arch. Neurol.*, 31:295–307.
10. Lima, L., and Coutinho, P. (1980): Clinical criteria for diagnosis of Machado-Joseph disease: Report of a non-Azorean Portuguese family. *Neurology*, 30:319–322.
11. Nakano, K. K., Dawson, D. M., and Spence, A. (1972): Machado disease: A hereditary ataxia in Portuguese immigrants to Massachusetts. *Neurology*, 22:49–55.
12. Perry, T., Currier, O., Hansen, S., and MacLean, J. (1977): Aspartate-taurine imbalance in dominantly inherited olivopontocerebellar atrophy. *Neurology*, 27:257.

13. Romanul, F. C. A., Fowler, H. L., Radvany, J. (1977): Azorean disease of the nervous system. *N. Engl. J. Med.*, 296:1,505–1,508.
14. Rosenberg, R. N. (1978): Joseph's disease. In: *The Inherited Ataxias*, edited by P. Kark, R. N. Rosenberg, and L. Schut. *Adv. Neurol.*, 21:1–424. Raven Press, New York.
15. Rosenberg, R. N. (1979): Peroneal muscular atrophy. In: *Textbook of Neurology*, 6th edition, edited by H. H. Merritt, pp. 569–572.
16. Rosenberg, R. N. (1980): Genetic variation and neurological disease. *Trends Neurosci.*, 3:144–148.
17. Rosenberg, R. N. (1982): Inherited, congenital, and idiopathic degenerative diseases of the nervous system. In: *Textbook of Medicine*, 16th edition, edited by J. B. Wyngaarden and L. H. Smith. pp. 2,036–2,045. W. B. Saunders Co., Philadelphia.
18. Rosenberg, R. N., Ivy, N., Kirkpatrick, J., Bay, C., Nyhan, W. L., and Baskin, F. (1981): Joseph disease and Huntington disease. Protein patterns in fibroblasts and brain. II. *Neurology*, 31:1,003–1,014.
19. Rosenberg, R. N., Nyhan, W. L., Bay, C. (1976): Autosomal dominant striatonigral degeneration: A clinical, pathologic, and biochemical study of a new genetic disorder. *Neurology*, 26:703–714.
20. Rosenberg, R. N., Robinson, A., and Partridge, D. (1978): Urine vapor pattern for olivopontocerebellar degeneration. *Clin. Biochem.*, 8:365–368.
21. Rosenberg, R. N., Thomas, L., Baskin, F., Kirkpatrick, J., Bay, C., and Nyhan, W. L. (1979): Joseph disease: Protein patterns in fibroblasts and brain. *Neurology*, 29:917–926.
22. Rosenberg, R. N., Vance, C. K., and Prashad, N. (1978): Differentiation of neuroblastoma, glioma, and hybrid cells in culture as measured by the synthesis of specific protein species: Evidence for neuroblast-reciprocal genetic regulation. *J. Neurochem.*, 30:1,343–1,355.
23. Smith, N. J., Espir, M. L. E., and Matthews, U. B. (1978): Familial myoclonic epilepsy with ataxia and neuropathy with additional features of Friedrich's ataxia and peroneal muscular atrophy. *Brain*, 101:461–472.
24. Woods, B. I., and Schaumburg, H. H. (1972): Nigro-spino-dentatal degeneration with nuclear ophthalmoplegia: A unique and partially treatable clinicopathological entity. *J. Neurol. Sci.*, 17:149–166.

Genetic Heterogeneity of the Hexosaminidase Deficiency Diseases

William G. Johnson

Columbia University College of Physicians and Surgeons, New York, New York 10032

FOUR STAGES IN UNDERSTANDING HEXOSAMINIDASE DEFICIENCY DISEASES

Hexosaminidase deficiency disease was first recognized nearly a century ago by Drs. Warren Tay (85) and Bernard Sachs (72), who described an infantile degenerative disease that came to be known as amaurotic family idiocy, or Tay-Sachs disease. In the brain, abnormally swollen neurons were filled with PAS-positive material. This pathological finding became the hallmark of amaurotic family idiocy and, for decades, virtually defined the disorder. During this time several disorders were lumped into amaurotic family idiocy, and at least six types were recognized (78). One was classic infantile Tay-Sachs disease. A second was "Tay-Sachs disease with visceral involvement," now known as G_{M1}-gangliosidosis, or Landing's disease. A third was congenital lipidosis, first described by Norman and Wood, which remains unsolved at the biochemical level, though sialidase deficiency is suspected (38,39). The other three are forms of ceroid storage disease, also known as ceroid lipofuscinosis, or Battens disease: (a) the late-infantile form, Jansky-Bielschowsky disease; (b) the juvenile form, Spielmeyer-Sjögren disease; and (c) the adult form, Kufs disease. An infantile form (77) of ceroid lipofuscinosis, Santavouri's disease, is also known. All forms of ceroid storage disease remain unsolved at the biochemical level.

After the clinical and pathological descriptions of amaurotic family idiocy, a third stage of understanding was reached when a marked increase in brain ganglioside (43) was found in infantile Tay-Sachs disease. The structure of this "Tay-Sachs ganglioside" was determined (82) and the material was designated G_{M2}-ganglioside. This finding distinguished Tay-Sachs disease from the other forms of amaurotic family idiocy. In Landing's disease, G_{M1}-ganglioside was stored (46,62). In the other forms no specific increase in ganglioside was found.

Classic infantile Tay-Sachs disease was then regarded as a G_{M2}-gangliosidosis, that is, a disorder in which G_{M2}-ganglioside was stored. In addition, at least two other compounds were found stored in brain or viscera (76) of these patients: kidney globoside (GL-4) and asialo-G_{M2}-ganglioside (also known as G_{A2}, or "Tay-Sachs globoside"). It was not clear why these compounds were stored. Were they abnormal in structure so that they could not be degraded? Was there an increase in their synthesis? Or was there a block in degradation?

A fourth stage of understanding of these disorders was the finding that a degradative enzyme was missing. Hexosaminidase A (37), one of three major tissue isoenzymes of hexosaminidase (beta-D-N-acetylhexosaminidase) was found to be absent (63,73) in classic infantile Tay-Sachs disease as determined by the artificial fluorogenic substrate 4-methylumbelliferyl-beta-D-N-acetylglucosaminide (4MU-beta-glcNAc) or the galactose-containing analog, 4MU-beta-galNAc. Likewise, activity of hexosaminidase was absent in classic infantile Tay-Sachs disease if a natural substrate, radiolabeled G_{M2}-ganglioside, was used for assay (9,44,74).

HEXOSAMINIDASE SUBSTRATES

At least three glycosphingolipid substrates have been found in hexosaminidase deficiency diseases. All have in common the lipid moiety ceramide (Cer). This consists of the long chain aminoalcohol, sphingosine, with a fatty acid attached in amide linkage to the 2-amino group. The glycosphingolipids globoside (kidney globosides, GL-3) and G_{A2} (asialo-G_{M2}) consist of Cer to which is attached at the 1-position various hexoses and N-acetylated hexosamines in glycosidic linkage:

$$\text{GalNAc-Gal-Glc-Cer } (G_{A2})$$
$$\text{GalNAc-Gal-Gal-Glc-Cer (globoside)}$$

Gangliosides contain, in addition, one or more residues of neuraminic acid, usually N-acetylated and usually attached to galactose or to each other. G_{M2}-ganglioside resembles kidney globoside and G_{A2} in having a terminal beta-linked N-acetylgalactosaminide moiety:

$$\text{GalNAc-Gal-Glc-Cer } (G_{M2})$$
$$\phantom{\text{GalNAc-}}\backslash$$
$$\phantom{\text{GalNAc-Ga}}\text{NANA}$$

All of these compounds are substrates for hexosaminidase because they contain terminal beta-linked N-acetylgalactosaminide moieties. Globoside and G_{A2} are cleaved by both hexosaminidase A (hex A) and hexosaminidase B (hex B), but G_{M2} appears to be cleaved only by an acidic subfraction of hex A (2).

Some other natural compounds contain beta-linked N-acetylgalactosaminide or N-acetylglucosaminide residues and are substrates for hexosaminidase. Among these are oligosaccharides (79,88), glycoproteins, and mucopolysaccharides (10,87).

There are also artificial substrates (8,37,70) for hexosaminidase. Although many have been devised, only two types are widely used. Colormetric substrates use p-nitrophenol (pNp) attached in glycosidic linkage to N-acetylgalactosamine (GalNAc) or N-acetylglucosamine (GlcNAc). The resulting compounds, pNp-beta-D-GalNAc and pNp-beta-D-GlcNAc, are faintly colored. However, the action of hexosaminidase produces *p*NP that is strongly yellow-colored in basic solution and easily measured spectrophotometrically at around 400 nm. Analogous fluorometric substrates use 4-methylumbelliferone, which fluoresces strongly in basic solution. These artificial substrates are cleaved by all the relevant hexosaminidase isozymes.

Since ganglioside and other lipid substrates accumulate in large quantities within nerve cells, it is reasonable to suppose that this causes the clinical disease. It is also reasonable, however, to wonder if some or even most of the pathology might result from changes of glycoproteins or mucopolysaccharides, acting perhaps at the cell surface in the excitable membrane of the neuron. Gangliosides are found in high concentration in presynaptic terminals. Abnormally formed synaptic boutons and deformed dentritic trees have been found in gangliosidosis (68).

HEXOSAMINIDASE ENZYMES

At least three important isozymes (4,5,6) of hexosaminidase are known: hex A, hex B, and hex S. The last has very low activity even in normal tissues and was the last to be discovered. Hex A and hex B are the major tissue isozymes of hexosaminidase (70). They are referred to as hexosaminidases because they cleave either GalNAc or GlcNAc residues if these are terminally located. These enzymes are specified for the beta-linkage, and the term beta-hexosaminidase is sometimes used. However, the term hexosaminidase is commonly understood to mean beta-hexosaminidase, because alpha-N-acetylgalactosaminidase and alpha-N-acetylglucosaminidase are different enzymes, distinct from each other, and are not considered to be alpha-hexosaminidases.

Hex C is genetically unrelated (66) to the hex A, B, S system. There are, however, several other hexosaminidases with isoelectric points intermediate between hex A and B. These have been given the designation hex I (intermediate), hex I_1, hex I_2, and hex P (pregnancy related). The subunit structure of these and their relation to the hex A, B, S system has not been clarified.

A FIFTH STAGE OF UNDERSTANDING: MOLECULAR GENETICS OF HEXOSAMINIDASE DEFICIENCY DISEASES

A fifth stage in understanding hexosaminidase deficiency disorders was reached when the relationship between the three major hexosaminidase isozymes measured by artificial substrates, A, B, and S, was understood. In classic infantile Tay-Sachs disease, hex A alone appeared absent (63,73). This form of the disorder was nearly always found in Ashkenazi Jewish infants. However, both the A and B forms of hexosaminidase were deficient in the variant form of Tay-Sachs disease with visceral storage of globoside, a form now known as Sandhoff's disease (75). To make matters more complex, a third form of infantile G_{M2}-gangliosidosis, the AB variant, lacked neither A nor B forms of hexosaminidase (73,76). Yet hexosaminidase as measured by the natural substrate G_{M2}-ganglioside was deficient in all three forms of the disease (44,76).

The explanation for these puzzling findings was that the hexosaminidase isoenzymes A, B, and S were composed of at least two kinds of subunits, alpha and beta (5,86), coded for, respectively, by the alpha-locus on chromosome 15 and the beta-locus on chromosome 5 (45). Hex A was composed of alpha- and beta-subunits in a structure of the form $(\text{alpha-beta})_n$. Hex B was composed of beta-subunits only in a structure of the form $(\text{beta-beta})_n$. Hexosaminidase was the alpha-homopolymer with structure $(\text{alpha-alpha})_n$. The two lines of work leading up to this conclusion were somatic cell hybridization and genetic complementation (86) (Fig. 1) and direct isozyme mixing with subunit exchange (5).

Sandhoff's disease, with deficiency of hex A and B, was clearly a beta-locus disorder, since both A and B contain the beta-subunit. The major residual hexosaminidase in Sandhoff's disease, designated hex S, contained no beta-subunits (6,17). Classic infantile Tay-Sachs disease with deficient hex A was clearly an alpha-locus disorder, since hex A was deficient but B was not. When hex S $(\text{alpha-alpha})_n$ was found in very small quantities in normal individuals, it was possible to show that S was also missing in Tay-Sachs disease (6), confirming the alpha-locus as the site of mutation in this disorder.

Study of the "AB variant" of infantile Tay-Sachs disease, with deficiency of neither hex A nor B, led to the finding that a heat-stable protein cofactor (23,25,48,49,55) was required for rapid cleavage of G_{M2} ganglioside. This factor was deficient in the AB variant, giving the clinical picture of infantile G_{M2}-gangliosidosis even though hex A and B were present (12,13). This factor must be coded for by yet another gene locus or loci. Thus, at least three gene loci are required for full hexosaminidase activity.

This heat-stable protein cofactor appears to bind to G_{M2}-ganglioside (K. Sandhoff, *personal communication*) so that a ganglioside-cofactor complex is the true substrate for the ganglioside-cleaving action of hex A. Although G_{M2}-ganglioside is rapidly cleaved only by hex A or an acidic subfraction of hex A, artificial substrates and nonganglioside glycolipids (kidney globoside and G_{A2}) are cleaved by both hex A and B (2).

Full hexosaminidase activity requires a complex group of at least three isoenzymes (A, B, and S), built up from at least two kinds of subunits (alpha and beta), requiring at lease one kind of activator (heat-stable protein cofactor), and coded for by at least three gene loci (alpha-locus, beta-locus, and activator locus). In addition to ganglioside and glycolipid substrates, hexosaminidase appears to cleave mucopolysaccharide (10,87), oligosaccharide (88), and possibly glycoprotein substrates as well.

THE CLINICAL SPECTRUM OF HEXOSAMINIDASE DEFICIENCY DISEASES

It is not surprising that the infantile encephalopathy of amaurotic family idiocy was the first clinical phenotype to be related to hexosaminidase deficiency, because the clinical picture is striking and the enzyme deficiencies are severe and clear-cut.

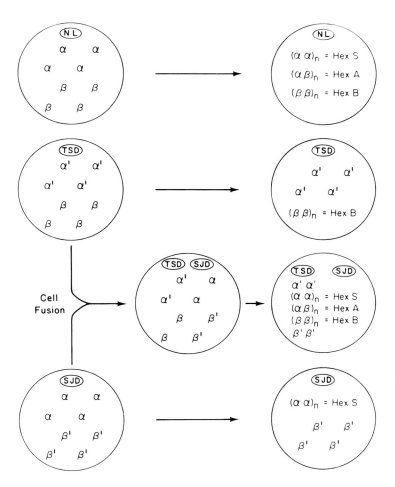

FIG. 1. Genetic complementation after cell fusion. *(Top)* A cell with a normal (NL) nucleus produces a normal alpha- and beta-subunit of hexosaminidase, and hex S, hex A, and hex B are found. *(Below)* A cell with a Tay-Sachs disease (TSD) nucleus produces normal beta-subunits and abnormal alpha-subunits, α'. In this cell, only hex B is found. *(Bottom)* A cell with a Sandhoff-Jatzkewitz disease (SJD) nucleus produces normal alpha-subunits and abnormal beta-subunits, β'. In this cell, only hex S is found. After cell fusion, the cell with both TSD and SJD nuclei produces both normal and abnormal alpha- and beta-subunits. In this cell hex S, hex A, and hex B are found.

An increasing variety of clinical disorders is now being associated with hexosaminidase deficiency, however. Some have infantile onset, some late-infantile, juvenile, adolescent, or adult onset. Some cause death in 1 to 2 years, others may not shorten the lifespan. Some (Table 1) affect the cerebral hemispheres, others the cerebellum, basal ganglia, spinal cord, or anterior horn cells. In general, the hexosaminidase deficiency disorders seem to affect neurons rather than myelin (Table 2).

TABLE 1. *Phenotypes of hexosaminidase deficiency diseases*

1. Infantile encephalopathy
2. Late-infantile or juvenile encephalopathy
3. (Basal ganglia disorder)[a]
4. Cerebellar or spinocerebellar ataxia
5. Motor neuron disease
6. Adult-onset encephalopathy
7. (Psychosis)[a]

[a]Phenotype was not the presenting or primary feature of the disorder.

TABLE 2. *Phenotypes not yet seen in hexosaminidase deficiency diseases*

1. Demyelinating neuropathy
2. Myopathy
3. Chorea
4. Parkinsonism
5. Familial spastic paraplegia
6. Primary generalized epilepsy (isolated)
7. Dementia (isolated)

TABLE 3. *Genetic classification of hexosaminidase deficiency diseases*

	Biochemical phenotype	Genotype
Normal	HEX A1, HEX B1	$HEX_{alpha}^1, HEX_{alpha}^1, HEX_{beta}^1, HEX_{beta}^1$
Tay-Sachs (T-S)	HEX A2, HEX B1	$HEX_{alpha}^2, HEX_{alpha}^2, HEX_{beta}^1, HEX_{beta}^1$
Sandhoff's	HEX A1, HEX B2	$HEX_{alpha}^1, HEX_{alpha}^1, HEX_{beta}^2, HEX_{beta}^2$
T-S carrier	HEX A2-1, HEX B1	$HEX_{alpha}^2, HEX_{alpha}^1, HEX_{beta}^1, HEX_{beta}^1$
Alpha-locus compound	HEX A2-3, HEX B1	$HEX_{alpha}^2, HEX_{alpha}^3, HEX_{beta}^1, HEX_{beta}^1$

Classification of alleles is according to O'Brien's terminology (57).

The number of hexosaminidase deficiency diseases is increasing. To stem the inevitable confusion arising from naming these, a rational classification based on which gene locus and which allele at that locus is involved would be helpful. Such a system (Table 3) has been put forward by O'Brien (57) and will be used here in conjunction with a phenotypic classification (Table 4). Biochemical phenotypes are designated: HEX A 1, HEX B 1 (normal phenotype); HEX A 2, HEX B 1 (Tay-Sachs disease); HEX A 1-2, HEX B 1 (Tay-Sachs carrier); HEX A 1, HEX B 2 (Sandhoff's disease); etc. Genotypes are designated HEX_{alpha}^1, HEX_{alpha}^1, HEX_{beta}^1, HEX_{beta}^1 (normal genotype); HEX_{alpha}^2, HEX_{alpha}^2, HEX_{beta}^1, HEX_{beta}^1 (Tay-Sachs disease); HEX_{alpha}^1, HEX_{alpha}^1, HEX_{beta}^2, HEX_{beta}^2 (Sandhoff's disease); HEX_{alpha}^2, HEX_{alpha}^1, HEX_{beta}^1, HEX_{beta}^1 (Tay-Sachs carriers); etc. Genetic compounds (Fig.

2 and Table 3) (See ref. 34 for review) are also being found and may ultimately prove to be more common than true homozygotes among cases in which the parents are unrelated and do not belong to a defined population. A number of disorders remain unclassified (1,19,52,90). Numerous attempts have been made to treat hexosaminidase deficiency diseases (35). This has not been successful for lysosomal disorders involving the nervous system, and remains possible, but of uncertain clinical benefit, for those involving only peripheral organs. The various clinical phenotypes can now be discussed.

Infantile Encephalopathy with Cherry-Red Spots

Alpha-Locus Disorders

Tay-Sachs disease

Genotype: HEX_{alpha}^2, HEX_{alpha}^2, HEX_{beta}^1, HEX_{beta}^1

TABLE 4. *Clinical-genetic classification of hexosaminidase deficiencies*

Phenotype
  Alpha-locus disorders
  Beta-locus disorders
  Activator-locus disorders

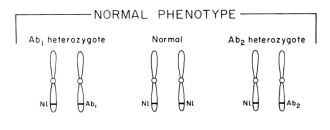

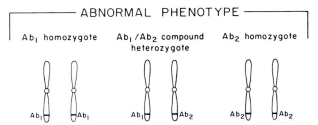

FIG. 2. The compound heterozygote or genetic compound.

The well-known clinical picture of classic infantile Tay-Sachs disease (92) begins at about 6 months with the onset in a previously healthy child of slowing of motor and mental development, myoclonic seizures, auditory-evoked myoclonus (exaggerated startle reflex to sound), and progressive visual deterioration with the appearance of the macular cherry-red spot (perimacular lipid accumulation). Later, mental and motor milestones are lost, the child enters a vegetative state, head circumference increases, and the child dies in the 3rd or 4th year of life or later. The disease is most common in infants of Ashkenazi Jewish families.

The diagnosis is made by demonstration of absent hex A in serum, leukocytes, and cultured skin fibroblasts (63,73). Heterozygous carriers can be detected (59) as with other lipid storage diseases (8). The conventional heat denaturation assay (59) accurately makes the diagnosis of this disease but cannot document the complete absence of hex A. Hex A should be studied by a method such as electrophoresis, which physically separates the isozymes, to confirm the diagnosis. Otherwise, subsequent prenatal diagnosis in that family may be inaccurate.

Tay-Sachs disease with residual Hex A

Alpha-Locus compound (see Fig. 2)
Genotype: HEX_{alpha}^2, HEX_{alpha}^4, HEX_{beta}^1, HEX_{beta}^1

This disorder was reported in a single patient from a non-Jewish family (60) who had a neurodegenerative disorder similar to Tay-Sachs disease. Motor weakness was first noted at 4 months and was followed by progressive muscular wasting, dulling of the sensorium, decrease in spontaneous movements, hyperacusis, and seizures. At 17 months she was hypertonic, hyperreflexic, with decerebrate posturing and bilateral optic atrophy with macular cherry-red spots. She died at 3 years. 26% residual hex A was found. The father had hex A in the "Tay-Sachs carrier" range, whereas the mother appeared normal. It was concluded that the child was a genetic compound.

Although this disorder is rare, it is important to recognize it, because such a couple could be falsely reassured at prenatal diagnosis that a fetus would be unaffected because some hex A was present.

Beta-Locus Disorders

Sandhoff's disease

Genotype: HEX_{alpha}^1, HEX_{alpha}^1, HEX_{beta}^2, HEX_{beta}^2

The clinical picture of this disorder (67) closely resembles that of classic infantile Tay-Sachs disease. It has no ethnic predisposition. Some patients with unusual clinical factors such as congestive heart failure have been reported (7). Biochemically, the disorder is different because of visceral storage of globoside (75). The diagnosis is readily made by finding nearly total absence of hexosaminidase in serum, leukocytes, and cultured skin fibroblasts (76). The small amount of residual enzyme is almost entirely hex S, which is heat labile (6).

Sandhoff's disease compound

>Beta-locus genetic compounds (see Fig. 2)
>Genotype: Not yet defined

A single case with the clinical and biochemical picture of infantile Sandhoff's disease was shown to be a genetic compound by study of the parents (47). The parents were shown to carry different beta-locus mutations by means of differences in heat denaturation and kinetic studies.

Activator Locus Disorder

AB variant

>Genotype: HEX_{alpha}^1, HEX_{alpha}^1, HEX_{beta}^1, HEX_{beta}^1

This disorder (14) also resembles classic infantile Tay-Sachs disease with progressive encephalopathy and macular cherry-red spots. However, hex A and B are normal or increased in amount by artificial substrate assay, although G_{M2}-ganglioside and other lipids are stored and G_{M2}-ganglioside hexosaminidase (natural substrate assay) is deficient (76). The explanation appears to be, as discussed earlier, that the protein activator of hexosaminidase is deficient and G_{M2}-gangliosode cannot be cleaved (12). Diagnosis of this disorder requires ganglioside analysis of brain or use of natural substrate assay for G_{M2}-ganglioside hexosaminidase.

Late-Infantile or Juvenile Encephalopathy with or without Cherry-Red Spots

Alpha-Locus Disorders

Late-infantile or juvenile Hex A deficiency

>Genotype: HEX_{alpha}^3, HEX_{alpha}^3, HEX_{beta}^1, HEX_{beta}^1

At least 24 patients in 18 families have been reported to have a neurodegenerative disorder beginning after the 1st year of life with partial deficiency of hex A (see ref. 34 for review). The clinical picture is heterogeneous and several disorders are probably represented in spite of the single genotype listed. The clinical features often include dementia, ataxia, muscle atrophy due to anterior horn cell disease, spasticity, and seizures. Cherry-red spot was noted in 2 patients, both from the same family, and was not as striking as in the infantile cases. G_{M2}-ganglioside accumulated in the brain in the cases in which this was measured, but to a lesser extent than in infantile cases. Hex A deficiency is quite variable by artificial substrate assay, varying from apparent absence of hex A to levels similar to those in the low

"Tay-Sachs carrier" range. Moreover, the level of residual artificial substrate hex A correlates badly with severity of symptoms. G_{M2}-ganglioside hexosaminidase (natural substrate assay) was measured in a few cases (58,84) and correlated better with the severity of symptoms. In general, these levels were only slightly higher than those in infantile cases.

Diagnosis is made by demonstration of decreased hex A in serum, leukocytes, and cultured skin fibroblasts. Natural substrate assay should be performed, if possible, for confirmation, because this may be required for accurate prenatal diagnosis.

Juvenile G_{M2}-gangliosidosis genetic compound

Alpha-locus genetic compound (see Fig. 2)
Genotype: HEX_{alpha}^{2}, HEX_{alpha}^{7}, HEX_{beta}^{1}, HEX_{beta}^{1}

A single case of this disorder is known (34), although a patient with late-infantile G_{M2}-gangliosidosis resulting from a donor insemination pregnancy was probably also a compound and falls into this classification (11). This child presented, with the onset at age 2, with ataxia and dementia without seizures. This was followed by visual deterioration and loss of speech and motor milestones. Bilateral macular cherry-red spots and myoclonic jerks, occurring 1 to 5 per min, were found on examination at 3 years. The diagnosis was made by finding partial deficiency of hex A in serum, leukocytes, and cultured skin fibroblasts. The paternal kindred contained 1 Ashkenazi great-grandparent and 4 individuals who had typical biochemical findings of the Tay-Sachs carrier state. The large maternal kindred contained no typical Tay-Sachs carriers, but several individuals related to the maternal grandfather had mild hex A deficiencies. The patient was therefore a genetic compound of two alpha-locus alleles.

Such disorders are rare (although careful family studies were not done in other juvenile G_{M2}-gangliosidoses) but it is important to document them when they occur. Such a couple would probably have been missed in Tay-Sachs population screening. Had prenatal diagnosis been done, they would have been falsely reassured that the fetus would be unaffected. Prenatal diagnosis would require natural substrate assay or a method of distinguishing qualitatively the abnormal residual enzyme in the compound from the normal residual enzyme in the HEX_{alpha}^{2} heterozygote.

Beta-Locus Disorder

Juvenile Sandhoff's disease

Genotype: Not yet defined

At least 2 patients (18,50) were described with the biochemical findings of Sandhoff's disease (total hexosaminidase level less than 10% of control values) but with juvenile-onset neurological disorders that were very slowly progressive. The clinical features included spasticity, ataxia, dystonic posturing, and mild mental deficiency. Such a disorder might be suspected in patients with apparently static encephalopathy and mental retardation.

Activator Locus Deficiency

None known.

Cerebellar Ataxia

Alpha-Locus Disorder

Atypical spinocerebellar ataxia with Hex A deficiency

Genotype: Not yet defined

This disorder was first described in a single sibship with 3 affected individuals (69). Falling and clumsiness began at age 3 to 4, followed by slowly progressive spasticity, ataxia, dysarthria, and muscle atrophy. Other findings were pes cavus and foot-drop, dystonic features, normal intelligence, and survival into (at least) the 4th decade. Multilamellar bodies were found in neurons. G_{M2}-ganglioside was increased in brain, and hex A was diminished in serum and leukocytes. Other similar families have been described (20,53,94).

Beta-Locus Disorders

Juvenile cerebellar ataxia with cherry-red spots

Genotype: Not yet defined

A single case was reported (32) with onset at 2½ years of slowly progressive cerebellar outflow tremor and ataxia similar to dyssynergia cerebellaris progressiva of J. Ramsey Hunt (27a). Macular cherry-red spots were also seen. Biochemical findings were similar to those of Sandhoff's disease in serum, leukocytes, and fibroblasts. On electrofocusing of urine (26) and fibroblasts (36), however, the major residual hexosaminidase resembled hex A except for a slightly more acidic isoelectric point. Hex S was present in lesser amount and hex B was absent. The carrier state was readily detected (31) and traced through four generations. Carriers had increased hex S (26), suggesting that the basic defect in beta-subunits was absent beta-beta binding and deficient beta-alpha binding.

Genetic complementation (Fig. 1) with Sandhoff's disease did not occur, thus confirming this as a beta-locus disorder (33).

Adult-onset spinocerebellar degeneration with Hex A and B deficiency

Genotype: Not yet defined

A single family was reported (65) in which 2 sisters developed sphinocerebellar degeneration after the age of 20. The clinical features were gait and limb ataxia, dysarthria, head titubation, tremor of the hands and fingers, facial grimacing, and chorea of the limbs, decreased vibratory sense, slightly decreased joint position sense, and hyperreflexia with equivocal plantar responses. Total hexosaminidase in serum and leukocytes was diminished to less than 10% of control values. Natural substrate activity using G_{A2} was reduced to about the same extent.

Activator Locus Deficiency

None known.

Motor Neuron Disease

Alpha-Locus Disorder

Amyotrophic lateral sclerosis-like syndrome with Hex A deficiency

Genotype: Not defined

A single example (40,41,96) of this disorder occurred in a 22-year-old Ashkenazi Jewish man with a 2-year history of neurological illness in whom total hex A deficiency was found during a Tay-Sachs carrier screening program. Clinical features included weakness, cramps, proximal muscle wasting, fasciculations, hyperreflexia, extensor plantar responses, and emotional depression. Other findings included anterior horn cell disease on electromyography, membranous cytoplasmic bodies on electron microscopy of rectal ganglion cells, and absent G_{M2}-ganglioside hexosaminidase in cultured skin fibroblasts. The patient was probably a genetic compound, although this has not been documented.

Adult-Onset Dementia

Alpha-Locus Disorder

None known.

Beta-Locus Disorder

None known.

Activator Locus Deficiency

AB variant with adult-onset dementia and normal pressure hydrocephalus

A single case was reported (64) with onset at age 18 of intellectual and behavioral deterioration with seizures and normal pressure hydrocephalus. Autopsy at age 22 showed storage of intracellular PAS-positive material in brain and spleen. Multilamellar cytoplasmic inclusions were seen by electron microscopy in the brain, spleen, and arachnoid granulations. G_{M2}-ganglioside was elevated in cerebral cortex. Hex A and hex B were both increased, leading to the conclusion that this disorder represented a form of AB variant. However, G_{M2}-ganglioside hexosaminidase was not measured and this conclusion must be considered tentative.

Asymptomatic Individuals with Hexosaminidase Deficiency—Genetic Compounds

A genetic compound (Fig. 2 and Table 3) is an individual who has no normal alleles of a particular gene because both available loci (one on each chromosome of the appropriate pair) are occupied by different abnormal alleles. Such an individual is not a homozygote because the two abnormal alleles are different.

The three known symptomatic genetic compounds were discussed in the last section.

At least 10 individuals in 6 families are known (15,16,24,42,54,56,60,91) to have apparently severe or total hex A, B, or A and B deficiencies without any symptoms at all. In general, they were discovered during population screening for Tay-Sachs carriers or because a relative had infantile Tay-Sachs or Sandhoff's disease. Obviously, asymptomatic adults cannot have biochemical findings identical to those of severely affected infants. In fact, when G_{M2}-ganglioside hexosaminidase (natural substrate assay) was measured in some of these individuals, about half normal activities were found (15,16,83). It seems likely that these individuals are compounds of two alpha-locus alleles (or two beta-locus alleles), one of which (for example, HEX_{alpha}^2) codes for enzyme inactive to both artificial and natural substrates and other of which (HEX_{alpha}^5 or HEX_{alpha}^6) codes for an enzyme inactive with artificial substrates but active with natural substrates. Two types of this phenomenon are recognized. Asymptomatic individuals with absent hex A have the genotype HEX_{alpha}^2, HEX_{alpha}^5, HEX_{beta}^1, HEX_{beta}^1. Asymptomatic individuals with deficient but not absent hex A have the genotype HEX_{alpha}^2, HEX_{alpha}^6, HEX_{beta}^1, HEX_{beta}^1. Although these individuals are asymptomatic, they could be "presymptomatic" and later develop symptoms of neurological disease. In at least one family such individuals have become symptomatic (53).

LUMPING AND SPLITTING—GENETIC HETEROGENEITY OF HEXOSAMINIDASE DEFICIENCY DISEASES

The foregoing discussion has been devoted to splitting: splitting of different types of amaurotic family idiocy, splitting hexosaminidase functions into enzyme protein and activator protein, splitting the hexosaminidase activity into isoenzymes, splitting each isoenzyme into subunits coded for by separate cistrons, and splitting the different clinical disorders into distinct syndromes.

It is of interest to ask how many different hexosaminidase deficiency disorders may potentially be found and whether this growing multitude of disorders can be organized into clinically, genetically, and biochemically useful categories.

Genetic Classification

The best model for genetic classification of the hexosaminidase deficiency disorders is the hemoglobinopathies. These disorders, classified according to a multiple loci-multiple allele system, now number nearly 400 (51). Mutations affecting the structural gene (cistron) of a hemoglobin polypeptide chain are classified by the number of the amino acid residue (beginning at the N-terminal end) affected by the mutation. Different mutations of the same codon may cause several different kinds of substitutions at the same amino acid residue; these are called euallelic mutations. Mutations affecting different amino acid residues of the same polypeptide chain are called heteroallelic. The number of different mutations affecting a single protein is astronomical, but nonetheless these can be readily classified by such a multiple loci-multiple allele system.

In the case of the hexosaminidase, at least three loci are known: the alpha-locus on chromosome 15, the beta-locus on chromosome 5, and the activator locus whose chromosomal localization is unknown. There may be other undiscovered hexosaminidase subunits required for some of the less well-studied hexosaminidase isoenzymes. There may be other activators, and the existing activator may be coded for by more than one locus. The hexosaminidase system may be more complex than is now appreciated (29). For example, hexosaminidase polymorphisms may exist in the normal population. Deficient regulatory genes (Fig. 3, Table 5) may cause some forms of hexosaminidase deficiency disease either through decreased

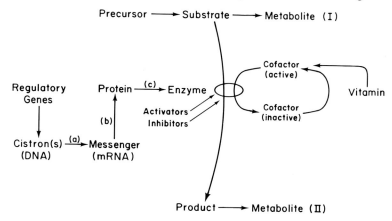

FIG. 3. The genetic mechanism for a hypothetical enzyme pathway. *(a)* Transcription; *(b)* translation; and *(c)* posttranslational modification.

TABLE 5. *Gene effects on hexosaminidase enzyme activity*

1. Defective regulation of enzyme synthesis
2. Mutation affecting developmental program
3. Inhibitor substance produced
4. Defect in posttranslational modification (glycosylation)
5. Mutation affecting gene splicing (intron)
6. Mutation involving signal sequence or "proenzyme" portion of the polypeptide chain
7. Mutation affecting hexosaminidase structural gene portion, giving rise to final enzyme protein

(or increased) rate of enzyme synthesis or through alteration of the developmental program for hexosaminidase. Hexosaminidase inhibitors (Fig. 3, Table 5) could cause hexosaminidase deficiency, though inheritance of such a disorder should be dominant rather than recessive. Hexosaminidase subunits could interact harmfully, as postulated in the metabolic interference hypothesis (28,30) (Table 5). Hexosaminidase is known to be subject to posttranslational modifications (21,22) (Table 5). Since hexosaminidases are glycoproteins, mechanisms must exist for their glycosylation. In mucolipidoses II and III, recessively inherited disorders, this mechanism is defective (22,27,71,89) and hexosaminidase, though enzymatically active, is not properly packaged in lysosomes. The defect may be a phosphotransferase which phosphorylates a glycoprotein-mannose residue (3); multiple lysosomal enzymes are affected.

It is appropriate now to classify hexosaminidase deficiency disorders as to which of the three known alleles is affected (Tables 3 and 4). Unfortunately, hexosaminidase has not yet been sequenced and it is not yet possible to say which amino acid residue has been substituted at which position on which subunit. Nonetheless, there is evidence that at least some (and probably all or nearly all) of the known hexosaminidase deficiency diseases result from mutations affecting structural genes (cistrons) (Table V). Cross-reacting material (enzymatically inactive hexosaminidase protein detected immunologically) has been found in both classic Tay-Sachs and Sandhoff's diseases (80,81). It is possible from biochemical studies to decide which subunit of hexosaminidase has been affected. Deficiency or defect of hex A and S but not B implicates the alpha-subunit and alpha-locus. Deficiency or defect of hex A and B but not S implicates the beta-subunit and beta-locus. Deficiency or defect of neither hex A, B, nor S in the presence of deficient G_{M2}-ganglioside hexosaminidase implicates a third locus such as the activator locus. Complementation (Fig. 1) with Tay-Sachs but not Sandhoff's fibroblasts implicates the beta-locus (33); complementation with Sandhoff's but not Tay-Sachs fibroblasts implicates the alpha-locus. The finding of any of the kinetic abnormalities of hexosaminidase described in the section on biochemical classification also implicates a structural gene mutation as the cause of hexosaminidase deficiency in the same way as the finding of cross-reacting material (16).

Useful as the genetic classification is, however, it does not immediately explain how the mutation affects the function of the hexosaminidase enzyme protein and why the patient has one clinical syndrome rather than another.

Biochemical Classification

Biochemical studies of hexosaminidases altered by mutations suggest that, in general, the same sorts of abnormalities may result from alpha-locus or beta-locus mutations and activator locus mutations as well. Although there are some differences, all three cause similar light and electron microscopic abnormalities in neurons. All three types of disorders cause decreased G_{M2}-ganglioside hexosaminidase. All three show neuronal accumulation of G_{M2}-ganglioside and other glycoconjugates. All three can cause infantile encephalopathy with cherry-red spots.

Later-onset activator locus disorders are not well characterized (if they, in fact, have yet been found), but both alpha-locus and beta-locus disorders can cause juvenile encephalopathy and cerebellar syndromes. Motor neuron disease has been reported for alpha-locus disorders only, but the number of cases is small.

Asymptomatic (or presymptomatic) adults have been reported with severely deficient hexosaminidase (by artificial substitute assay) referable to both alpha-locus and beta-locus mutations. How, then, can the striking difference between the clinical syndromes be explained?

Again, an answer is suggested by comparison with the hemoglobinopathies (51). Mutations affecting the same hemoglobin beta-chain can cause a wide variety of clinical disorders ranging from severe sickle diseases to asymptomatic polymorphisms. The differences result from the specific effect of the amino acid substitution on the function of the protein. In fact, the study of mutations affecting hemoglobin is one of the main avenues of determining how normal hemoglobin functions. Several classes of mutations can be defined by their effect on the protein structure and function (Table 6).

Mutations affecting the active site of the hexosaminidase enzyme may have a direct and profound effect on enzyme activity. Damage to the active site commonly causes decreased affinity of enzyme for substrate (and increase of K_m, the Michaelis constant) or a decrease in the maximum enzyme activity (V_{max}). Increased affinity for substrate (and decreased K_m) may also decrease maximum reaction velocity. These changes may be seen with artificial substrates, natural substrates, or both. Active site involvement may reasonably be suspected in any of the alpha-locus or beta-locus hexosaminidase deficiencies described to date, but it has not been documented.

Another effect of mutational change on an enzyme is altered charge. This happens when the substituted amino acid and the normally occurring amino acid come from

TABLE 6. *Effect of mutational change on hexosaminidase enzyme protein*

1. Active site effect
    A. Increased K_m
    B. Decreased V_{max}
    C. Both
2. Altered pI: protein charge affected
3. Subunit binding affected
    A. Homopolymer binding
    B. Heteropolymer binding
4. Cofactor or activator binding affected
5. Unstable enzyme protein produced
6. Substrate specificities affected
7. Subunits interact harmfully (metabolic interference)
8. Small inactive subunit produced by terminator mutation
9. Mutation affecting "signal sequence" or "proenzyme" portion of the polypeptide

different groups rather than the same group of neutral, acidic, or basic amino acids. Altered charge of an enzyme need not reflect a change in amino acid structure. For example, in the sialidoses (deficiencies of alpha-L-N-acetylneuraminidase), hexosaminidase isozymes (and other lysosomal enzymes) may be "oversialated." That is, the protein has an excess of sialic acid residues on its carbohydrate portion. Since these carry a full negative charge each, the enzyme's isoelectric point (pI) is lowered and its electrophoretic migration is changed. However, an altered pI of hex A and B only or A and S only would implicate, respectively, a beta-locus or alpha-locus disorder. Apparent acid shift of pI of residual hex A-like enzyme but not hex S has been found in one beta-locus hexosaminidase deficiency (26,36). Unfortunately, hex B activity was not detectable in that disorder and its pI could not be determined.

Another effect of mutation may be to affect the site at which subunits bind to each other. A mutation may affect binding of subunits to like subunits (alpha-alpha or beta-beta) or to unlike subunits (alpha-beta) or both. In one hexosaminidase deficiency disorder, such a defect has been suggested (26,36), affecting beta-beta binding severely and beta-alpha binding somewhat less severely. In that beta-locus disorder, hex B appeared totally deficient, but some residual A-like enzyme was found along with hex S. Further evidence for a subunit binding defect was the observation that carriers had greatly increased hex S. Apparently their defective beta-subunits failed to bind to normal beta- or alpha-subunits and the excess unbound alpha-subunits combined to form large amounts of the alpha-homopolymer, hex S. In another beta-locus variant in an asymptomatic adult, a binding defect has been postulated (15).

A fourth effect of mutation may be to affect the site of an enzyme to which a cofactor, activator, or inhibitor binds. The heat-stabile activator protein is thought to bind to ganglioside, so that the true substrate for ganglioside hexosaminidase is the ganglioside-activator protein complex. It is probable, however, that the activator interacts in some way with the hexosaminidase enzyme also, so that a mutation affecting the alpha-subunit or beta-subunit may prevent binding of the activator-ganglioside complex to the enzyme.

A fifth effect of mutation is to produce an unstable enzyme protein. Such a protein may be enzymatically active but the instability may greatly reduce the enzyme activity of tissues. At least two examples of unstable hexosaminidases are known, both with mutations affecting the beta-subunit (16,47).

A sixth effect of mutation on an enzyme protein is to affect substrate specificities. Substrates for hexosaminidase include G_{M2}-ganglioside, nonganglioside glycolipids (kidney globoside and asialo-Tay-Sachs ganglioside), mucopolysaccharides, oligosaccharides, and possibly glycoproteins. Some mutations may prevent hexosaminidase action on all of these. Other mutations may affect some substrates more than others. Such mechanisms may have a profound effect on the clinical syndrome produced. The cells of the nervous system, especially the neurons, differ greatly in size, shape, and metabolism from region to region. Some neuronal populations differ in phylogenetic origin, migration, and developmental program. Consequently, greater involvement of one substrate may affect one group of neurons for which that substrate is especially important; it may affect migration of cells, altering one class of neurons more than others; it may affect certain neurons at different times during development.

A seventh way in which mutation may affect protein functions is to give rise to harmful interactions between subunits in a multisubunit enzyme (or structural protein). This is the metabolic interference hypothesis, described in detail elsewhere (28,30).

An eighth effect of a mutation may be to cause termination of the growing polypeptide chain before it is complete. This would give rise to a subunit smaller than normal, probably abnormal in tertiary (three-dimensional) structure, and possibly missing regions important for function.

Finally, some mutations of the structural gene might never be seen in the finished protein. A mutation in an intervening sequence (intron) necessary for splicing of the message might not be seen as an amino acid substitution in the finished protein but could have devastating effects (61). Likewise, a mutation affecting the "signal sequence" of an enzyme precursor protein (21,93) might not be seen in the finished protein because of removal during posttranslational processing, but it could have devastating effects on the packaging of enzymes into the proper subcellular compartment.

When hexosaminidase deficiency diseases have been classified according to the effect of the mutation on the enzyme protein, the correlation of the clinical syndromes with enzyme protein change may be closer than with the locus affected.

Clinical Classification

It is often stated that clinical classification of diseases is a preliminary step, but ultimate classification must await the work of the biochemist or geneticist (or, in earlier times, the pathologist). In a sense, this is true. However, careful clinical classification must not be overlooked as an important tool for distinguishing different genetic entities (29). For example, the Hurler and Scheie syndromes were easily distinguished at a time when the biochemical changes in each (alpha-L-iduronidase deficiency, apparently total) were indistinguishable.

For this reason, it is important to carry out careful clinical study of patients with hexosaminidase deficiency diseases (and other recessive genetic diseases), to study affected sibs equally carefully, and to compare the clinical and biochemical findings with those of other apparently identically affected families. In this way, consistent clinical differences between patients in different families may define genetically different entities. In the same way, studies of the carrier state in family members other than sibs is important because consistent biochemical difference between the maternal and paternal relatives is one important way to document genetic compounds (34).

Finally, a word should be said about demonstrating the causal relationship between a biochemical defect and a genetic disease. Some biochemical defects may

TABLE 7. *Criteria for accepting an enzyme deficiency as the cause of an associated autosomal recessive disease*

---
1. The enzyme deficiency is found in multiple similarly affected patients in a single family or in other families.
2. The enzyme deficiency is lacking in (older) unaffected siblings and other unaffected individuals.
3. Parents of the patients, though unaffected, have a partial deficiency of the enzyme.
4. Some unaffected family members (patient's sibs, parents' sibs, grandparents, etc.) have partial deficiency of the enzyme, in a pattern consistent with the pedigree.
5. The enzyme is deficient in all tissues in which it is normally present.
6. Genetically unrelated enzymes are not deficient.
7. A substrate abnormality appropriate to the enzyme deficiency can be demonstrated.

TABLE 8. *Criteria for accepting a structural gene mutation as the cause of deficiency of an enzyme activity*

---
1. The enzyme deficiency does not result from presence of an inhibitor.
2. Cross-reacting material and/or residual enzyme activity is present.
3. The residual enzyme has abnormal properties (for example, altered K_m or increased lability).
4. The enzyme or one of its subunits has an altered amino acid sequence.
5. Messenger RNA has a corresponding codon alteration.
6. DNA has an appropriate codon alteration.
---

be unrelated to any clinical disease but may nonetheless be found by chance in a patient with genetic disease. An example is fucosidase deficiency in serum which occurs in 6 to 7% of apparently normal individuals. Certain criteria are useful (Table 7) for accepting an enzyme deficiency as the cause of an associated autosomal recessive disorder. Some of these may be fulfilled in the case of a genetically determined enzyme deficiency unrelated to the clinical syndrome [for example, #3 and #4 for the serum fucosidase polymorphism (96)]. Many well-established disorders have not yet been shown to meet all of these criteria. However, these are useful as guidelines for studying patients and families. Other criteria are useful (Table 8) for accepting a structural gene mutation as the cause of deficiency of an enzyme activity. None of the hexosaminidase deficiencies has yet met all of these criteria. They are, nonetheless, useful as goals for future studies.

ACKNOWLEDGMENTS

This work was supported in part by grants 1-RO1-NS-15281 from the National Institutes of Health and by the March of Dimes Birth Defects Foundation, The Muscular Dystrophy Association (H. Houston Merritt Clinical Research Center for Muscular Dystrophy and Related Diseases), and a Career Scientist Award (W.G.J.) from the Irma T. Hirschl Foundation.

REFERENCES

1. Andermann, F., Andermann, E., Carpenter, S., Karpati, G., and Wolfe, L. (1974): Late onset G_{M2}-gangliosidosis in two Lebanese families. *Am. J. Hum. Genet. (abstract)*, 26:10A.
2. Bach, G., and Suzuki, K. (1975): Heterogeneity of human hepatic N-acetyl-beta-D-hexosaminidase A activity toward natural glycosphingolipid substrates. *J. Biol. Chem.*, 250:1,328–1,332.
3. Ben-Yoseph, Y., Hahn, L. C., DeFranco, C. L., and Nadler, H. L. (1980): Deficiency of mannan kinase in mucolipidoses types II and III. *Am. J. Hum. Genet. (abstract)*, 32:35A.
4. Beutler, E. (1979): The biochemical genetics of the hexosaminidase system in man. *Am. J. Hum. Genet.*, 31:95–105.
5. Beutler, E., and Kuhl, W. (1975): Subunit structure of human hexosaminidase verified: Interconvertibility of hexosaminidase isozymes. *Nature*, 258:262–264.
6. Beutler, E., Kuhl, W., and Comings, D. (1975): Hexosaminidase isozymes in type O G_{M2}-gangliosidosis (Sandhoff-Jatzkewitz disease). *Am. J. Hum. Genet.*, 27:628–638.
7. Blieden, L. C., Desnick, R. J., Carter, J. B., Krivit, W., Moller, J. H., and Sharp, H. L. (1974): Cardiac involvement in Sandhoff's disease. *Am. J. Cardiol.*, 34:83–88.
8. Brady, R. O., Johnson, W. G., and Uhlendorf, B. W. (1971): Identification of heterozygous carriers of lipid storage diseases. *Am. J. Med.*, 51:423–431.
9. Brady, R. O., Tallman, J. F., Johnson, W. G., and Quirk, J. M. (1972): An investigation of the metabolism of Tay-Sachs ganglioside specifically labeled in critical portions of the molecule. In: *Sphingolipids, Sphingolipidoses, and Allied Disorders*, 3rd edition, edited by B. W. Volk, and S. M. Aronson, pp. 277–285. Plenum Press, New York.
10. Cantz, M., and Kresse, H. (1974): Sandhoff disease: Defective glycosaminoglycan catabolism in cultured fibroblasts and its correction by beta-N-acetyl-hexosaminidase. *Eur. J. Biochem.*, 47:581–590.
11. Chutorian, A. M., Johnson, W. G., and Schwartz, R. C. (1980): Donor insemination: Who should be screened for what? Tay-Sachs disease resulting from donor insemination. *Am. J. Hum. Genet. (abstract)*, 32:102A.
12. Conzelmann, E., and Sandhoff, K. (1978): AB variant of infantile G_{M2}-gangliosidosis: Deficiency of a factor necessary for stimulation of hexosaminidase A-catalyzed degradation of ganglioside G_{M2} and glycolipid G_{A2}. *Proc. Natl. Acad. Sci. USA*, 75:3,979–3,983.
13. Conzelmann, E., Sandhoff, K., Nehrkorn, H., Geiger, B., and Arnon, R. (1978): Purification, biochemical and immunological characterization of hexosaminidase A from variant AB of infantile G_{M2}-gangliosidosis. *Eur. J. Biochem.*, 84:27–33.
14. De Baecque, C. M., Suzuki, K., Rapin, I., Johnson, A. B., Wethers, D. L., and Suzuki, K. (1975): G_{M2}-gangliosidose, AB variant. *Acta Neuropathol.*, 33:207–226.
15. Dreyfus, J. C., Poenaru, L., and Svennerholm, L. (1975): Absence of hexosaminidase A and B in a normal adult. *N. Engl. J. Med.*, 292:61–63.
16. Dreyfus, J. C., Peonaru, L., Vibert, M., Rairse, N., and Boje, J. (1977): Characterization of a variant of beta-hexosaminidase: "Hexosaminidase Paris." *Am. J. Hum. Genet.*, 28:281–283.
17. Geiger, B., Arnon, R., and Sandhoff, K. (1977): Immunochemical and biochemical investigation of hexosaminidase S. *Am. J. Hum. Genet.*, 29:508–522.
18. Goldie, W. D., Holtzman, D., and Suzuki, K. (1977): Chronic hexosaminidase deficiency. *Ann. Neurol.*, 2:156–158.
19. Gordon, B. A., Pozsony, J., Geeling, S., Kaufmann, J. C. E., and Haust, M. D. (1974): An unusual variant of Tay-Sachs disease. *Clin. Res. (abstract)*, 22:740a.
20. Grabowski, G. A., Willner, J. P., Bender, A., Gordon, R. E., and Desnick, R. J. (1980): Chronic beta-hexosaminidase A deficiency—Clinical and biochemical studies of a new neuromuscular disease. *Pediatr. Res. (abstract)*, 14:632.
21. Haselik, A., and Neufeld, E. F. (1980a): Biosynthesis of lysosomal enzymes in fibroblasts. Synthesis as precursors of higher molecular weight. *J. Biol. Chem.*, 255:4,937–4,945.
22. Haselik, A., and Neufeld, E. F. (1980b): Biosynthesis of lysosomal enzymes in fibroblasts. Phosphorylation of mannose residues. *J. Biol. Chem.*, 255:4,946–4,950.
23. Hechtman, P., and LeBlanc, D. (1977): Purification and properties of the hexosaminidase A-activating protein for human liver. *Biochem. J.*, 167:693–701.
24. Hechtman, P., and Rowlands, A. (1979): Apparent hexosaminidase B deficiency in two healthy members of a kindred. *Am. J. Hum. Genet.*, 31:428–438.
25. Hechtman, P., and Kachra, Z. (1980): Interaction of activating protein and surfactants with human liver hexosaminidase A and G_{M2}-ganglioside. *Biochem. J.*, 185:583–591.

26. Hiatt, A., and Johnson, W. G. (1977): A new hexosaminidase deficiency disease: Isoelectric focusing of the residual enzyme from urine. *American Society of Human Genetics, 28th Annual Meeting*, p. 53A *(abstract)*.
27. Hickman, S., and Neufeld, E. F. (1972): A hypothesis for I-cell disease: Defective hydrolases that do not enter lysosomes. *Biochem. Biophys. Res. Commun.*, 49:992–999.
27a. Hunt, J. Ramsey (1915): Dyssynergia cerebellaris progressiva—a chronic progressive form of cerebellar tremor. *Brain*, 37:247–268.
28. Johnson, W. G. (1978): The isolated compound: A new form of simple inheritance. *Am. J. Hum. Genet. (abstract)*, 30:124A.
29. Johnson, W. G. (1979): Principles of genetics in neuromuscular disease. In: *Practice of Pediatrics*, Vol. 4, edited by V. C. Kelley. Harper and Row, New York.
30. Johnson, W. G. (1980): Metabolic interference and the + / − heterozygote. A hypothetical form of simple inheritance which is neither dominant nor recessive. *Am. J. Hum. Genet.*, 32:374–386.
31. Johnson, W. G., and Chutorian, A. M. (1978): Inheritance of the enzyme defect in a hexosaminidase deficiency disease. *Ann. Neurol.*, 4:399–403.
32. Johnson, W. G., Chutorian, A., and Miranda, A. (1977): A new juvenile hexosaminidase deficiency disease presenting as cerebellar ataxia—Clinical and biochemical studies. *Neurology*, 27:1,012–1,018.
33. Johnson, W. G., Chutorian, A., Miranda, A., and Nette, G. (1977): Clinical, genetic, and biochemical studies in a new juvenile hexosaminidase deficiency disease. *Neurology (abstract)*, 27:384.
34. Johnson, W. G., Cohen, C. S., Miranda, A. F., Waran, S. P., and Chutorian, A. M. (1980): Alpha-locus hexosaminidase genetic compound with juvenile gangliosidosis phenotype: Clinical, genetic, and biochemical studies. *Am. J. Hum. Genet.*, 32:508–518.
35. Johnson, W. G., Desnick, R. J., Long, D. M., Sharp, H. L., Krivit, W., Brady, B., and Brady, R. O. (1973): Intravenous injection of purified hexosaminidase A into a patient with Tay-Sachs disease. In: *Enzyme Therapy in Genetic Diseases*, edited by R. J. Desnick, R. W. Bernlohr, and W. Krivit, pp. 120–124. The National Foundation, New York.
36. Johnson, W. G., and Miranda, A. (1978): A new hexosaminidase deficiency: The residual enzyme in fibroblasts. *Trans. Am. Soc. Neurochem (abstract)*, 9:205.
37. Johnson, W. G., Mook, G., and Brady, R. O. (1972): Beta-hexosaminidase A from human placenta. *Methods Enzymol.*, 28:857–861.
38. Johnson, W. G., Thomas, G. H., Miranda, A. F., and Driscoll, J. M. (1980): Congenital sialidosis, a new form of alpha-L-neuraminidase deficiency—Its possible relation to hydrops fetalis. *Neurology (abstract)*, 30:377.
39. Johnson, W. G., Thomas, G. H., Miranda, A. F., Driscoll, J. M., Wigger, J. H., Yeh, M. N., Schwartz, R. C., Cohen, C. S., Berdon, W. E., and Koenigsberger, M. R. (1980): Prenatal diagnosis in two pregnancies at risk for congenital sialidosis. *Ann. Neurol. (abstract)*, 8:216.
40. Kaback, M., Miles, J., Yaffe, M., Itabashi, H., McIntyre, H., Goldberg, M., and Mohandas, T. (1978): Hexosaminidase A (HEX A) deficiency: A new type of G_{M2}-gangliosidosis. *Am. J. Hum. Genet. (abstract)*, 30:31A.
41. Kaback, M., Miles, J., Yaffe, M., Itabashi, H., McIntyre, H., Goldberg, M., Mohandas, T., and O'Brien, J. S. (1979): Type VI G_{M2}-gangliosidosis: An amyotrophic lateral sclerosis phenocopy. *Clin. Res. (abstract)*, 27:121A.
42. Kelly, T. E., Reynolds, L. W., and O'Brien, J. S. (1976): Segregation within a family of two mutant alleles for hexosaminidase A. *Clin. Genet.*, 9:540–543.
43. Klenk, E. (1939/1940): Beitrage zur Chemie der Lipoidosen. Niemann-Pick'sche Krankheit und amaurotische Idiotie. *Hoppe-Seylers Z. Physiol. Chem.*, 262:128.
44. Kolodny, E. H., Brady, R. O., and Volk, B. W. (1969): Demonstration of an alteration of ganglioside metabolism in Tay-Sachs disease. *Biochem. Biophys. Res. Commun.*, 37:526.
45. Lalley, P. A., and Shows, T. B. (1976): Expression of human hexosaminidase A phenotype depends on genes assigned to chromosome 5 and 15. *Cytogenet. Cell Genet.*, 16:192–196.
46. Landing, B. H., Silverman, F. N., Craig, J. M., Jacoby, M. D., Lahey, M. D., and Chadwick, D. L. (1964): Familial neurovisceral lipidosis. *Am. J. Dis. Child*, 108:503–522.
47. Lane, A. B., and Jenkins, T. (1978): Two variant beta-chain alleles segregating in a South African family. *Clin. Chim. Acta*, 87:219–228.
48. Li, S. C., Nakamura, T., Ogamo, A., and Li, Y. T. (1979): Evidence for the presence of two separate protein activators for the enzymatic hydrolysis of G_{M1}- and G_{M2}-gangliosides. *J. Biol. Chem.*, 254:10,592–10,595.

49. Li, Y. T., Mazzotta, M. Y., Wan, C. C., Orth, R., and Li, S. C. (1973): Hydrolysis of Tay-Sachs ganglioside by beta-hexosaminidase A of human liver and urine. *J. Biol. Chem.*, 248:7,512–7,515.
50. MacLeod, P. M., Wood, S., Jan, J. E., Applegarth, D. A., and Dolman, C. L. (1977): Progressive cerebellar ataxia, spasticity, psychomotor retardation, and hexosaminidase deficiency in a 10-year-old child: Juvenile Sandhoff disease. *Neurology*, 27:571–573.
51. McKusick, V. A. (1978): *Mendelian Inheritance in Man*, 5th edition. Johns Hopkins University Press, Baltimore.
52. Momoi, T., Suco, M., Taniaka, K., and Nakas, Y. (1978): Tay-Sachs disease with altered beta-hexosaminidase B; a new variant? *Pediatr. Res.*, 12:77–81.
53. Navon, R., Brand, N., and Sandbank, U. (1980): Adult (G_{M2}) gangliosidosis: Neurological and biochemical findings in an apparently new type. *Neurology (abstract)*, 30:449–450.
54. Navon, R., Geiger, B., Ben-Yoseph, Y., and Rattazzi, M. (1976): Low levels of hexosaminidase A in healthy individuals with apparent deficiency of this enzyme. *Am. J. Hum. Genet.*, 28:339–349.
55. Novak, A., and Lowden, J. A. (1980): Kinetics of rat liver G_{M2}-ganglioside-beta-D-N-acetylhexosaminidase. *Can. J. Biochem.*, 58:82–88.
56. Navon, R., Padeh, B., and Adam, A. (1973): Apparent deficiency of hexosaminidase A in healthy members of a family with Tay-Sachs disease. *Am. J. Hum. Genet.*, 25:287–293.
57. O'Brien, J. S. (1978): Suggestions for a nomenclature for the G_{M2}-gangliosidoses making certain (possibly unwarrantable) assumptions. *Am. J. Hum. Genet.*, 30:672–675.
58. O'Brien, J. S., Norden, A. G. W., Miller, A. L., Frost, R. G., and Kelly, T. E. (1977): Ganglioside G_{M2} N-acetyl-beta-D-galactosaminidase and G_{M2} (G_{A2}) N-acetyl-beta-D-galactosaminidase: Studies in human skin fibroblasts. *Clin. Genet.*, 11:171–183.
59. O'Brien, J. S., Okada, S., Chen, A., and Fillerup, D. (1970): Tay-Sachs disease: Detection of heterozygotes and homozygotes by serum hexosaminidase assay. *N. Engl. J. Med.*, 283:15.
60. O'Brien, J. S., Tennant, L., Veath, M. L., Scott, C. R., and Bucknall, W. E. (1978): Characterization of unusual hexosaminidase A-deficient human mutants. *Am. J. Hum. Genet.*, 30:602–608.
61. Ohno, S. (1980): Origin of intervening sequences within mammalian genes and the universal signal for their removal. *Differentiation*, 17:1–15.
62. Okada, S., and O'Brien, J. S. (1968): Generalized gangliosidosis: Beta-galactosidase deficiency. *Science*, 160:1,002–1,004.
63. Okada, S., and O'Brien, J. S. (1969): Generalized absence of a beta-D-N-acetylhexosaminidase component. *Science*, 165:698–700.
64. O'Neill, B., Butler, A. B., Young, E., Falk, P. M., and Bass, N. H. (1978): Adult-onset G_{M2}-gangliosidosis. *Neurology*, 28:1,117–1,123.
65. Oonk, J. G. W., van der Helm, H. J., and Martin, J. J. (1979): Spinocerebellar degeneration: Hexosaminidase A and B deficiency in two adult sisters. *Neurology*, 29:380–383.
66. Penton, E., Poenaru, L., and Dreyfus, J. C. (1975): Hexosaminidase C in Tay-Sachs and Sandhoff disease. *Biochim. Biophys. Acta*, 391:162–169.
67. Pilz, H., Muller, D., Sandhoff, K., and ter Meulen, V. (1968): Tay-Sachssche krankheit mit hexosaminidase-defekt. *Deutsch. Med. Wochenschr.*, 93:1,833–1,839.
68. Purpura, D. P., Highstein, S. M., Karabelas, A. B., and Walkley, S. U. (1980): Intracellular recording and HRP staining of cortical neurones in feline ganglioside storage disease. *Brain Res.*, 181:446–449.
69. Rapin, I., Suzuki, K., Suzuki, K., and Valsamis, M. P. (1976): Adult (chronic) G_{M2}-gangliosidosis—Atypical spinocerebellar degeneration in a Jewish sibship. *Arch. Neurol.*, 33:120–130.
70. Robinson, D., and Stirling, J. L. (1968): N-Acetyl-beta-glucosaminidases in human spleen. *Biochem. J.*, 107:321–327.
71. Rousson, R., Ben-Yoseph, Y., Fiddler, M. B., and Nadler, H. L. (1979): Demonstration of altered acid hydrolases in fibroblasts from patients with mucolipidosis II by lectin titration. *Biochem. J.*, 180:501–505.
72. Sachs, B. (1887): On arrested cerebral development with special reference to its cortical development. *J. Nerv. Ment. Dis.*, 14:541–553.
73. Sandhoff, K. (1969): Variation of beta-N-acetylhexosaminidase patterns in Tay-Sachs disease. *F.E.B.S. Lett.*, 4:351–354.
74. Sandhoff, K. (1970): The hydrolysis of Tay-Sachs ganglioside (TSG) by human N-acetyl-beta-D-hexosaminidase A. *F.E.B.S. Lett.*, 11:342.

75. Sandhoff, K., Andreae, U., and Jatzkewitz, H. (1968): Deficient hexosaminidase activity in an exceptional case of Tay-Sachs disease with additional storage of kidney globoside in visceral organs. *Life Sci.*, 7:283–288.
76. Sandhoff, K., Harzer, K., Wassle, W., and Jatzkewitz, H. (1971): Enzyme alterations and lipid storage in three variants of Tay-Sachs disease. *J. Neurochem.*, 18:2,469–2,489.
77. Santavouri, P., Haltia, M., Rapola, J., and Raitta, C. (1977): Infantile type of so-called neuronal ceroid lipofuscinosis. Part I. A clinical study of 15 patients. *J. Neurol. Sci.*, 18:257–267.
78. Schettler, G., and Kahlke, W. (1967): Gangliosidoses. In: *Lipids and Lipidoses*, edited by G. Schettler, pp. 213–259. Springer-Verlag, New York.
79. Seyama, Y., and Yamakawa, T. (1974): Multiple components of beta-N-acetylhexosaminidase from equine kidney. *J. Biol. Chem.*, 75:495–507.
80. Srivastava, S. K., and Ansari, N. H. (1978): Altered alpha-subunits in Tay-Sachs disease. *Nature*, 273:245–246.
81. Srivastava, S. K., and Beutler, E. (1974): Studies on human beta-D-N-acetylhexosaminidases. III. Biochemical genetics of Tay-Sachs and Sandhoff's diseases. *J. Biol. Chem.*, 249:2,054–2,057.
82. Svennerholm, L. (1962): The chemical structure of normal brain and Tay-Sachs gangliosides. *Biochem. Biophys. Res. Commun.*, 9:436.
83. Tallman, J. F., Brady, R. O., Navon, R., and Padeh, B. (1974): Ganglioside catabolism in hexosaminidase A-deficient adults. *Nature*, 252:254–255.
84. Tallman, J. F., Johnson, W. G., and Brady, R. O. (1972): The metabolism of Tay-Sachs ganglioside: Catabolic studies with lysosomal enzymes from normal and Tay-Sachs brain tissue. *J. Clin. Invest.*, 51:2,339–2,345.
85. Tay, W. (1881): Symmetrical changes in the region of the yellow spot in each eye of an infant. *Trans. Ophthal. Soc. U.K.*, 1:55.
86. Thomas, G. H., Taylor, H. A., Jr., Miller, C. S., Axelman, J., and Migon, B. R. (1974): Genetic complementation after fusion of Tay-Sachs and Sandhoff cells. *Nature*, 250:580–582.
87. Thompson, J. N., Stoolmiller, A. C., Matalon, R., and Dorfman, A. (1973): N-Acetyl-beta-hexosaminidase: Role in the degradation of glycosaminoglycans. *Science*, 181:866–876.
88. Tsay, G. C., and Dawson, G. (1976): Oligosaccharide storage in brains from patients with fucosidosis, G_{M1}-gangliosidosis, and G_{M2}-gangliosidosis (Sandhoff's disease). *J. Neurochem.*, 27:733–740.
89. Ullrich, K., Basner, R., Gieselmann, V., and Von Figura, K. (1979): Recognition of human alpha-N-acetyl-glucosaminidase by rat hepatocytes. Involvement of receptors specific for galactose, mannose-6-phosphate, and mannose. *Biochem. J.*, 180:413–419.
90. Van Hoof, F., Evrard, P., and Hers, H. G. (1972): An unusual case of G_{M2}-gangliosidosis with deficiency of hexosaminidase A and B. In: *Sphingolipids, Sphingolipidoses, and Allied Diseases*, pp. 343–350. Plenum Press, New York.
91. Vidgoff, J., Buist, N. R. M., and O'Brien, J. S. (1973): Absence of N-acetyl-D-hexosaminidase A activity in a healthy woman. *Am. J. Hum. Genet.*, 25:372–381.
92. Volk, B. W., Schneck, L., and Adachi, M. (1970): Clinic, pathology, and biochemistry of Tay-Sachs disease. In: *Leucodystrophies and Poliodystrophies*. In: *Handbook of Clinical Neurology*, Vol. 10, edited by P. J. Vinken and G. Bruyn, pp. 385–426. American Elsevier Publishing Co., New York.
93. Wickner, W. (1980): Assembly of proteins into membranes. *Science*, 210:861–868.
94. Willner, J. P., Bender, A. N., Straus, L., Yahr, M., and Desnick, R. J. (1979): Total beta-hexosaminidase A deficiency in two adult Ashkenazi Jewish siblings: Report of a new clinical variant. *Am. J. Hum. Genet. (abstract)*, 31:86A.
95. Wood, S. (1979): Human alpha-fucosidase: A common polymorphic variant for low serum enzyme activity—Studies of serum and leukocyte enzyme. *Hum. Hered.*, 29:226–229.
96. Yaffe, M. G., Kaback, M., Goldberg, M., Miles, J., Itabashi, H., McIntyre, H., and Mohandas, T. (1979): An amyotrophic lateral sclerosis-like syndrome with hexosaminidase A deficiency: A new type of G_{M2}-gangliosidosis. *Neurology (abstract)*, 29:611.

Genetics of Neurological and Psychiatric Disorders, edited by Seymour S. Kety, Lewis P. Rowland, Richard L. Sidman, and Steven W. Matthysse. Raven Press, New York © 1983.

Glycogen-Storage Diseases of Muscle: Genetic Problems

Lewis P. Rowland and Salvatore DiMauro

Department of Neurology and the H. Houston Merritt Clinical Research Center for Muscular Dystrophy and Related Diseases, Columbia-Presbyterian Medical Center, New York, New York 10032

Of the hereditary diseases of muscle, the glycogen-storage diseases are understood most thoroughly, and there has been considerable progress in elucidating the enzymatic abnormalities, as recorded in several reviews (20,21,45,87). Yet, as we reported a decade ago (89), there have been several genetic conundrums. It is useful to highlight some of the current problems because the glycogen diseases have often served as examples in the analysis of other heritable diseases.

Our attention is now focused on the evidence of heterogenity at several levels (20). At the simplest level, the same clinical syndrome may be caused by lack of different enzymes. For instance, the syndrome of cramps and myoglobinuria may be caused by lack of either phosphorylase or phosphofructokinase. The syndrome of progressive proximal limb weakness may be caused by lack of acid maltase or the debrancher enzyme system.

There is another kind of clinical heterogeneity. Lack of the same enzyme may be associated with different clinical syndromes. For each enzyme deficiency, there is more than one clinical phenotype. For each of the disorders, there is a typical syndrome, but atypical syndromes have also been described.

There is also evidence of molecular heterogeneity. For each enzyme deficiency, there are cases in which the enzyme protein seems to be totally lacking and others in which a mutant protein has been identified by demonstrating immunologically cross-reacting material (CRM). For some syndromes, the isoenzyme that characterizes muscle may be affected but the enzyme is spared in heart or liver, whereas in other cases the enzyme may be affected in other organs as well as skeletal muscle.

For ease of description, we shall consider each of the main enzyme categories separately. We shall not consider type I glycogenosis (lack of glucose-6-phosphatase) or type VI (lack of liver phosphorylase) in which muscle is not affected, or

---

This chapter is dedicated to Professor David Shemin in honor of his 70th birthday.

type IV disease (lack of brancher enzyme) in which muscle may be affected but the myopathy is much less significant than the overwhelming liver disorder.

TYPE II GLYCOGEN-STORAGE DISEASES (LACK OF ACID MALTASE)

Clinical Variations

This disease has played an important part in the history of human biochemical genetics. It was one of the first glycogen-storage diseases to be identified, when Pompe described the first case in 1932 (79), and recognition of the lack of acid α-glucosidase (acid maltase) in 1963 by Hers (42) established the general concept of lysosomal storage diseases.

The typical syndrome (Pompe disease) is evident in the first weeks of life, and almost all affected infants die before age 2 years (19). Flaccid quadriplegia is due to glycogen storage in both muscle and motor neurons. Glossomegaly, cardiomegaly, and congestive heart failure are other major clinical manifestations, and the liver is also involved; it is a generalized disease. Autosomal recessive inheritance is implied by high incidence of affected siblings without appearance of the disease in other generations and by disproportionate frequency of consanguinity (19).

Soon after the enzymatic abnormality in a typical infantile case was identified, late-onset syndromes were recognized (31,33,40,44,80,85,95,98,100). Some begin before age 2 years, others in later childhood, and still others in adult years, even in the fifth decade (19,44). These cases are manifest primarily by limb weakness, but respiratory muscles are often affected (85). The course is usually one of slow progression with varying degrees of incapacity, or the respiratory disorder may ultimately be fatal. Cardiomegaly and congestive heart failure or hepatomegaly are exceptional in these forms. There have been cases in siblings, but many of the reported late-onset cases were sporadic. It is therefore uncertain how many distinct late disorders there are; this category is likely to be of heterogeneous pathogenesis.

Chemical Pathology

In the infantile disease, glycogen accumulates and acid maltase activity cannot be detected in many organs, including skeletal muscle, heart, motor neurons, and liver (19). In skeletal muscle, glycogen content is five to 10 times more than the normal limit of 1.5 g/100-g wet weight of muscle. The enzyme is absent in white blood cells, cultured skin fibroblasts, and cultured muscle (4,14,19,24,37,71,77, 81,101). An apparent exception is the kidney, in which total acid maltase may be normal by conventional assays (19,78,97), although glycogen accumulation was shown morphologically (110).

In the late-onset forms, glycogen also accumulates in skeletal muscle. Other organs are spared clinically and glycogen does not accumulate excessively in them, but the enzyme activity is much reduced in the same organs affected by the infantile disorder (19). There are important differences, however. The extent of glycogen accumulation in muscle is rarely more than five times normal and may even be normal. In all cases there is some residual acid maltase activity, usually between 5 and 10% of the normal mean (19). There have been only two biochemical studies of postmortem cases of late-onset disease; in both, enzyme activity was less than 10% of normal in heart and brain but glycogen did not accumulate in these organs (25,63). In fibroblasts from late-onset cases, as in skeletal muscle, there is some residual enzyme activity, whereas no activity is found in fibroblasts from infantile cases. Glycogen accumulation, however, is approximately the same in fibroblasts from infantile, childhood, or adult cases despite these differences in enzyme activity (81). Failure to find more severe glycogen accumulation in fibroblasts from infantile cases has been attributed to the artificial environment of the culture or to the rapid turnover of cells (81).

In both infantile and late-onset forms, the heterozygous state of parents has been identified by intermediate levels of enzyme activity in leucocytes, cultured skin fibroblasts, muscle, or urine (19,32,66,100). Sensitive methods have been developed to measure enzyme activity in amniotic fluid cells so that antenatal diagnosis is possible (77).

Molecular Pathology

In biochemical assays of renal tissue from infantile cases, acid maltase activity may be normal although, as in other tissues, there is no immunological evidence of the normal protein (66). This seeming paradox occurs because in addition to an acid maltase similar to that of other tissues, the kidney of normal individuals also expresses a separate "renal" isoenzyme active at acid pH (19). This kidney enzyme differs in heat-stability, electrophocusing, electrophoresis, and antibody reactions (19,81) and seems to be under separate genetic control.

It has been difficult to ascertain the difference between infantile and late-onset forms. At first, it seemed possible that a related enzyme, neutral maltase, might be present in amounts sufficient to compensate for the lack of acid maltase in late-onset cases, thus accounting for the less malignant course (2,3). Subsequent studies, however, showed that this could not explain the differences because neutral maltase activity was normal in both early and late cases (3,19,66,67,81).

In infantile cases, catalytic enzyme activity is always lacking and, in the first 5 cases so studied, there was no evidence of an immunologically reacting protein (9,18,54,57,76). Beratis et al. (7, and *personal communication*, 1980) found evidence of CRM in 1 of 10 infantile cases; in that case there was no enzyme activity in the material precipitated by the antibody. In cultured muscle, acid maltase is lacking and glycogen accumulates (4,81). The infantile disease, therefore, seems to be due either to total lack of enzyme protein or to the presence of a catalytically inactive mutant protein.

In the late-onset forms, the residual enzyme seems to have normal characteristics for pH optimum, kinetics, electrophoresis, reactions with inhibitors, and immunoprecipitation (7,9,18,54,57,66,67,76, and N. G. Beratis, *personal communica-*

tion, 1980). Even after precipitation by antibody, the enzyme retains catalytic activity (54). Therefore, in late cases, there seems to be a quantitative reduction in the amount of a normal enzyme protein, but it is not clear whether this is due to retarded synthesis or accelerated degradation.

These detailed studies have excluded the possibility that the late-onset cases could be due to a mutant enzyme with abnormally decreased catalytic activity, a theory that has also been considered (28). In normal individuals, however, there is electrophoretic evidence of polymorphism. Instead of the normal protein, some individuals have a variant form of acid maltase that occurs in 1 of 16 Europeans (99) and 1 of 32 New Yorkers (8). The variant enzyme has much less activity with glycogen as substrate than with maltose or a widely used fluorogenic substrate (8). It has been suggested that individuals who are homozygous for the variant enzyme might be at special risk of developing late-onset disease (8).

Problems

Coding for acid maltase has been assigned to chromosome 17 (15,96), but among the several problems that remain to be elucidated with regard to acid maltase deficiency are the normal structure and function of the enzyme itself. The protein may be a tetramer, but subunit structure and other aspects are not known. The role of this enzyme in the normal economy of the cell is uncertain, and the glycogen-storage diseases as a group highlight this uncertainty. Acid maltase activity must be essential in some regard or its absence would not be so clearly associated with excessive glycogen accumulation and clinical disease. However, if it does function in the normal degradation of glycogen, as these syndromes attest, why does acid maltase activity not prevent accumulation of glycogen in other glycogen-storage diseases that result from lack of phosphorylase, phosphofructokinase, or debrancher? In these diseases, the amount of acid maltase activity is normal, but glycogen accumulates nevertheless. Alternatively, it is not clear why extralysosomal glycogen should be found in acid maltase deficiency when the other enzymes are normal.

Similarly, it is not clear why, in the late-onset forms, glycogen accumulates in muscle but not in heart, brain, or liver, even though the activity of the enzyme in these organs is only approximately 10% of normal, as it is in skeletal muscle. If lack of this enzyme is responsible for the excess glycogen in muscle, why not in other organs?

The clinical heterogeneity of late-onset cases is also a problem. It is difficult to believe that a disease starting before age 2 years in siblings is the same as a sporadic disease beginning after age 40. Some of the adult-onset cases could be acquired rather than inherited.

If two different clinical syndromes are associated with lack of the same enzyme, the disorders may be allelic, involving mutations of the same cistron, or nonallelic, involving mutations at different gene loci (46). A commonly used test for allelism is to look for complementation by fusing fibroblasts taken from individuals with different forms of the disease; when cultured alone, the fibroblasts from each type lack enzyme activity, but enzyme activity found in fused cells is evidence of complementation, implying that each cell originally contained a subunit that was inactive alone although the two, together, function effectively. If complementation studies imply activity of more than one subunit, each must be under separate genetic control and more than one allele must be involved. Failure to show complementation suggests that the disorders are allelic, whereas complementation may be taken nearly always as evidence of nonallelism of the two disorders. Nonallelic conditions may be due to loss of different subunit components of an enzyme; fused fibroblasts may regain the missing catalytic activity because of complementation, with each of the deficient cells providing one subunit that is insufficient to function catalytically alone (65). This kind of study was carried out by Reuser et al. (81) who fused fibroblasts from infantile, childhood, and adult cases of acid maltase deficiency in numerous combinations, but the enzyme activity was not regained.

The possibility that the early and late forms are allelic was strengthened by the observation of a Dutch family in which a girl had the typical infantile form and her grandfather had a late-onset form (10,53). Assuming that one allele is responsible for the infantile form and a second allele for the late form, there are at least three possible explanations for this state of affairs (62):

(a) The grandfather might be homozygous for the late form. If his son, necessarily heterozygous for this form, married a woman who was heterozygous for the infantile form, their child might be a genetic compound (two different mutant genes at the same locus) for the infantile form (from her mother) and the late form (from her father).

(b) The grandfather could have been a genetic compound for the two forms. Each of his children would be heterozygous for either one of the two forms. If the affected infant's father were heterozygous for the infantile disease and her mother were also heterozygous for the same form, the child would be homozygous for the infantile form.

(c) The two conditions might be nonallelic and have appeared in the same family by coincidence.

Resolution of this problem awaits more detailed knowledge about the structure of the normal enzyme.

A major practical problem is treatment, which has been notably unsuccessful although attempts have been made to replace the missing enzyme (19,109). Still another problem is presented by cases of lysosomal glycogen storage with normal acid maltase activity (17).

GLYCOGEN-STORAGE DISEASE TYPE III
(LACK OF DEBRANCHER ENZYME SYSTEM)

Clinical Variations

The major clinical variations of the glycogen-storage diseases associated with lack of the debrancher enzyme are concerned with the organs involved. In the typical form, all symptoms can be attributed to liver disease commencing in early childhood. The disorder is often mild, but it may be manifest by hepatomegaly, growth retardation, fasting hypoglycemia, ketonuria, or nonprogressive cirrhosis. In this hepatic form, limb weakness is usually not prominent and is overshadowed by the liver disorder. In 10 patients, however, there was prominent myopathy (22), with symptoms of either limb weakness or exercise intolerance. In some of these cases symptoms began in childhood; in others onset was delayed to adult years. None had episodes of myoglobinuria. The myopathy is less severe than that of acid maltase deficiency; involvement of respiratory muscles has not been reported and there have been no fatal cases yet, although the heart may be involved to some extent (22,86). The disorder seems to be inherited in autosomal recessive fashion.

Biochemical Pathology

Glycogen accumulation and lack of enzyme activity have been demonstrated in muscle, red blood cells, white blood cells, and liver (22), but there have been no autopsy studies. The extent of glycogen accumulation is less than in the infantile form of acid maltase deficiency, between 2.8 and 6.7 g per 100-g wet weight of muscle in debrancher cases with or without clinical myopathy. The stored glycogen has an abnormal structure, with outer chains of glycosyl units that are shorter than those of normal glycogen. In cases with myopathy, there is no rise in venous lactate after ischemic work, but contracture is not induced by this test (22). In all cases blood glucose does not increase after administration of epinephrine or glucagon, but there is a normal rise after a galactose load, reflecting the block of glycolysis in liver. Glycogen accumulation and lack of enzyme activity have been demonstrated in cultured fibroblasts and cultured muscle (24,47,69).

Molecular Pathology

Although little is known of the subunit structure of the debrancher enzyme, the mechanism of action has been studied thoroughly and several biochemical variants of the disorder were identified by Van Hoof and Hers (105). In the normal degradation of glycogen, the outer chains are first removed by the action of phosphorylase, which stops as the branch points are neared. Debranching then requires two sequential reactions: a maltotriosyl unit is first transferred from a donor chain to an acceptor chain of the dextrin, then the residual amylo-1,6-linked glycosyl unit is removed. These two activities, transferase and amylo-1,6-glucosidase, are apparently catalyzed by the same enzyme and can be measured by suitably designed assays. Van Hoof and Hers (105) found varying combinations of loss of one or both activities in either or both liver and muscle. In the most common form, accounting for 75% of 45 cases studied, both enzyme activities were lacking in both tissues, but in the other cases there were different patterns of enzyme deficiency. The different enzyme patterns were all associated with similar clinical disorders.

Problems

With meager knowledge of this enzyme, it is not surprising that there are genetic problems. It is not clear why the liver disorder dominates in some cases and myopathy in others, especially because the extent of glycogen accumulation in muscle may be the same whether or not there is overt weakness. Among the patients with symptomatic myopathy, it is not clear why weakness is the dominant symptom in most, but exercise intolerance predominates in others. Because the actions of phosphorylase and debrancher are closely linked, and because ischemic work is not followed by normal lactate formation if either enzyme is missing, it is not clear why myoglobinuria has not been seen in debrancher deficiency. The clinical difference from lack of phosphorylase may be due to the availability of the short outer chains of the accumulated glycogen or to the chronically increased content of fatty acids and ketones in blood, which could provide an alternate source of energy for working muscle (22). It is not known why there is no hemolytic anemia, since the red blood cells contain excessive glycogen and lack the enzyme. Finally, it is not known why the heart is involved clinically in some cases and not others.

GLYCOGEN-STORAGE DISEASE TYPE V (LACK OF MUSCLE PHOSPHORYLASE)

Clinical Variations

The essential clinical features of this condition were recognized by McArdle in 1951 (64) and the enzymatic abnormality was identified 8 years later (75,92). In the next 20 years, 62 cases were reported (19). In typical cases of muscle phosphorylase deficiency (McArdle disease), symptoms are restricted to exercise intolerance and attacks of myoglobinuria. The amount of work that can be done varies from day to day, but attacks of myoglobinuria may be precipitated by either brief strenuous activity or prolonged but less vigorous exercise; normal walking and activities of daily living are rarely curtailed. Many patients recognize the second-wind phenomenon; after exercising to the point of muscle aches, they rest and then can exercise for a longer time. Although symptoms usually start in childhood, attacks of myoglobinuria do not appear until late adolescence or adult years. Weakness of limb muscles is rarely prominent in typical cases, but some weakness has been reported in approximately 20% of the cases (19).

Within the limits of exercise intolerance, patients may lead a relatively normal life, but renal failure may follow attacks of myoglobinuria (5,38). Heart and liver are never affected; a few patients have had convulsions for reasons not known (19), but there is no other evidence of cerebral disease. The uterus is not involved, and affected women may expect normal pregnancy (13).

Two atypical forms have been reported. In one, exercise intolerance is not prominent and symptoms of limb weakness appear in late life. In 3 of these cases, lack of enzyme activity was demonstrated by biochemical assay of muscle homogenates (34,52,83). In 2 others, however, biochemical assay was not carried out and diagnosis rested solely on lack of phosphorylase activity in histochemical reactions (43,51). To verify an atypical form of a metabolic disease, it would seem desirable to confirm the diagnosis by biochemical assay and to provide biochemical evidence of the metabolic block.

In the other atypical form, congenital weakness caused death at age 13 and 16 weeks, respectively, in two sisters (21,70). In one of these siblings, the enzyme was lacking by biochemical assay and immunodiffusion, and the metabolic block was demonstrated by studies of anerobic glycolysis (21).

The typical form is presumed to be inherited as an autosomal recessive trait, as witnessed by high frequency of affected siblings and consanguinity (19). Men are affected three times more often than women, however (19). In two reports, autosomal dominant inheritance was suggested. In one of these there was insufficient detail for evaluation (94). In the other (12), the propositus had a syndrome of exercise intolerance but no myoglobinuria; there was no rise in venous lactate after ischemic work, and phosphorylase activity was lacking in muscle on biochemical assay. The patient's son had similar symptoms, a flat lactate curve and no histochemical evidence of phosphorylase, but glycogen content was normal histochemically. The patient's mother also had similar symptoms and flat lactate curve, but phosphorylase activity was normal histochemically. Because the studies were not complete, it is difficult to evaluate this family, but even if phosphorylase deficiency affected two generations, "pseudodominance" was not excluded; that is, the patient (possibly homozygous) could have married a heterozygous carrier of the same recessive trait and the son could then have been homozygous. This could be tested because heterozygotes have approximately half-normal activity of phosphorylase in biopsies (6).

Biochemical Pathology

Both exercise intolerance and attacks of myoglobinuria are attributed to lack of availability of glycogen to supply fuel for working muscle, and the second-wind phenomenon is attributed to mobilization of lipid in the blood as an alternate fuel. The failure of venous lactate content to rise after ischemic work is accounted for in the same manner, as originally suggested by McArdle himself (64). Symptoms are restricted to skeletal muscle. The muscle content of glycogen is increased above the normal 1.5% but almost always below 5 g % (19). The structure of the glycogen is normal. When incubated anaerobically with endogenous or added glycogen, lactate is not produced, but lactate is formed normally if glucose-1-phosphate or other glycolytic intermediates are added.

Molecular Pathology

The restriction of symptoms to skeletal muscle can be attributed to separate genetic control for the phosphorylase isoenzymes of heart, liver, and brain. This has been documented by the appearance of heritable liver disease due to lack of phosphorylase (glycogen-storage disease type VI), which does not affect muscle (46). Normal heart contains three isoenzymes: a specific cardiac isoenzyme, the muscle isoenzyme, and a hybrid of the two (70). In 1 of the 2 fatal infantile cases of muscle phosphorylase deficiency, both muscle and hybrid forms were lacking in the heart, and only the cardiac isoenzyme was present (70). In normal people, the cardiac isoenzyme greatly predominates over the muscle form, which may explain the lack of heart disease in McArdle patients.

In the first reported cases to be tested for immunoreactive protein, none was found in muscle (27,38,82,90). Later, however, several patients were found to have antigenic CRM when tested with antibodies to normal muscle phosphorylase (5,6,11,27,36,38,56), and it is now uncertain which form is more common, that with or without CRM. Additionally, catalytically inactive phosphorylase protein has been shown by electrophoresis in gels containing sodium dodecyl sulfate in some patients but not others (36,56); in one case, there was electrophoretic evidence of catalytically inactive phosphorylase but there was no CRM (56). The inactive protein may regain catalytic activity if experimentally phosphorylated by incubation with protein kinase or phosphorylase kinase (11), but there is no evidence that these phosphorylating enzymes are abnormal in living patients.

An apparent mystery arose when muscle was cultured from patients lacking muscle phosphorylase and enzyme activity was found in these cells that presumably lacked the genetic program for this protein (84). The mystery deepened when it was reported that the enzyme protein was the same isoenzyme found in normal mature muscle (68). However, workers in several laboratories found that the isoenzymes in skeletal muscle, liver, and heart were all different (20,26,55,70) and that the enzyme in cultured muscle (from either normals or McArdle patients) had the properties of a fetal enzyme (20,91). The presence of a fetal enzyme could also explain the histochemical appearance of phosphorylase activity in regenerating fibers of patients with the disease (73). Phosphorylase activity is also normal in cultured fibroblasts and white blood cells (24), again presumably because the isoenzymes differ from that in skeletal muscle.

Problems

Among the several problems that remain to be elucidated in phosphorylase deficiency is a molecular explanation for the several forms: typical McArdle disease, late-onset weakness, and the fatal infantile variety. In the typical form, the link between the enzymatic defect and the symptoms of cramp and myoglobinuria is still fuzzy. It is assumed that myoglobinuria results because ATP production fails, but this has not been proved (30,88). The apparently different patterns of genetic transmission also await molecular explanation, and treatment remains unsatisfactory. The predominance of affected males could be due to more strenuous activity, but it might also have genetic implications of some sort of sex-linkage.

GLYCOGEN-STORAGE DISEASE TYPE VII
(LACK OF PHOSPHOFRUCTOKINASE)

Clinical Variations

This syndrome was first identified by Tarui et al. in 1965 (104); it seems to be the least common of the glycogen-storage diseases of muscle, and only 9 cases have been reported (19). The typical syndrome is clinically indistinguishable from McArdle disease except that evidence of hemolytic anemia was found in all 6 patients who were studied appropriately, and overt jaundice was found in 2 cases (19). The pattern of exercise intolerance, cramps, and myoglobinuria does not differ from that in McArdle disease. As in phosphorylase deficiency, males have been predominantly affected, accounting for 7 of the 9 cases (19). The pattern of inheritance is otherwise compatible with an autosomal recessive trait. In 5 other cases, there was evidence of hemolytic anemia without myopathy (29,35,74,108).

Atypical forms have also been reported. In 3 cases (16,39) the enzyme was lacking in infants with severe congenital weakness, leading to death within a few weeks or 4 years; one of these infants also had fixation of multiple joints in a pattern compatible with arthrogryposis multiplex congenita (18). A syndrome of progressive limb weakness, without cramps or myoglobinuria, has been reported in 2 cases. In one, weakness began in adolescence (93); in the other, overt weakness was not apparent until age 56 (41).

Chemical Pathology

In typical cases, several findings are indistinguishable from McArdle disease, including flat lactate response and contracture after ischemic work, and histochemical appearance of stored glycogen in muscle. Biochemically, the amount of glycogen in muscle is also similar (1.6 to 4.4 g glycogen/100-g muscle wet weight). The differential diagnosis can therefore be made only by biochemical assay of the enzymes, or by histochemical stains for the two enzymes. In contrast to phosphorylase deficiency, glucose-6-phosphate and fructose-6-phosphate concentrations in muscle increase because of the block in glycolysis (78,104). The structure of glycogen in muscle is normal, but an abnormal polysaccharide has also been identified in some cases (1,41).

Evidence of hemolytic anemia also identifies phosphofructokinase deficiency. There may or may not be overt anemia or bilirubinemia, but reticulocyte counts are usually increased and the life-span of labeled red blood cells is reduced (29).

Molecular Pathology

In the original cases (78,104), the hemolytic anemia in patients with myopathy suggested that the enzymatic abnormality affected two organs. In these cases, phosphofructokinase activity was essentially absent in muscle and approximately 50% of normal in erythrocytes. Antibodies to normal muscle phosphofructokinase did not react with either muscle or red blood cells from an affected patient (60,78). It was therefore suggested that there were muscle and red cell isoenzymes and that the red cell enzyme included subunits of muscle type (102–104). In a series of investigations, Layzer et al. (50,58–61) concluded that the pattern could be explained if the enzyme were a tetramer (49), if the muscle enzyme had four identical subunits, and if red cells had a combination of two muscle and two red cell subunits. In patients with both myopathy and anemia, the red cell enzyme would be devoid of muscle subunits. In patients with anemia alone, a different mutation of the muscle subunit has been suggested, causing unstable rather than absent enzyme activity (35,48). Continuous *de novo* synthesis of these labile subunits would keep the enzyme activity normal in muscle, but lack of synthesis in erythrocytes would cause a partial enzyme defect (35).

With improved chromatographic and electrophoretic techniques, Vora et al. (106) showed that the pattern suggested by Layzer et al. was essentially correct and that the erythrocyte subunit was the same as that found in liver. In conformity with standard nomenclature, the subunits were called M (muscle) and L (liver). Normal muscle phosphofructokinase is therefore composed of identical subunits (M_4); liver is composed only of L_4 units; and erythrocyte phosphofructokinase is composed of five isoenzymes (M_4, M_3L, M_2L_2, ML_3, L_4). As predicted, red cells from a patient had only the L_4 band (106).

Platelets and white blood cells do not normally contain the M units and therefore are not affected in the disease (106). Normal fibroblast phosphofructokinase seems to be composed of L, M, and P (platelet) subunits (107), but fibroblasts from a patient have not yet been studied. In muscle cultured from patients with myopathic phosphofructokinase deficiency, enzyme activity appears, as in phosphorylase deficiency (72).

Problems

The major genetic problem is an explanation of the differences among the typical form of phosphofructokinase deficiency, the severe infantile form, and the form manifest only by limb weakness. Accumulation of an unidentified polysaccharide in some cases (1,41) remains to be explained, and treatment, again, is unsatisfactory. The enzymatic abnormality probably bears the same relationship to muscle symptoms that phosphorylase deficiency does, but neither is understood.

CONCLUSIONS

Analysis of the clinical features and biochemistry of the glycogen-storage diseases of muscle indicates considerable progress and different kinds of heterogeneity. Some of the heterogeneous aspects can be explained in molecular terms; for instance, some organs are affected and others are spared in different diseases because the isoenzymes in different organs are under separate genetic control. In the typical form of all these diseases, the molecular basis for lack of enzyme activity is uncertain and could be due to gene deletion, mutation of a structural gene, or mutation in another locus affecting expression of the gene. In cases in which there is residual activity, it is uncertain whether synthesis is retarded or degradation accelerated. Other manifestations of heterogeneity may be explained by allelic variation, as in acid maltase deficiency, but this is not yet proved. Some variations of clinical syndromes are not understood, as in myopathies owing to lack of phosphorylase or phosphofructokinase, which could also be allelic disorders. In others, as in the pure hemolytic disorder of phosphofructokinase deficiency, the variations cannot be allelic. We do not understand the male predominance in McArdle disease or Tarui-Layzer disease, nor apparently autosomal dominant transmission of what are usually recessive traits. In all of these diseases, it is difficult to explain symptoms on the basis of the known biochemical abnormality, and in all of them treatment remains unsatisfactory.

ACKNOWLEDGMENT

This work was supported by Center Grants from the Muscular Dystrophy Association and the National Institute of Neurological and Communicative Disorders and Stroke (NS-11766).

REFERENCES

1. Agamanolis, D., Askari, A., DiMauro, S., Hays, A., Kumar, K., Lipton, M., and Raynor, A. (1980): Muscle phosphofructokinase deficiency: Two cases with unusual polysaccharide accumulation and immunologically active enzyme protein. *Muscle Nerve*, 3:456–467.
2. Angelini, C., and Engel, A. G. (1972): Comparative study of acid maltase deficiency: Biochemical differences between infantile, childhood, and adult types. *Arch. Neurol.*, 26:344–349.
3. Angelini, C., and Engel, A. G. (1973): Subcellular distribution of acid and neutral alpha-glucosidases in normal, acid maltase deficient, and myophosphorylase deficient human skeletal muscle. *Arch. Biochem. Biophys.*, 156:350–355.
4. Askanas, V., Engel, W. K., DiMauro, S., Brooks, B. R., and Mehler, M. (1976): Adult-onset acid maltase deficiency; morphologic and biochemical abnormalities reproduced in cultured muscle. *N. Engl. J. Med.*, 294:573–578.
5. Bank, W. J., DiMauro, S., and Rowland, L. P. (1972): Renal failure in McArdle's disease. *N. Engl. J. Med.*, 287:1102.
6. Bank, W. J., DiMauro, S., Rowland, L. P., and Milestone, R. (1972): Heterozygotes in muscle phosphorylase deficiency. *Trans. Am. Neurol. Assoc.*, 97:179–183.
7. Beratis, N. G., LaBadie, G. U., and Hirschhorn, K. (1978): Characterization of the molecular defect in infantile and adult acid alpha-glucosidase deficiency fibroblasts. *J. Clin. Invest.*, 62:1264–1274.
8. Beratis, N. G., LaBadie, G. U., and Hirschhorn, K. (1980): An isozyme of acid α-glucosidase with reduced catalytic activity for glycogen. *Am. J. Hum. Genet.*, 32:137–149.
9. Brown, B. I., Murray, A. K., and Brown, D. H. (1975): The lysosomal α-glucosidases of mammalian tissues. In: *Physiological Effects of Food Carbohydrates, American Chemical Society Symposium Series*, edited by A. Jeanes and J. Hodge. American Chemical Society, Washington, D. C., 15:223–234.
10. Busch, H. F. M., Koster, J. F., and van Weerden, T. W. (1979): Infantile and adult-onset acid maltase deficiency occurring in the same family. *Neurology*, 29:415–416.
11. Cerri, C., and Willner, J. H. (1980): Phosphorylation of McArdle phosphorylase induces activity. *Neurology*, 30:369.
12. Chui, L. A., and Munsat, T. L. (1976): Dominant inheritance of McArdle syndrome. *Arch. Neurol.*, 33:636–641.
13. Cochrane, P., and Alderman, B. (1973): Normal pregnancy and successful delivery in myophosphorylase deficiency (McArdle's disease). *J. Neurol. Neurosurg. Psychiatry*, 36:225–227.

14. Dancis, J., Hutzler, J., Lynfield, J., and Cox, R. P. (1969): Absence of acid maltase in glycogenosis type 2 (Pompe's disease) in tissue culture. *Am. J. Dis. Child.*, 117:108–111.
15. D'Ancona, G. G., Wurm, J., and Croce, C. M. (1979): Genetics of type II glycogenosis: Assignment of the gene for acid α-glucosidase to chromosome 17. *Proc. Natl. Acad. Sci. U.S.A.*, 76:4526–4529.
16. Danon, M. J., Carpenter, S., Manaligod, J. R., and Schliselfeld, L. H. (1981): Fatal infantile glycogen storage disease: Deficiency of phosphofructokinase and phosphorylase B-kinase. *Neurology*, 31:1303–1310.
17. Danon, M. J., Oh, S. J., DiMauro, S., Manaligod, J. R., Eastwood, A., Naidu, S., and Schliselfeld, L. S. Lysosomal glycogen storage diseases with normal acid maltase. *Neurology*, 31:51–57.
18. de Barsy, T., Jacquemin, P., Devos, P., and Hers, H. G. (1972): Rodent and human acid-glucosidase: Purification, properties and inhibition by antibodies. Investigation in type II glycogenosis. *Eur. J. Biochem.*, 31:156–165.
19. DiMauro, S. (1979): Metabolic myopathies. *Handb. Clin. Neurol.*, 41:175–234.
20. DiMauro, S., Arnold, S., Miranda, A., and Rowland, L. P. (1978): McArdle disease. The mystery of reappearing phosphorylase activity in muscle culture. A fetal isoenzyme. *Ann. Neurol.*, 3:60–66.
21. DiMauro, S., and Hartlage, P. L. (1978): Fatal infantile form of muscle phosphorylase deficiency. *Neurology*, 28:1124–1128.
22. DiMauro, S., Hartwig, G. B., Hays, A., Eastwood, A. B., Franco, R., Olarte, M., Chang, M., Roses, A. D., Fetell, M., Schoenfeldt, R. S., and Stern, L. Z. (1979): Debrancher deficiency: Neuromuscular disorder in five adults. *Ann. Neurol.*, 5:422–436.
23. DiMauro, S., Mehler, M., Arnold, S., and Miranda, A. (1977): Genetic heterogeneity of glycogen diseases. In: *Pathogenesis of Human Muscular Dystrophies*, edited by L. P. Rowland. Excerpta Medica, Amsterdam, pp. 506–514.
24. DiMauro, S., Rowland, L. P., and Mellman, W. J. (1973): Glycogen metabolism of human diploid fibroblast cells in culture. I. Studies of cells from patients with glycogenosis type II, III and V. *Pediatr. Res.*, 7:739–744.
25. DiMauro, S., Stern, L. Z., Mehler, M., Nagle, R. B., and Payne, C. (1978): Adult-onset acid maltase deficiency. A postmortem study. *Muscle Nerve*, 1:27–36.
26. Dreyfus, J. C. (1972): Phosphorylases and phosphorylase regulation in diseases. In: *Clinical Studies in Myology*, edited by B. Kakulas. Excerpta Medica, Amsterdam, pp. 62–69.
27. Dreyfus, J. C., and Alexandre, Y. (1971): Immunological studies on glycogen storage diseases Type III and V: Demonstration of an immunoreactive protein in one case of muscle phosphorylase activity. *Biochem. Biophys. Res. Commun.*, 44:1364–1370.
28. Dreyfus, J. C., Proux, D., and Alexandre, Y. (1974): Molecular studies on glycogen storage diseases. *Enzyme (Basel)*, 18:60–72.
29. Dupond, J. L., Robert, M., Carbillet, J. P., and Lecont des Floris, R. (1977): Glycogenose musculaire et anémie hémolytique par déficit enzymatique chez deux germains. Forme familiale de maladie de Tarui, par déficit en phosphofructokinase musculaire et erythrocytaire. *Nouv. Presse Med.*, 6:2665–2668.
30. Edwards, R. H., Young, A., and Wiles, M. (1980): Needle biopsy of skeletal muscle in the diagnosis of myopathy and the clinical study of muscle function and repair. *N. Engl. J. Med.*, 302:261–271.
31. Engel, A. G. (1970): Acid maltase deficiency in adults: Studies in four cases of a syndrome which may mimic muscular dystrophy and other myopathies. *Brain*, 93:599–616.
32. Engel, A. G., and Gomez, M. R. (1970): Acid maltase levels in muscle in heterozygous acid maltase deficiency and in non-weak and neuromuscular disease controls. *J. Neurol. Neurosurg. Psychiatry*, 33:801–804.
33. Engel, A. G., Gomez, M. R., Seybold, M. E., and Lambert, E. H. (1973): The spectrum and diagnosis of acid maltase deficiency. *Neurology*, 23:95–106.
34. Engel, W. K., Eyerman, E. L., and Williams, H. E. (1963): Late-onset type of skeletal muscle phosphorylase deficiency. A new family with completely and partially affected subjects. *N. Engl. J. Med.*, 282:697–704.
35. Etiemble, J., Kahn, A., Boivin, P., Bernard, J. F., and Goudemand, M. (1976): Hereditary hemolytic anemia with erythrocyte phosphofructokinase deficiency. Studies of some properties of erythrocyte and muscle enzymes. *Hum. Genet.*, 31:83–91.

36. Feit, H., and Brooke, M. H. (1976): Myophosphorylase deficiency: Two different molecular etiologies. *Neurology*, 26:963–987.
37. Fujimoto, A., Fluharty, A. L., Stevens, R. L., Kihara, H., and Wilson, M. G. (1976): Two alpha-glucosidases in cultured amniotic cells and their differentiation in the prenatal diagnosis of Pompe's disease. *Clin. Chim. Acta*, 68:177–186.
38. Grunfeld, J. P., Ganeval, D., Chanard, J., Fardeau, M., and Dreyfus, J. C. (1972): Acute renal failure in McArdle's disease: Report of 2 cases. *N. Engl. J. Med.*, 286:1237–1241.
39. Guibaud, P., Carrier, H., Mathieu, M., Dorche, C., Parchoux, B., Bethenod, M., and Larbe, F. (1978): Observation familape de dystrophie musculaire congénitale par déficit en phosphofructokinase. *Arch. Fr. Pediatr.*, 35:1105–1115.
40. Gullotta, F., Stefan, H., and Mattern, H. (1976): Pseudodystrophische Muskelglykogenose im Erwachsenalter (Saure-Maltase-Mangel-Syndrome). *J. Neurol.*, 213:199–216.
41. Hays, A. P., Hallett, M., Delfs, J., Morris, J., Sotrel, A., Shevchuk, M. M., and DiMauro, S. (1981): Muscle phosphofructokinase deficiency: Abnormal polysaccharide in a case of late-onset myopathy. *Neurology*, 31:1077–1086.
42. Hers, H. G. (1963): a-Glucosidase deficiency in generalized glycogen storage disease (Pompe's disease). *Biochem. J.*, 86:11–16.
43. Hewlett, R. H., and Gardner-Thorpe, C. (1978): McArdle's disease. What limit to age of onset? *S. Afr. Med. J.*, 1:60–62.
44. Hudgson, R., Gardner-Medwin, D., Worsfold, M., Pennington, R. J. T., and Walton, J. N. (1968): Adult myopathy from glycogen storage disease due to adult acid maltase deficiency. *Brain*, 91:435–460.
45. Huijing, F. (1975): Glycogen metabolism and glycogen storage diseases. *Physiol. Rev.*, 55:609–658.
46. Johnson, W. G. (1979): Principles of genetics in neuromuscular disease. In: *Practice of Pediatrics*, edited by V. C. Kelley, Harper and Row, Hagerstown, Md. pp. 73–95.
47. Justice, P., Ryan, C., Hsia, D. Y. Y., and Krmpotik, E. (1970): Amylo-1,6-glucosidase in human fibroblasts: studies of type III glycogen storage disease. *Biochem. Biophys. Res. Commun.*, 39:135–141.
48. Kahn, A., Etiemble, J., Meinhoefer, M. C., and Bovin, P. (1975): Erythrocyte phosphofructokinase deficiency associated with an unstable variant of muscle phosphofructokinase. *Clin. Chim. Acta*, 61:415–419.
49. Karadsheh, N. S., Uyeda, Y., and Oliver, R. M. (1977): Studies on the structure of human erythrocyte phosphofructokinase. *J. Biol. Chem.*, 252:3515–3524.
50. Kaur, J., and Layzer, R. B. (1977): Nonidentical subunits of human erythrocyte phosphofructokinase. *Biochem. Genet.*, 15:1133–1142.
51. Korenyi-Both, A., Smith, B. H., and Baruah, J. K. (1977): McArdle's syndrome. Fine structural changes in muscle. *Acta Neuropathol. (Berl)*, 40:11–19.
52. Kost, G. J., and Verity, M. A. (1980): A new variant of late-onset myophosphorylase deficiency. *Muscle Nerve*, 3:195–201.
53. Koster, J. F., Busch, H. F. M., and Slee, R. C. (1978): Glycogenosis type II: Infantile and late onset acid maltase deficiency observed in one family. *Clin. Chim. Acta*, 87:451–453.
54. Koster, J. F., and Slee, R. G. (1977): Some properties of human liver acid α-glucosidase. *Biochim. Biophys. Acta*, 482:89–97.
55. Koster, J. F., Slee, R. G., Daegelen, D., Meienhofer, M. C., Dreyfus, J. C., Niermeyer, M. F., and Fernandes, J. (1976): Isoenzyme pattern of phosphorylase in white blood cells and fibroblasts from patients with liver phosphorylase deficiency. *Clin. Chim. Acta*, 69:121–125.
56. Koster, J. F., Slee, R. G., Jennekens, G. I., Wintzen, A. R., and van Berkel, T. J. C. (1979): McArdle's disease: A study of the molecular basis of two different etiologies of myophosphorylase deficiency. *Clin. Chim. Acta*, 94:229–235.
57. Koster, J. F., Slee, R. G., Van der Klei-Van Moorsel, J. M., Rietra, P. J. G. M., and Lucas, C. J. (1976): Physico-chemical and immunological properties of acid α-glucosidase from various human tissues in relation to glycogenosis type II (Pompe's disease). *Clin. Chim. Acta*, 68:49–58.
58. Layzer, R. B., and Conway, M. M. (1970): Multiple isoenzymes of human phosphofructokinase. *Biochem. Biophys. Res. Commun.*, 40:1259–1265.
59. Layzer, R. B., and Rasmussen, J. The molecular basis of muscle phosphofructokinase deficiency. *Arch. Neurol.*, 31:411–417.
60. Layzer, R. B., Rowland, L. P., and Bank, W. J. (1969): Physical and kinetic properties of human phosphofructokinase from skeletal muscle and erythrocytes. *J. Biol. Chem.*, 244:3823–3831.

61. Layzer, R. B., Rowland, L. P., and Ranney, H. M. (1967): Muscle phosphofructokinase deficiency. *Arch. Neurol.*, 17:512–523.
62. Loonen, M. C. B., Busch, H. F. M., Koster, J. F., Martin, J. J., Niermeijer, M. F., Schram, A. W., Brouwer-Kelder, B., Mekes, W., Slee, R. G., and Tager, J. M. (1981): A family with different clinical forms of acid maltase deficiency (glycogenosis type II): Biochemical and genetic studies. *Neurology*, 31:1209–1316.
63. Martin, J. J., deBarsy, H., and DenTandt, W. R. (1976): Acid maltase deficiency in non-identical twins. A morphological and biochemical study. *J. Neurol.*, 213:265–275.
64. McArdle, B. (1951): Myopathy due to a defect in muscle glycogen breakdown. *Clin. Sci.*, 10:13–33.
65. McKusick, V. A., Howell, R. R., and Hussels, I. E. (1972): Allelism, non-allelism and genetic compound among the mucopolysaccharidoses. *Lancet*, 1:993–996.
66. Mehler, M., and DiMauro, S. (1976): Late-onset acid maltase deficiency. *Arch. Neurol.*, 33:692–695.
67. Mehler, M., and DiMauro, S. (1977): Residual acid maltase activity in late-onset acid maltase deficiency. *Neurology*, 27:178–184.
68. Meienhofer, M. C., Askanas, V., Proux-Daegelen, D., Dreyfus, J. C., and Engel, W. K. (1977): Muscle-type phosphorylase activity present in muscle cells cultured from three patients with myophosphorylase deficiency. *Arch. Neurol.*, 34:779–781.
69. Miranda, A. F., DiMauro, S., Antler, A., Stern, L. Z., and Rowland, L. P. (1981): Glycogen debrancher deficiency is reproduced in muscle culture. *Ann. Neurol.*, 9:283–288.
70. Miranda, A. F., Nette, G., Hartlage, P. L., and DiMauro, S. (1979): Phosphorylase isoenzymes in normal and myophosphorylase deficient heart. *Neurology*, 29:1538–1541.
71. Miranda, A. F., Somer, H., and DiMauro, S. (1979): Isoenzymes as markers of differentiation. In: *Muscle Regeneration*, edited by A. Mauro. Raven Press, New York, pp. 453–473.
72. Miranda, A., Trevisan, C., Shanske, S., Antler, A., and DiMauro, S. (1980): The expression of genetic defects in muscle cultures. Now you see it, now you don't. *Neurology*, 30:367.
73. Mitsumoto, H. (1979): McArdle disease: Phosphorylase activity in regenerating muscle fibers. *Neurology*, 29:258–262.
74. Miva, S., Sato, T., and Murao, H. (1972): A new type of phosphofructokinase deficiency: Hereditary nonspherocytic hemolytic anemia. *Acta Haematol. Jpn.*, 35:113–118.
75. Mommerts, W. F. H. M., Illingworth, B., Pearson, C. M., Guillory, R. J., and Seraydarian, K. (1959): A functional disorder of muscle associated with the absence of phosphorylase. *Proc. Natl. Acad. Sci. U.S.A.*, 45:791–797.
76. Murray, A. K., and Brown, D. H. (1979): The molecular heterogeneity of purified human liver lysosomal α-glucosidase (acid α-glucosidase). *Arch. Biochem. Biophys.*, 185:511–524.
77. Niermeijer, M. F., Koster, J. F., Jahodova, M., Fernandes, J., Heukels-Dully, M. J., and Galjaard, H. (1975): Prenatal diagnosis of type II glycogenosis (Pompe's disease) using microchemical analysis. *Pediatr. Res.*, 9:498–503.
78. Pilz, H., Goebel, H. H., Stefan, H., Seidel, D., and Kohlschutter, A. (1977): Biochemischdiagnostische Kreitien bei Glykogenose Typ II (Saure-Maltase-Defekt). *J. Clin. Chem. Clin. Biochem.*, 15:705–708.
79. Pompe, J. C. (1932): Over idiopatische hypertrophie van het hart. *Ned Tijdschr. Geneedskd.*, 76:304–311.
80. Pongratz, D., Schlossmacher, I., Koppenwallner, C., and Hubner, G. (1976): An especially mild myopathic form of glycogenosis type II. Problems of clinical and light microscopic diagnosis. *Pathol. Eur.*, 11:39–44.
81. Reuser, A. J. J., Koster, J. F., Hoogeveen, A., and Galjaard, H. (1978): Biochemical, immunological, and cell genetic studies in glycogenosis type II. *Am. J. Hum. Genet.*, 30:132–143.
82. Robbins, P. W. (1960): Immunological study of human muscle lacking phosphorylase. *Fed. Proc.*, 19:193.
83. Roelofs, R. I., Corbin, J., Peter, J. B., and Infants, E. (1974): A new variant of McArdle's disease. *Neurology*, 24:397.
84. Roelofs, R. I., Engel, W. K., and Chauvin, P. B. (1972): Histochemical demonstration of phosphorylase activity in regenerating skeletal muscle fibers from phosphorylase deficiency patients. *Science*, 177:795–797.
85. Rosenow, E. C., and Engel, A. C. (1978): Acid maltase deficiency in adults presenting as respiratory failure. *Am. J. Med.*, 64:485–491.

86. Rossignol, A. M., Meyer, M., Rossignol, R., Palcoux, M. P., Raynaud, E. J., and Bost, M. (1979): La myocardiopathie de la glycogenose type III. *Arch. Fr. Pediatr.*, 36:303–308.
87. Rowland, L. P. (1977): Glycogen storage disease of muscle. In: *Scientific Approaches to Clinical Neurology*, edited by E. S. Goldensohn and S. H. Appel. Lea and Febiger, Philadelphia, pp. 1692–1714.
88. Rowland, L. P., Araki, S., and Carmel, P. (1965): Contracture in McArdle's disease. *Arch. Neurol.*, 13:541–544.
89. Rowland, L. P., DiMauro, S., and Bank, W. J. (1973): Glycogen storage diseases of muscle. Problems in biochemical genetics. *Birth Defects*, 7:43–51.
90. Rowland, L. P., Fahn, S., and Schotland, D. L. (1963): McArdle's disease. *Arch. Neurol.*, 9:325–342.
91. Sato, K., Imai, F., Hatayama, I., and Roelofs, R. (1977): Characterization of glycogen phosphorylase isoenzymes present in cultured skeletal muscle from patients with McArdle's disease. *Biochem. Biophys. Res. Commun.*, 78:663–668.
92. Schmid, R., and Mahler, R. (1959): Chronic progressive myopathy with myoglobinuria: Demonstration of a glycogenolytic defect in muscle. *J. Clin. Invest.*, 38:2044–2058.
93. Serratrice, G., Monges, A., Roux, H., Aquaron, R., and Gambarelli, D. (1969): Forme myopathique du déficit en phosphofructokinase. *Rev. Neurol. (Paris)*, 120:271–277.
94. Shevchenko, A. M. (1977): McArdle's disease (a family observation). *Zh. Nevropathol. Psikhiatr. Korsakov.*, 76:655–659, 1976. *Muscular Dystrophy Abstr.*, 21:213.
95. Smith, J., Zellweger, H., and Afifi, A. K. (1967): Muscular form of glycogenosis type II (Pompe): Report of a case with unusual features. *Neurology*, 17:537–549.
96. Solomon, E., Swallow, D., Burgess, S., and Evans, L. (1979): Assignment of the human acid alpha-glucosidase (aGLU) to chromosome 17 using somatic cell hybrids. *Ann. Hum. Genet. (London)*, 42:273–281.
97. Soyama, K., Ono, E., Shimada, N., Tanaka, K., Oya, N., and Kusunoki, T. (1977): Properties of the alpha-glucosidase from various human tissues in relation to glycogenosis type II (Pompe's disease). *Clin. Chim. Acta*, 78:473–478.
98. Swaiman, K. F., Kennedy, W. R., and Sauls, H. S. (1968): Late infantile acid maltase deficiency. *Arch. Neurol.*, 18:642–648.
99. Swallow, D. M., Corney, G., Harris, H., and Hirschhorn, R. (1975): Acid α-glucosidase: A new polymorphism in man demonstrated by affinity electrophoresis. *Ann. Hum. Genet. (London)*, 38:391–406.
100. Tanaka, K., Shimazu, S., Oya, N., Romisawa, M., Kusunoki, T., Soyama, K., and Ono, E. (1979): Muscular form of glycogenosis type II (Pompe's disease). *Pediatrics*, 63:124–129.
101. Taniguchi, N., Kato, E., Yoshida, H., Iwaki, S., Ohki, T., and Koisumi, S. (1978): Alpha-glucosidase activity in human leucocytes: Choice of lymphocytes for the diagnosis of Pompe's disease and the carrier state. *Clin. Chim. Acta*, 89:293–299.
102. Tarui, S., Kono, N., Kuwajima, M., and Ikura, Y. (1978): Type VII glycogenosis. Muscle and erythrocyte phosphofructokinase deficiency. *Monogr. Hum. Genet.*, 9:42–47.
103. Tarui, S., Kono, N., Nasu, T., and Nishikawa, M. (1969): Enzymatic basis for the coexistence of myopathy and hemolytic disease in inherited muscle phosphofructokinase deficiency. *Biochem. Biophys. Res. Commun.*, 34:77–83.
104. Tarui, S., Okuno, G., Ikua, Y., Tanaka, T., Suda, M., and Nishikawa, M. (1965): Phosphofructokinase deficiency in skeletal muscle. A new type of glycogenosis. *Biochem. Biophys. Res. Commun.*, 19:517–523.
105. Van Hoof, F., and Hers, H. G. (1967): The subgroups of type III glycogenosis. *Eur. J. Biochem.*, 2:271–274.
106. Vora, S., Corash, L., Engel, W. K., Durham, S., Seaman, C., and Piomelli, S. (1980): The molecular mechanism of the inherited phosphofructokinase deficiency associated with hemolysis and myopathy. *Blood*, 55:629–635.
107. Vora, S., Durham, S., and Piomelli, S. (1979): Isozymes of phosphofructokinase in human blood cells: Molecular and genetic evidence for a multigenic system. *Blood*, 54:35a.
108. Waterbury, L., and Frenkel, L. E. P. (1972): Hereditary nonspherocytic hemolysis with erythrocyte phosphofructokinase deficiency. *Blood*, 39:415–425.
109. Williams, J. C., and Murray, A. K. (1980): Enzyme replacement in Pompe disease with an a-glucosidase-low density lipoprotein complex. *Birth Defects: Original Article Series*, 16:415–423.
110. Witzleben, C. L. (1969): Renal cortical tubular glycogen localization in glycogenosis type II (Pompe's disease). *Lab. Invest.*, 20:424–429.

Genetics of Neurological and Psychiatric Disorders, edited by Seymour S. Kety, Lewis P. Rowland, Richard L. Sidman, and Steven W. Matthysse. Raven Press, New York © 1983.

Genetic Predisposition to Environmental Factors

Jack P. Antel and Barry G. W. Arnason

Department of Neurology, University of Chicago Pritzker School of Medicine and Division of Biological Sciences, Chicago, Illinois 60637

The environment may be defined as the sum of all the conditions and elements that surround and act on an individual. In this chapter, we shall discuss interactions between genetically determined traits and selected environmental factors that affect the expression of diseases of brain, nerve, and muscle. Such interactions can be either positive or negative and are frequently multifactorial. They include both the influence of environmental factors on the expression of genetically determined defects and the role of gene products in conferring not only susceptibility, but also resistance to environmentally determined diseases. Examples for discussion have been chosen to illustrate general principles of gene-environment interplay. Special emphasis has been placed on the genetics of the immune system and on the role of immunogenetic endowment in modulating response to pathogens and immunogens that cause nervous system disease.

INFLUENCE OF ENVIRONMENTAL FACTORS ON EXPRESSIONS OF GENETICALLY DETERMINED DEFECTS

Physical Factors

Stress, fever, and exercise can provoke clinical flare-ups of several muscle disorders due to defined or suspected inborn metabolic errors. Classic examples include the muscle necrosis and myoglobinuria in McArdle's disease (glycogen phosphorylase deficiency) and in disorders of fatty acid metabolism or transport (e.g., carnitine palmityl transferase deficiency) (61). In these disorders, the association between physical factors and disease is obvious because of the explosively sudden and dramatic clinical presentation. In other conditions, the clinical picture evolves slowly or subtly or only becomes apparent as many minor events are summed over time, and the relationship between physical factors in the environment and clinical disease may not be appreciated. Thus, the notion that mental disorders, for example, may be provoked by stress should be given the attention it deserves. Stress, possibly mediated by sympathetic innervation of the lymphoid organs, exerts substantial effects on immune response and could impose an ancillary or adjuvant effect in diseases of the nervous system in which immune regulatory mechanisms have a role, as discussed below (45).

Drugs

One person in 3,000 exhibits a bizarre response to the short-acting, depolarizing neuromuscular blocking agent succinyl choline (21). The response consists of profound muscular relaxation with apnea and lasts for several hr. Succinyl choline is degraded by plasma cholinesterase, and the basis of the abnormality lies in an atypical plasma cholinesterase that hydrolyses succinyl choline 100 times more slowly than normal. The defect can be detected by the so-called dibucaine test; the abnormal protein is inhibited much less than normal. The condition is inherited in an autosomal recessive manner. Heterozygotes produce both normal and abnormal cholinesterase—the normal product suffices to hydrolyse succinyl choline at a normal or near normal rate. The abnormal gene is present in about 2% of population groups studied throughout the world. This finding implies a total lack of selective pressures related to geographic or ethnic differences for this allelic defect. This rare condition illustrates that totally silent genetic defects that are inconsequential under ordinary life circumstances can be "unmasked" by medical intervention with potentially lethal results.

A second syndrome that can be unmasked by succinyl choline, halothane, and other anesthetic agents is malignant hyperthermia (25). During anesthesia, muscular rigidity occurs with massive production of lactic acid, an accompanying steep drop in blood pH level, high fever, tachycardia, ventricular arrhythmia, hyperventilation, and hypotension. The syndrome carries a mortality of 60%. Several distinct conditions predispose to malignant hyperthermia. The most common is a dominantly inherited myopathy that is usually silent clinically (30). Nonetheless, carriers of the trait often have elevated serum creatine phosphokinase (CPK) levels and many show myopathic changes in muscle on biopsy (46). Abnormally enhanced contractility of biopsied muscle to caffeine has been employed as a diagnostic test to detect family members at risk for malignant hyperthermia. The defect appears to involve calcium release mechanisms triggered by anesthetic agents with excess intracellular calcium ions. Caffeine increases myoplasmic calcium content.

The malignant hyperthermia syndrome has also been described in dominantly inherited myotonia, in central core disease, and in boys of short stature with cryptorchism, lumbar lordosis, thoracic kyphosis, and a myopathy. This last entity appears to be a sex-linked recessive. A model disease occurs in Landrace pigs (49). Malignant hyperthermia illustrates the conversion of a mild myopathic defect into a lethal entity when the appropriate environmental agents are given.

Acute intermittent porphyria (AIP), an autosomal dominant disorder with variable penetration, can present as a subacute peripheral neuropathy, as autonomic dysfunction, or as a cerebral disorder, including aberrant behavior and hypothalamic disturbance (4). Although how the neurologic dysfunction arises is unknown, the basic inherited defect involves the enzyme uroporphyrinogen-I-synthetase. This enzyme defect results ultimately in decreased heme production. Secondary to this reduction, the gene coding for the enzyme δ-amino levulinic acid (ALA) synthetase, an early step in heme synthesis, is derepressed. The resultant increased δ-ALA synthetase activity is associated with the clinical syndrome in an as yet undetermined manner. Phenobarbital and other drugs that further induce this enzyme activity exacerbate the clinical disease. The potentiating effects of phenobarbital on AIP illustrate that usually benign drugs in the susceptible host can induce serious clinical disorders and emphasize the need for caution in the use of pharmacologic agents.

Dietary Factors

The autosomal recessive disorder of phenylketonuria (PKU) (Type 1 hyperphenylalaninemia) illustrates not only the deleterious effect of a normal dietary constituent on a genetically predisposed individual, but also the great clinical benefit to be derived from proper dietary management (65). Before the advent of neonatal detection and treatment of PKU, most individuals with PKU exhibited profound retardation, seizures, behavior disorders, and growth retardation. Restricted phenylalanine diet for infants with classic PKU can prevent the neurologic sequelae. Mental development is inversely related to the age at which diet management is initiated, with the best outcome predicted if therapy is begun before age 2 months. Dietary therapy has now extended to other disorders of amino acid metabolism.

Wilson's disease (hepatolenticular degeneration) is another example in which dietary control contributes to effective disease management. In Wilson's disease, an autosomal recessive disorder, development of the neurologic syndrome is associated with deposition of excess copper in extrahepatic sites, including the brain, and invariably in Descemet's membrane of the eye (Kayser-Fleischer rings). Although the precise role for copper in inducing the classic extrapyramidal syndrome is not yet defined, effective clinical control is possible using copper-chelating agents, usually D-penicillamine and a low-copper diet (76). Wilson's disease and amino acid disorders exemplify how controlling an environmental (nutritional) factor can permit successful management of some inherited disorders.

ROLE OF GENE PRODUCTS ON CONFERRING SUSCEPTIBILITY OR RESISTANCE TO ENVIRONMENTALLY DETERMINED DISEASES

Nutritional Deficiencies

Blass et al. (11) demonstrated a role for genetic factors in determining susceptibility to development of Wernicke-Korsakoff's (W-K) syndrome, which arises secondary to thiamine deficiency and is most commonly found in malnourished alcoholics. Blass et al. (11) found that the enzyme transketolase, measured in cultured fibroblasts from 4 chronic alcoholics with W-K syndrome, had genetically determined reduced avidity for the coenzyme thiamine pyrophosphate. This genetic defect results in a greater than expected decline in enzyme activity as thiamine levels fall in malnutrition and makes the individuals more likely to develop the clinical features of W-K syndrome. It is not known whether these findings can be extended either to other neurologic disorders arising in malnourished alcoholics, such as central white matter destruction (Marchiafava-Bignami, central pontine myelinolysis), peripheral neuropathy, or myopathy, or to other vitamin deficiency disorders associated with neuropsychiatric sequelae, such as pellagra (niacin) and subacute combined degeneration of the spinal cord (vitamin B-12).

Drug Reaction and Toxicity

Several genetic mechanisms underlie an individual's abnormal reactivity to a particular drug, i.e., an idiosyncratic reaction. Drug-provoked clinical flare-up of genetic defects has been previously discussed. Drug toxicity can appear in individuals with genetic incapacity to metabolize effectively particular drugs. For example, isoniazid (INH) is acetylated in the liver (56). Individuals with genetically reduced amounts of liver acetyl transferase, when treated with the usual doses of INH, accumulate toxic levels of the drug and become prone to clinical toxicity, including peripheral neuropathy. Certain immune-mediated drug reactions dependent on immunogenetic factors will be discussed later. Individuals also vary in sensitivity to particular drugs, again on a genetic basis. For example, sensitivity to mydriatic agents varies markedly among racial groups (13). Such genetic considerations further indicate the complexity of factors influencing drug responses and the need for careful monitoring of drug therapy.

IMMUNOGENETIC FACTORS

Major Histocompatibility Complex

The first evidence that specific genetically determined immune traits modulate immune function was provided by studies showing that allogeneic graft rejection and susceptibility to particular oncogenic viruses in mice were controlled by specific gene loci contained within a defined chromosomal region, subsequently termed the major histocompatibility complex (MHC) (42). The MHC, often termed a "supergene," contains multiple gene loci subserving multiple immune-related functions. In mice, the MHC is termed the H_2 region and is found on chromosome 17; in humans, it is termed the HLA (human leukocyte antigen) region and is located on the short arm of chromosome 6.

MHC Antigens

Among the MHC genes are those encoding for the MHC antigens, which are glycoproteins expressed on the cell surface of lymphocytes and in variable amounts on other nucleated cells (including neural cells). In mice, genes within the H and K regions of the H_2 complex encode for these antigens. In humans, the currently defined loci encoding for HLA antigens include the A, B, and C loci which control antigens that are defined by serologic techniques using antisera specific for each

HLA antigen. D locus antigens are detected on T lymphocytes by means of an operationally cumbersome mixed lymphocyte reaction. The human DR locus antigens, akin to the antigens determined by the Ia region of the mouse, are expressed on B cells and activated T cells and can be serologically defined.

Multiple alleles are found at each of these loci. Standardized nomenclature has been adopted for each determinant. A given HLA antigen is identified by the locus of origin and an arbitrary number. It is likely that many allelic determinants at known gene loci and perhaps other loci remain to be defined.

The several MHC loci are in close proximity on the chromosome and the specific determinants at each locus are only infrequently separated by crossover between homologous chromosomes during meiosis. This close approximation is termed genetic linkage. The HLA-D and -DR loci are the most tightly linked loci. The HLA genes (loci) are in genetic linkage with other gene loci within the MHC that code for numerous immune traits, so that HLA antigens act as genetic markers for these traits. The combination of HLA antigens determined by one chromosome constitutes a haplotype. Every individual inherits one haplotype from each parent. In routine HLA typing of unrelated individuals, we cannot determine which HLA antigens comprise the haplotype; family analysis is required to obtain such data.

There are many questions about the frequency of particular HLA antigens and haplotypes within different populations. The frequency with which specific HLA antigens occur in different racial groups varies markedly. The frequency with which an antigen of one locus is associated with a specific antigen of a second locus is not predicted by chance, and specific combinations occur more frequently than expected—i.e., linkage disequilibrium exists. The explanation of these findings is not precisely known. Perhaps certain HLA haplotypes only recently have entered the population and have not yet been diluted out by further crossovers. Alternately, a particular HLA antigen or haplotype may offer selective genetic advantage or disadvantage, possibly related to resistance to a given disease. For instance, members of a Dutch colony that had emigrated to South America in the 19th century had only 1% HLA-B8-positive individuals, compared to 26% in the native Dutch population. It is possible that, during the known epidemics occurring among the emigrants, B8-positive individuals were selected against. HLA frequency variations among different racial and geographic populations could reflect similar selective pressures.

The explanation for the persistence of polymorphism within the HLA system is also unknown. Polymorphism may prevent viruses or other pathogens from adapting either to become optimally suited to interact with a universal HLA antigen that acts as a receptor or to escape immune surveillance by MHC region-determined immune reactions.

Mechanisms of MHC-Determined Control of Immune Response to External Antigens

A major aim of the immune system is to recognize self from nonself and to destroy foreign antigens in a controlled fashion, sparing self-antigens. How MHC-determined gene products control immune reactivity continues to be defined as our knowledge of the overall organization of the immune system expands.

MHC antigens as receptors

The role for MHC antigens as receptors for external antigens, particularly viruses, is under study. Helenius et al. (27) demonstrated that HLA-A and -B loci antigens in humans and H-2K and -D loci antigens in mice can act as cell surface receptors for Semliki forest virus. Haspel et al. (26), however, demonstrated that, for measles virus, HLA antigens neither serve as receptors nor are incorporated into the maturing viral particle. Furthermore, measles virus infection of human lymphoblastoid cells did not diminish the concentration of HLA antigens; in contrast, cell infection with poliovirus and vesicular stomatitis virus did reduce the concentration by inhibiting host protein synthesis (26). Pellegrino et al. (54) suggested that virus infection (vesicular stomatitis virus) may alter expression of HLA antigens (i.e., appearance of different or altered HLA determinants). Reduction or alteration of HLA antigens on the cell surface may interfere with virus-cell interaction and alter host susceptibility to infection by specific agents.

HLA antigens may themselves cross-react with external antigens and become the target of an immune response. Cross-reactivity between Klebsiella antigen and B27 positive lymphocytes from patients with ankylosing spondylitis (AS) (20) is suggested because spondyloarthropathies occur in HLA-B27-positive individuals after dysentery epidemics and because anti-Klebsiella antiserum selectively lyses/B27-positive cells. Filtrates from Klebsiella cultures can modify B27-positive lymphocytes from patients without AS so that they are lysed by the antiserum. These results suggest that environmental agents can modify HLA antigens, thus influencing susceptibility to disease. In addition to such distinct properties (private determinants) of each HLA antigen, some HLA antigens share certain common structures (public determinants—e.g., B5, 15,18,35,21). These common structures may also become the target of both humoral and cellular immunity. There are also structural similarities among HLA determinants, B2 microglobulin, and immunoglobulins.

MHC Restriction of the immune response

MHC gene products play a central role in determining that an immune effector cell response is restricted to a specific target cell. *In vitro*, cytotoxic cells that develop after exposure to viral-infected cells are specific for both the virus (e.g., influenza) (44) and the HLA haplotype of the cell. That the HLA antigen itself is a recognition site on the target cell has been established by showing that monoclonal anti-HLA antibody blocks lysis of the target cell. Some HLA antigens (e.g., B7 and B8) result in consistent recognition (i.e., recognized as self), whereas others, A2 and 3, do not (44). These differences could influence the specificity and magnitude of a cytotoxic response. The MHC gene products that most influence cytotoxic specificity are probably linked more closely to the HLA-A and -B (H2-K and -D) regions than to the DR region (Ia region).

Several hypotheses have been proposed to explain how MHC-mediated restriction of T cell responses occurs. For instance, there could be a dual recognition system in which both the MHC antigen and foreign antigen are recognized independently. Alternately, Bodmer (12) suggested that after antigen binding to the T cell receptor, the receptor undergoes a conformational change. An interaction could then occur between HLA antigens and the T cell receptor, permitting a response such as killing the target cell or triggering it to divide. If certain T cell receptors interact with only certain HLA antigens, this results in restriction of response without requiring specific recognition of HLA specificities by cells (12).

Immune response genes

Immune response genes are autosomal dominant genes that control the level of immune response to specific protein antigens. Initial observations, made with well-defined simple antigens and inbred strains of animals, indicated that the magnitude of antibody response was determined by specific gene loci on the I region of the MHC (7). With more complex antigens, multiple immune response genes may determine the ultimate response. Immune response genes probably promote antigen recognition and presentation of antigens to T cells by encoding for helper cells and by augmenting cell-cell interaction within the immune system (see later section). As mentioned, HLA antigens are in genetic linkage with MHC-immune response genes; this linkage probably accounts for most associations between HLA antigens and both immune reactivity and disease susceptibility.

Immune suppressor genes

One mechanism of immune nonresponse to external antigens is mediated via discrete subsets of cells termed suppressor cells. These cells also exhibit antigen specificity. Passive transfer of suppressor cells can induce immune unresponsiveness in the recipient. The MHC region (Ij subregion in the mouse) is probably the major site of gene loci encoding for suppressor cells. The molecules mediating the suppressor effect appear themselves to be encoded by loci within the Ij subregion (73).

Cell-cell interaction

An effective immune response requires interaction among several cell subsets. Required interactions include macrophage-T cells during antigen presentation, macrophage-B cells, and T regulator cells with B and T effector cells. The Ia (DR in humans) MHC antigens (found on macrophages, B cells, and some T cells) play particularly vital roles in these processes. For example, the interaction between human macrophages and T cells, in terms of antigen presentation, as studied with herpes simplex virus, requires HLA-D (DR) region (equivalent to Ia region) homology (24). The role of MHC antigens in restricting the immune response has already been discussed.

Non-MHC Immunogenetic Factors

MHC-encoded products do not entirely determine immune recognition and immune reactivity. Antigen-specific non-MHC-specific cytotoxicity responses do occur (such as to the hapten trinitrophenyl), suggesting there are important non-MHC cell surface determinants (10). Similarly, T cells infected by Epstein-Barr (EB) virus can lyse HLA-mismatched EB virus-infected targets. Non-MHC loci also influence nonresponsiveness, such as the nonresponse to lipopolysaccharide in C3H/HeJ mice that results from a determinant on chromosome 4 (9).

Caution also needs to be raised that all inherited traits are truly such. For instance, Gorczynski et al. (22) demonstrated that acquired tolerance to foreign H_2 antigens in mice could be transmitted to the F1 generation.

This role of genetically determined MHC traits in immune function may account for differences among individuals in response to environmental agents and susceptibility to disease. MHC-determined properties influence recognition of antigens and initiation of immune responses. Differences in cell surface MHC antigens may determine whether the particular pathogen can utilize the antigen as a receptor or whether immune cross-sensitivity may occur between them. Some native or altered surface antigens could predispose to defective recognition of self versus nonself, leading to autoimmune phenomena. As discussed, some MHC antigens are more strongly recognized as self-antigens than others. MHC-determined differences in cell-cell interactions and in the balance of helper and suppressor cells determine immune capacity and efficiency. Heterogeneity of all these MHC-encoded factors would result in heterogeneity among individuals in terms of immune response to external agents.

HLA-Linked Immune Responses in Humans

The levels of immune reactivity to defined external antigens vary among people. In at least some cases, these variations are linked to HLA phenotype. The immune response of individuals to several bacterial antigens (including mycobacterium leprae, tetanus toxoid, and streptococcus) and allergens are HLA linked. Whether exposure to M. leprae results in a lepromatous or tuberculoid response may be determined by the HLA type of the individual (75). Japanese people with A9-B5-DHO haplotype showed a low response to tetanus toxoid, even though most were not homozygous for this haplotype (63). This finding in heterozygotes suggests that the reduced response is due to an HLA-linked immune suppressor gene rather than altered interactions of the antigen with the cell. Greenberg et al.'s (23) analysis of responses to a purified extracellular antigen from a group A streptococcus indicated that either (a) complementation between at least two HLA-linked immune response genes was required for the response, a finding analogous with previous studies in mice, or (b) the response depended on one HLA-linked locus and one non-HLA-linked locus (23). The response to some potent allergens, e.g., Ra5 antigen, of ragweed pollen also link in a complex manner to HLA type (41).

Immune reactivity to some viral antigens also correlates with particular HLA antigens. The hemagglutination inhibition antibody response to intranasal inoculation with live attenuated influenza A virus in individuals bearing HLA-BW16 is less than in W16-negative individuals (69). W16-positive individuals vaccinated intramuscularly with inactivated virus show a normal response, indicating that the reduced response to intranasal inoculation reflects HLA-linked interference with cellular attachment of the virus. In this regard, Dausset et al. (16) found that fibroblasts from HLA-identical siblings showed similar patterns of infectivity to a panel of viruses, whereas an HLA-nonidentical sibling's fibroblasts differed markedly. DeVries et al. (17) reported that the responses of CW3-positive individuals to vaccinia immunization measured in terms of both antibody and cellular immunity (lymphocyte transformation) were less than CW3-negative individuals. Several possible explanations can be proposed for these differences: (a) altered interaction between virus and its receptors in CW3-positive individuals; (b) molecular memory—i.e., if the external antigen and the HLA antigen have the same specificities, it is unlikely that an immune response would be raised against the external antigen; (c) presence of immune suppressor genes in CW3-positive individuals; and (d) enhanced response in CW3-negative individuals. The antibody response to herpes simplex virus I (HSVI) and the incidence of herpes labialis may be reduced in DR-W2-positive individuals (51).

The antibody response to measles virus as a function of HLA constitution of the individual has been studied extensively (3,8,19). Arnason et al. (3) found that serum antibody titers were increased in HLA-A3, -B7 individuals. These antigens are the ones overrepresented in patients with multiple sclerosis (MS) and may account for the elevated measles serum [and cerebrospinal fluid (CSF)] antibody titers characteristic of MS patients.

HLA-linked differences in immunoreactivity also occur in response to nonpathogens and nonspecific immune activators. The levels of antibovine serum albumin antibodies are linked with specific HLA types (66). Hyporeactivity to the predominantly T cell mitogen concanavalin A (Con A) is suggestively linked with HLA-B7 and may partially account for the reduced mitogenic response in MS patients. Paty et al. (52) reported that antibody response to pokeweed mitogen, as measured by a plaque-forming cell assay, was reduced in DR-W2 individuals. The mechanism of these altered responses is also speculative.

HLA-linked Disease Susceptibility and Resistance

Most clearly HLA-associated diseases share certain features: inflammation is common; familial incidence is increased, but not in a simple Mendelian inheritance pattern; and most individuals bearing the HLA type do not develop the disease. Most neurologic disorders associated with HLA types are of known or suspected viral or autoimmune etiology. Among "viral" diseases, HLA associations have been reported for poliomyelitis (HLA-A3 and -B7), subacute sclerosing panencephalitis, and Jakob-Creutzfeldt disease (Table 1). Variable association has also been reported in leprosy. Among diseases of probable immune-mediated etiology that show linkage with HLA antigens are MS, myasthenia gravis (MG), relapsing polyneuritis (but not the usual form of Guillain-Barre syndrome), and adult and childhood forms of dermatomyosistis, or polymyositis. Such findings of HLA linkage with immune-

TABLE 1. *HLA-associated neurologic and psychiatric disorders*

Disease	HLA association	References
MS	A3-B7-D2-DR-W2	Reviewed in 47
MG		14,18,62
young adult	A1-B8-DR-W3	
elderly	A2, A3	
Alzheimer's disease	CW3[a], B15, ? none	28,60,72
Schizophrenia	A9[a], A28	43
paranoid	A28[b]	
manic-depressive disease	B5[a], B13	70
SSPE[e]	AW29[a]	35,36
Jakob-Creutzfeldt	A2[a], A28	34
paralytic poliomyelitis	A3[a], B7	37,57
ALS	A3, B35, B40 (B12)[c]	2,5,29,32,33,53
Guillain-Barre syndrome	—	38
relapsing polyneuritis	A1-B8-DR-W3[a]	1
polymyositis—adult	B8, B14[d]	6,15
dermatomyositis—juvenile	B14[a]	48
leprosy		59,75
lepromatous	B8[a],B14	
tuberculoid	A9[a], BW-21	

[a] Findings restricted to a single study.
[b] Data pooled from four studies.
[c] More than one study report each of these findings; negative studies also reported.
[d] ? if B8 and B14 show cross-reactivity.
[e] SSPE-Subacute sclerosing panencephalitis.

mediated disease parallel those linking susceptibility to experimental allergic encephalomyelitis and lymphocytic choriomeningitis virus with MHC traits of animals.

The association of these viral and immune-mediated diseases with HLA phenotypes has resulted in attempts to link other nervous and mental diseases of unknown etiology with HLA type, which would provide evidence of viral-immune etiology. Among the diseases studied are schizophrenia and manic-depressive disease, but with no consistent results yet found (43,70). Alzheimer's disease has been only inconsistently linked with HLA type (28). No data imply that specific HLA types are associated with longevity in general.

Establishing both the existence and significance of linkage between a disease and HLA antigens can be difficult. In numerous diseases, results have been inconsistent, and in no disease is the association 100%. These variations probably reflect statistical, methodological, and biological variables. Since multiple variables (HLA antigens) are being measured in limited numbers of patients in disease-related HLA studies, chance association may reach statistical significance. Diseases of unknown etiology may be heterogenous entities and strong linkage with a single HLA type would not be expected. Conversely, numerous immune response traits may predispose to the development of a single disease, as will be discussed in MS.

Weak linkage between an HLA type and disease may result from inadequacy of the HLA-typing sera used. As typing sera that define new alleles at known loci or at new loci become available, stronger associations between HLA types and disease can be expected. This postulate is demonstrated by the findings in MS. Initially, linkage was found with the A(A3) and B(B7) locus antigens; however, much stronger linkage exists with the DR locus antigen DR-W2. The currently used typing sera may not be absolutely specific for a single HLA antigen, but may detect closely related alleles. As better-defined typing sera become available (e.g., monoclonal antibody), one can anticipate finding clearer associations.

Development of disease may depend not only on susceptibility-inducing gene products, but also on genetically determined resistant factors. Analysis of haplotypes in multiple diseases (74), including AS, MS, and amyotrophic lateral sclerosis (ALS), indicated that some haplotypes containing a given HLA antigen are associated with increased incidence of a disease, whereas other haplotypes with the same antigen are associated with the reduced incidence of the disease. In ALS, HLA-B12 is associated with slow progression of disease, whereas in more rapid forms of ALS, we and others have found overrepresentation of HLA-A3 (2). In MS, the haplotype from the sixth chromosome (which does not contain the putative disease-predisposing immune response genes) may also modulate the development and course of the disease (47).

Among neurologic diseases, the most consistent associations between HLA antigens and disease are found in MG and MS. The HLA-related studies in these two entities indicate the complexities involved in performing HLA analysis on heterogenous populations with possible heterogenous diseases.

At least two subgroups of patients can be identified among MG patients, namely young women without thymoma and elderly adults, more frequently men. Both groups show circulating antiacetylcholine receptor (AchR) antibody, but in our experience, only the latter groups show cell-mediated immunity to AChR (62). In the younger group, HLA-B8 and DR-W3, which is in genetic disequilibrium with A8, are overrepresented, whereas in the older group, either HLA-A2 or -A3, or both, may be overrepresented. HLA-B8 is also overrepresented in other autoimmune disorders that affect young women, such as rheumatoid arthritis. These findings are consistent with the postulate that MG is comprised of two discrete diseases or that the immunogenetic constitution modulates the time of appearance and course of disease. In addition, patients with restricted ocular MG do not have the overrepresentation of some HLA types seen in generalized MG. We (62) found no correlation between HLA type and the presence of cell-mediated immunity to AChR, but Compston et al. (5) found that antibody levels were highest in HLA-B8-positive patients. How immunogenetic factors contribute to the development of MG is unknown, but susceptibility to experimental acute MG in mice is linked with the H_2 constitution of the animal.

In MS, some HLA antigens are overrepresented in selected populations and some antigens are underrepresented. HLA-A3, -B7, and DW2 are consistently found to be overrepresented in the North European caucasian and North American MS populations. The increased frequency of DW2 and DR-W2 exceeds that of A3 and B7, suggesting that the disease susceptibility gene is more closely linked with the DR locus. HLA-B8 is also somewhat overrepresented, but HLA-B12 is underrepresented. Oger et al. (47) and Stendall-Broden et al. (71) reported that MS associated with B8 is frequently more benign and less associated with an increase in CSF immunoglobulin than in B7-positive individuals. Madigan et al. (40) found that B7 correlated with the relapsing form of the disease but that B8 tended to correlate with the chronic progressive form of the disease. Arnason et al. (3) found that patients with optic neuritis, but without evidence of MS, did not show an overrepresentation of HLA-A3 or -B7. In non-Northern European caucasians (e.g., Israeli Jews), there was no association with A3-B7-DW2, while BW40 was increased. In Jordanian MS patients, DW4 was increased. In the Japanese, the incidence of MS linked most strongly with DR-W1. These data indicate that both disease susceptibility and resistance factors are determined by immunogenetic factors and that, in different populations, either different immune response genes are operative or genetic linkage patterns differ.

Studies of families in which several members have MS are informative in immunogenetic studies because haplotype-identical and -nonidentical individuals can be recognized; haplotype-identical individuals are likely to share immune response traits along the entire MHC region. These studies are hindered by the lack of precision in diagnosis, particularly in "unaffected" family members. Overall, familial studies (47) have failed to establish any consistent segregation of HLA types in affected and unaffected members, suggesting that no single gene in the HLA complex determines susceptibility to MS. Immunogenetic predisposition to MS probably depends on the interplay of multiple "susceptibility" and "resistance" factors.

HLA antigens can also serve as genetic markers for nonviral, nonimmune-mediated diseases. In some, but not all, families with autosomal dominant spinocerebellar degenerations, there is strong linkage between disease inheritance and HLA types (31). Linkage with HLA type is also suggested in neural tube defects (58). In mice, the MHC (H_2) region is in close linkage with developmental genes located at the T locus, and MHC gene products serve as genetic markers for developmental genes. Similar considerations probably apply in humans.

HLA-Linked Therapeutic and Toxic Drug Responses

Both therapeutic and toxic reactions to drugs have been linked with specific HLA antigens. Smeraldi et al. (67,68) found a positive correlation between HLA-A1 in schizophrenic patients and a positive response to chlorpromazine, whereas the presence of HLA-A2 was perhaps inversely correlated. They suggested that specific HLA antigens modify monoaminergic cell receptors, increasing their avidity for binding of chlorpromazine. Perres et al. (55) observed an increased relapse rate in HLA-A3-positive patients with major affective disorders treated with lithium, suggesting that A3 may interfere with lithium cell binding and transport. These studies implicate a direct role for HLA antigens in determining therapeutic effects of drugs.

HLA-linked drug toxicity has been reported with levamisole (64), a drug unsuccessfully utilized in MS and with gold salt or D-penicillamine therapy for rheumatic diseases. Panayi et al. (50) found that DR-W2 or -W3-positive rheumatoid patients, taking one or both of the latter drugs, had toxic reactions more frequently than did DR-W2 and -W3-negative patients. In our experience (Fig. 1), the risk of toxicity from gold salts was markedly increased in HLA-B12-positive patients (39). The toxic manifestations of gold and D-penicillamine are believed to be immune mediated. The HLA antigens linked to toxicity are probably genetic markers for specific immune response genes. Potentially, linkage between HLA antigens and response to therapeutic agents could aid in selecting appropriate drug regimens.

CONCLUSIONS

The interactions between genetic properties of an individual and environmental factors determine the capacity of the individual to function within the environment. These interactions may be either positive or negative and often involve multiple factors. Environmental factors can cause known or unsuspected genetic defects to express themselves. Genetic factors modulate both susceptibility and resistance to the effects of external agents. The complexity of genetic-environmental interactions would predict heterogeneity of clinical expression of diseases in which both sets of factors are operative. The more precisely the nature of these interactions are

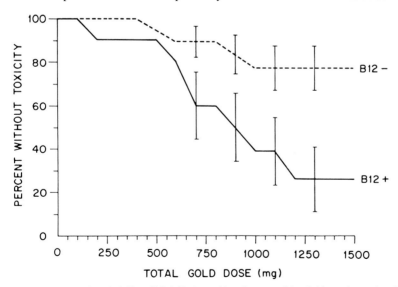

FIG. 1. The increased probability of HLA-B12-positive rheumatoid arthritis patients developing adverse responses to gold salt therapy compared to B12 negative patients, as a function of cumulative drug dosage *(abscissa)*, is shown. Data are presented as a Kaplan-Meier survival plot.

defined for a specific clinical entity, the greater the potential to develop effective therapeutic measures.

ACKNOWLEDGMENTS

We wish to thank Ms. Ann Yarbrough and Dr. Joel J.-F. Oger for their help in preparation and review of the manuscript. This work is supported in part by Grant MS:RG 1130-C-16 from the National Multiple Sclerosis Society, the Muscular Dystrophy Association, and NIH Grant RO1-AGO1798 from the National Institutes of Health.

REFERENCES

1. Adams, D., Festenstein, H., Gibson, J. D., Hughes, R. A. C., Jaraquemada, J., Papasteriadis, C., Sachs, J., and Thomas, P. K. (1979): HLA antigens in chronic relapsing idiopathic inflammatory polyneuropathy. *J. Neurol. Neurosurg. Psychiatry*, 42:184–186.
2. Antel, J. P., Medof, M. E., Richman, D. P., and Arnason, B. G. W. (1979): Immunologic considerations in amyotrophic lateral sclerosis. In: *Clinical Neuroimmunology*, edited by F. C. Rose, pp. 227–244. Blackwell Scientific Publications, London.
3. Arnason, B. G. W., Fuller, T. C., Lehrich, J. R., and Wray, S. H. (1974): Histocompatibility types and measles antibodies in multiple sclerosis and optic neuritis. *J. Neurol. Sci.*, 22:419–428.
4. Becker, D. M., and Kramer, S. (1977): The neurological manifestations of porphyria: A review. *Medicine*, 56:411–423.
5. Behan, P. O., Durward, W. F., and Dick, H. (1976): Histocompatibility antigens associated with motor neuron disease. *Lancet*, 2:803.
6. Behan, W. M. H., Behan, P. O., and Dick, H. (1978): HLA-B8 in polymyositis. *N. Engl. J. Med.*, 298:1,260–1,261.
7. Benacerraf, B., and McDevitt, H. O. (1972): Histocompatibility-linked immune response genes. *Science*, 175:273–279.
8. Bertrams, F., von Fisenne, E., Hoher, P. G., and Kuwert, E. (1973): Lack of association between HLA antigens and measles antibody in multiple sclerosis. *Lancet*, 2:441.
9. Bick, P. H., Persson, U., Smith, E., Möller, E., and Hammarstrom, L. (1977): Genetic control of lymphocyte activation: Lack of response to low doses of concanavalin A in lipopolysaccharide-nonresponder mice. *J. Exp. Med.*, 146:1,146–1,151.
10. Biddison, W., Payne, S., Shearer, G., and Shaw, G. (1980): Human cytotoxic T cell responses to trinitrophenyl hapten and influenza virus. Diversity of restriction antigens and specificity of HLA-linked genetic regulation. *J. Exp. Med.*, 152:204s.
11. Blass, J. P., and Gibson, G. E. (1977): Abnormality of a thiamine-requiring enzyme in patients with Wernicke-Korsakoff syndrome. *N. Engl. J. Med.*, 297:1,367–1,370.
12. Bodmer, W. F. (1980): Models and mechanisms for HLA and disease associations. *J. Exp. Med.*, 152:353s–357s.
13. Chen, K. K., and Poth, E. J. (1929): Racial differences as illustrated by the mydriatic action of cocaine, euphthalmine, and ephidrine. *J. Pharmacol. Exp. Ther.*, 36:429.
14. Compston, D. A. S., Vincent, A., Newsome-Davis, J., and Batchelor, J. R. (1980): Clinical, pathological, HLA antigen, and immunological evidence for disease heterogeneity in myasthenia gravis. *Brain*, 103:579–602.
15. Cumming, W. J. K., Hudgson, P., and Wilcox, C. B. (1978): HLA antigens in adult polymyositis. *N. Engl. J. Med.*, 299:1,365.
16. Dausset, J., Florman, A. L., Bachvaroffe, R., Kanra, G. Y., Sasportes, M., and Rapaport, F. T. (1972): *In vitro* approach to a correlation of cell susceptibility to viral infection with HLA genotypes and other biological markers. *Proc. Soc. Exp. Biol. Med.*, 140:1,344–1,349.
17. DeVries, R. R. P., Kreeftenberg, H. G., Loggen, H. G., and Van Rood, J. J. (1977): *In vitro* immune responsiveness to vaccinia virus and HLA. *N. Engl. J. Med.*, 297:692–696.
18. Feltkamp, T. E. W., Den Berg-Loonen, P. M., Nijenhuis, L. E., Engelfriet, C. P., Van Rossum, A. L., Van Loghem, J. J., and Oosterhuis, H. J. (1974): Myasthenia gravis, autoantibodies, and HLA antigens. *Br. Med. J.*, 26:131–133.

19. Fewster, M., Ames, F., and Botha, M. C. (1979): Measles antibodies and histocompatibility types in multiple sclerosis. *J. Neurol. Sci.*, 43:19–26.
20. Geczy, A. F., Alexander, K., and Bashir, H. V. (1980): A factor(s) in *Klebsiella* culture filtrates specifically modifies an HLA-B27-associated cell surface component. *Nature*, 283:782–784.
21. Goedde, H. W., and Agarwal, D. P. (1978): Pseudocholinesterase variations. *Hum. Genet. (Suppl.)*, 1:45–55.
22. Gorczynski, R. M., and Steele, E. J. (1980): Inheritance of acquired immunological tolerance to foreign histocompatibility antigens in mice. *Proc. Natl. Acad. Sci.*, 77:2,871–2,875.
23. Greenberg, L. J., Chapyk, R. -L., Bradley, P. W., and Lalouel, J. -M. (1980): Immunogenetics of response to a purified antigen from Group A streptococci. *Immunogenetics*, 2:161–167.
24. Hansen, G. S., Rubin, B., Sorenson, S. F., and Svejgaard, A. (1978): Importance of HLA-D antigens for the cooperation between human monocytes and T lymphocytes. *Eur. J. Immunol.*, 8:520.
25. Harriman, D. G. F., Ellis, F. R., Franks, A. J., and Sumner, D. W. (1978): Malignant hyperthermia myopathy in man: An investigation of 75 families. In: *Malignant Hyperthermia*, edited by J. A. Aldrete and B. A. Britt, p. 67. Grune and Stratton, New York.
26. Haspel, M. V., Pellegrino, M., Lampert, P., and Oldstone, M. B. (1977): Human histocompatibility determinants and virus antigens: Effect of measles virus infection on HLA expression. *J. Exp. Med.*, 146:146–155.
27. Helenius, A., Morein, B., Fries, E., Simons, K., Robinson, P., Schirrmacher, V., Terhorst, C., and Strominger, J. L. (1978): Human (HLA-A and HLA-B) and murine (H-2K and H-2D) histocompatibility antigens are cell surface receptors for Semliki Forest virus. *Proc. Natl. Acad. Sci.*, 75:3,846–3,850.
28. Henschke, P. J., Bell, D. A., and Cape, R. D. T. (1978): Alzheimer's disease and HLA. *Tissue Antigens*, 12:132–135.
29. Hoffman, P. M., Robbins, D. S., and Nolte, M. T. (1978): Cellular immunity in Guamanians with amyotrophic lateral sclerosis. *N. Engl. J. Med.*, 299:680–685.
30. Isaacs, H. (1978): Myopathy and malignant hyperthermia. In: *Malignant Hyperthermia*, edited by J. A. Aldrete and B. A. Britt, p. 89. Grune and Stratton, New York.
31. Jackson, J. F., Currier, R. F., Terasaki, P. I., and Norton, N. E. (1977): Spinocerebellar ataxia and HLA linkage. *N. Engl. J. Med.*, 296:1,138–1,141.
32. Jokelainen, M., Tiilikainen, A., and Lapinleimu, K. (1977): Polio antibodies and HLA antigens in amyotrophic lateral sclerosis. *Tissue Antigens*, 10:259–266.
33. Kott, E., Livni, E., and Zamir, R. (1979): Cell-mediated immunity to polio and HLA antigens in amyotrophic lateral sclerosis. *Neurology*, 29:1,040–1,044.
34. Kovanen, J., Tiilikainen, A., and Haltia, M. (1980): Histocompatibility antigens in familial Creutzfeldt-Jakob disease. *J. Neurol. Sci.*, 45:317–321.
35. Kreth, H. W., ter Meulen, V., and Eckert, G. (1975): HLA and subacute sclerosing panencephalitis. *Lancet*, 2:415.
36. Kurant, J. E., Sever, J. L., and Terasaki, P. (1975): HLA-W29 and subacute sclerosing panencephalitis. *Lancet*, 1:927.
37. Lasch, E. E., Joshua, H., Gazit, E., El-Massri, M., Marcus, O., and Zamir, R. (1979): Study of the HLA antigen in Arab children with paralytic poliomyelitis. *Isr. J. Med. Sci.*, 15:12–13.
38. Latovitzki, N., Suciu-Foca, N., Penn, A., Olarte, M., and Chutorian, A. (1979): HLA typing and Guillain-Barre syndrome. *Neurology*, 29:743–745.
39. Latts, J. R., Antel, J. P., Levinson, D. J., Arnason, B. G. W., and Medof, M. E. (1980): Histocompatibility antigens and gold toxicity: A preliminary report. *J. Clin. Pharmacol.*, 20:206–209.
40. Madigand, M., Fauchet, R., Oger, J., and Sabouraud, O. (1981): Sclérose en plaque: Corrélation entre formes cliniques et groupe HLA. *Nelle Press Med. (Paris) (in press)*.
41. Marsh, D. G., Bias, W. B., and Hsu, S. H. (1973): HLA-7 cross-reacting group with a specific reaginic antibody response in allergic man. *Science*, 179:691–693.
42. McDevitt, H. O., and Bodmer, W. F. (1974): HLA immune response genes and disease. *Lancet*, 1:1,269–1,274.
43. McGuffin, P. (1979): Is schizophrenia an HLA-associated disease? *Psychol. Med.*, 9:721–728.
44. McMichael, A. J., Parham, P., Brodsky, F. M., and Pilch, J. R. (1980): Influenza virus-specific cytotoxic T lymphocytes recognize HLA molecules. *J. Exp. Med.*, 152:195–203.
45. Miles, K., Quintans, J., Chelmicka-Schorr, E., and Arnason, B. G. W. (1980): The sympathetic nervous system modulates antibody response. *J. Neuroimmunol.*, 1:101–105.

46. Nelson, T. E. (1978): Malignant hyperthermia susceptible muscle. In: *Malignant Hyperthermia*, edited by J. A. Aldrete and B. A. Britt, p. 23. Grune and Stratton, New York.
47. Oger, J., and Arnason, B. G. W. (1979): Immunogenetics of neurological diseases. *Trends Neurosci.*, 2:68–70.
48. Pachman, L. M., Jonasson, O., and Cannon, R. A. (1977): HLA-B8 in juvenile dermatomyositis. *Lancet*, 2:567.
49. Palmer, E. G., Topel, D. G., and Christian, L. L. (1978): Light and electron microscopy of skeletal muscle from malignant hyperthermia-susceptible pigs. In: *Malignant Hyperthermia*, edited by J. A. Aldrete and B. A. Britt, p. 103. Grune and Stratton, New York.
50. Panayi, G. S., Wooley, P., and Batchelor, J. R. (1978): Genetic basis of rheumatoid disease: HLA antigens, disease manifestations, and toxic reactions to drugs. *Br. Med. J.*, 2:1,326–1,328.
51. Paty, D. W., Cousin, H. K., Stiller, C. R., Boucher, D. W., Furesz, J., Warren, K. G., Marchuk, L., and Dossetor, J. B. (1977): HLA-D typing with an association of DW2 and absent immune responses towards herpes simplex (Type I) antigen in multiple sclerosis. *Trans. Proc.*, 9:1,845–1,848.
52. Paty, D. W., Cousin, H. K., Stiller, C. R., and Dossetor, J. B. (1978): An HLA-D-linked low response to polyclonal B cell activation in multiple sclerosis. *Transplant. Proc.*, 10:973–975.
53. Pedersen, L., Platz, P., and Jersild, C. (1977): HLA (SD and LD) in patients with amyotrophic lateral sclerosis (ALS). *J. Neurol. Sci.*, 31:313–318.
54. Pelligrino, M. A., Ferrone, S., Brautbar, C., and Hayflick, L. (1976): Changes in HLA antigen profiles on SV40 transformed human fibroblasts. *Exp. Cell Res.*, 97:340.
55. Perris, C., Strandman, E., and Wahlby, L. (1979): HLA antigens and the response to prophylactic lithium. *Neuropsychobiology*, 5:114–118.
56. Peters, J. H., Miller, K. S., and Brown, P. (1965): Studies on the metabolic basis for the genetically determined capacities for isoniazid inactivation in man. *J. Pharmacol. Exp. Ther.*, 150:298.
57. Pietch, M. C., and Morris, P. J. (1974): An association of HLA-3 and HLA-7 with paralytic poliomyelitis. *Tissue Antigens*, 4:50.
58. Pietrzyk, J. J., and Turowski, G. (1979): Immunogenetic basis of congenital malformations: Association of HLA-B27 with spina bifida. *Pediatr. Res.*, 13:879–883.
59. Rea, T. H., Levan, N. E., and Terasaki, P. I. (1976): Histocompatibility antigens in patients with leprosy. *J. Infect. Dis.*, 134:615–618.
60. Renvoize, E. B., Hambling, M. H., Pepper, M. D., and Rajah, S. M. (1979): Possible association of Alzheimer's disease with HLA-BW15 and cytomegalovirus infection. *Lancet*, 1:1,238.
61. Reza, M. J., Kar, N. C., Pearson, C. M., and Kark, R. A. P. (1978): Recurrent myoglobinuria due to muscle carnitine palmityl transferase deficiency. *Ann. Intern. Med.*, 88:610–615.
62. Richman, D. P., Antel, J. P., Patrick, J. W., and Arnason, B. G. W. (1979): Cellular immunity to acetylcholine receptor in myasthenia gravis: Relationship to histocompatibility type and antigenic site. *Neurology*, 29:291–296.
63. Sasazuki, T., Kohno, Y., and Iwamoto, I. (1978): Association between an HLA haplotype and low responsiveness to tetanus toxoid in man. *Nature*, 272:359–361.
64. Schmidt, K. L., and Mueller-Eckhardt, C. (1977): Agranulocytosis, levamisole, and HLA-B27. *Lancet*, 2:85.
65. Shear, C. S., Wellman, N. S., and Nyhan, W. L. (1974): Phenylketonuria: Experience with diagnosis and management. In: *Heritable Disorders of Amino Acid Metabolism*, edited by W. L. Nyhan, pp. 141–159. J. Wiley and Sons, New York.
66. Singal, D. P., Perets, A., and Dolovich, J. (1979): HLA antigens and serum antibodies to bovine serum albumin. *Transplant. Proc.*, 11:1,864–1,868.
67. Smeraldi, E., Bellodi, L., Saccheti, E., and Cazzullo, C. L. (1976): The HLA system and the clinical response to treatment with chlorpromazine. *Br. J. Psychiatry*, 129:486–489.
68. Smeraldi, E., and Scorza-Smeraldi, R. (1976): Interference between anti-HLA antibodies and chlorpromazine. *Nature*, 260:532–533.
69. Spencer, M. J., Cherry, J. D., and Terasaki, P. I. (1976): HLA antigens and antibody response after influenza A vaccination. *N. Engl. J. Med.*, 294:13–16.
70. Stember, R. H., and Fieve, R. R. (1977): Histocompatibility antigens in affective disorders. *Clin. Immunol. Immunopathol.*, 7:10–14.
71. Stendahl-Brodin, L., Link, H., Möller, E., and Norrby, E. (1979): Genetic basis of multiple sclerosis: HLA antigens, disease progression, and oligoclonal IgG in CSF. *Acta Neurol. Scand.*, 59:297–308.

72. Sulkava, R., Koskimies, S., Wikström, J., and Jorma, P. (1980): HLA antigens in Alzheimer's disease. *Tissue Antigens*, 16:191–194.
73. Tada, T., Taniguchi, M., and David, C. S. (1976): Properties of the antigen-specific suppressive T cell factor in the regulation of antibody response of the mouse. IV. Special subregion assignment of the gene(s) that codes for the suppressive T cell factor in the H-2 histocompatibility system. *J. Exp. Med.*, 121:2,241.
74. Terasaki, P. I., and Mickey, M. R. (1975): HLA haplotypes of 32 diseases. *Transplant. Rev.*, 22:105–119.
75. Thorsby, E., Godal, T., and Myrvang, B. (1973): HLA antigens and susceptibility to diseases. II. Leprosy. *Tissue Antigens*, 3:373–377.
76. Walshe, J. M. (1967): The physiology of copper in man and its relation to Wilson's disease. *Brain*, 90:149.

Genetics of Neurological and Psychiatric Disorders, edited by Seymour S. Kety, Lewis P. Rowland, Richard L. Sidman, and Steven W. Matthysse. Raven Press, New York © 1983.

Familial Spongiform Encephalopathies

David M. Asher, *Colin L. Masters, D. Carleton Gajdusek, and Clarence J. Gibbs, Jr.

Laboratory of Central Nervous System Studies, National Institute of Neurological and Communicative Disorders and Stroke, National Institutes of Health, Bethesda, Maryland 20205

The subacute spongiform encephalopathies, kuru and Creutzfeldt–Jakob disease (CJD) of humans and scrapie and mink encephalopathy of animals, are a group of noninflammatory degenerative diseases of the central nervous system. Each disease has a progressive, unremitting clinical course ending in death and a typical constellation of anatomic findings including vacuolation of neurons leading to spongiform changes of gray matter, proliferation and hypertrophy of glia, and sometimes amyloid plaques. The cerebral cortex, cerebellum, and other areas of the brain may be involved, but no pathological changes have been recognized outside the CNS. These encephalopathies are infections caused by filterable agents that we call unconventional viruses because of their very small size and because many of their physical properties, especially their resistance to inactivation and their failure to elicit a detectable immune response, are unlike those of known viruses. Their structure remains obscure. The diseases were described in earlier volumes in this series (42,50) and have been reviewed since (5,40,51).

Each of the spongiform encephalopathies sometimes occurs in siblings, progeny, and other relatives of affected individuals. The purpose of this chapter is to examine such familial clusters of disease and to review the evidence that genetic influence may be responsible.

SCRAPIE AND MINK ENCEPHALOPATHY

Although scrapie has been recognized for many years to be an infectious disease of sheep, most investigators remain convinced that the disease has an important hereditary component. As early as 1913, a hereditary etiology was considered by Stockman and dismissed in favor of infection (98). At the other extreme, Parry (90) went so far as to propose that scrapie is never naturally infectious and that the

---

*Present address: Department of Pathology, University of Western Australia, Perth, Australia

transmissible agent of scrapie is ubiquitous in sheep but produces disease only in animals of a particular genotype.

The infectious etiology of scrapie has been amply confirmed by repeated experimental transmission of disease from tissues of affected sheep to sheep from previously disease-free flocks as well as to other species of animals that never get spongiform encephalopathies naturally and by demonstration that the pathogen replicates in animals (70). However, there is evidence that heredity also plays an important role in scrapie; most striking is the observation that some breeds of sheep are highly resistant to the disease. For example, in a study of 24 breeds of sheep inoculated with the same dose of scrapie virus, the incidence of scrapie varied from 78% in Herdwicks to zero in Dorset Downs (56).

The nature of genetic control of susceptibility to scrapie in sheep is not clear. Parry (90) hypothesized an autosomal recessive gene for scrapie based on the observation that when both parents were naturally infected with scrapie they had virtually 100% affected progeny, whereas progeny of other mating combinations had lower rates of disease; the phenotypes of progeny predicted from genotypes deduced for their parents on the basis of Parry's hypothesis agreed reasonably well with those actually observed. Other investigators, however, found serious inconsistencies between phenotypes predicted from the simple recessive hypothesis and those observed. In five breeds of sheep, crosses between scrapie-susceptible parents produced many scrapie-resistant progeny (57); more than half of the Suffolk sheep from flocks totally free of scrapie for generations nevertheless gave birth to scrapie-affected lambs when bred to scrapied animals and had many fewer discordant twin pairs than predicted by the recessive hypothesis (36). Thus, the autosomal recessive hypothesis for scrapie in sheep has generally been discarded.

Evidence for genetic control of the incubation period of scrapie in sheep was investigated by Dickinson and co-workers. In a study of experimental scrapie in Cheviot sheep inoculated intracerebrally with high doses of scrapie virus, incubation periods tended to fall into two discontinuous groups, and in breeding experiments, the short incubation period seemed to segregate as if under the control of a single fully dominant gene called *"sip"* (35). When long- and short-incubation-period sheep were inoculated subcutaneously with high doses of scrapie virus, those deduced to have the *sip* gene became ill about 10 months later, whereas those presumed to carry only long-incubation-period genes *(lip)* never developed scrapie in 3 years of observation. So the *sip* gene seemed to confer susceptibility to subcutaneous infection with scrapie under one set of conditions. If the situation is analogous to that in mice with scrapie, animals with the *lip* gene may not have been absolutely resistant to subcutaneous inoculation with scrapie but may simply have had incubation periods so long that their lives ended before disease became manifest. The apparent resistance of those animals with the *lip* gene depended not only on the route of infection but also on the dose, strain, and passage history of the scrapie virus (30). A 12-year breeding program of Herdwick sheep that yielded flocks uniformly susceptible or largely resistant to the strain of scrapie virus used by Dickinson was also interpreted as fitting best the same hypothesis—that susceptibility to infection by the subcutaneous route is controlled by two alleles of a single autosomal gene, with susceptibility dominant to resistance (89). Minor differences in distribution of CNS lesions have been observed among breeds of sheep, although the basic histopathology was the same (60). When scrapie spread from infected sheep to goats with which they were in contact, there was a difference in attack rates among various breeds of goat (64), suggesting that genetic constitution might also affect their susceptibility; no resistance to intracerebral inoculation of scrapie agent has been observed in any breed of goat.

Dickinson also studied genetic control of experimental scrapie in mice. Mice are more convenient to study than sheep because many inbred lines are available, generation times are short, and the incubation period of scrapie in mice may be only a few months. Also, scrapie virus can be cloned by terminal dilution in mice to give strains of agent more homogeneous in biological properties than are uncloned viruses, which appear to be mixtures of strains (34). Two genes have been identified as significantly affecting the incubation period of scrapie in mice, although other genes may have minor effects (28). Two lines of mice differed markedly in incubation periods observed after inoculation with a strain of scrapie virus (ME7), and hybrids of the two lines had incubation periods intermediate to those of their parents; subsequent filial crosses and back crosses to parents yielded progeny with incubation periods consistent with a single-gene hypothesis. The locus influencing the incubation period of ME7 scrapie virus in mice was called *"sinc"* (scrapie incubation), and its two alleles designated *s7*, for "shortened incubation period," and *p7*, for "prolonged incubation period"; neither allele showed dominance (32). Other strains of scrapie virus were found to resemble ME7 in incubation periods elicited in mice, but the 22A strain of scrapie agent paradoxically produced incubation periods that were longer in *s7s7* and shorter in *p7p7* homozygous mice and still longer in *s7p7* heterozygotes (28,31).

The difference between short and long incubation periods mediated by the *sinc* gene was associated with a delay in the detectable replication of virus both in the spleen after intraperitoneal inoculation and in the brain after intracerebral inoculation; once replication of virus began in those tissues, the rate of increase in titers and the final titers achieved were the same in mice of all genotypes (26,28,33). When a low dose of scrapie virus was inoculated into a long-incubation-period mouse by the intraperitoneal route, virus was detected in the spleen only after a period of almost 2 years and was never found in the brain at all (29,31); although the mice were obviously not resistant to infection in a strict sense, the incubation after intraperitoneal inoculation with low doses of some strains of virus effectively exceeded their life-spans. It is possible that the *sinc* gene or linked genes may also have influenced the distribution and intensity of vacuolation and amyloid plaque formation throughout the brain in response to various strains of scrapie virus (16,28,38) as well as the tendency of some strains to undergo changes in incubation periods and in distribution of lesions after serial passages (17,25).

A gene for dominant hemimelia *(Dh)* also appears to affect the incubation period of scrapie in mice. *Dh* heterozygote mice have variable hind-limb malformations and asplenia, and homozygotes are usually not viable. In inbred mice homozygous for the *sinc* allele *s7*, those animals with hemimelia and asplenia had significantly longer incubation periods after intraperitoneal inoculation with scrapie virus than did mice homozygous for the wild-type allele at the *Dh* locus (27,28). Unlike the *sinc* gene, which influenced the incubation period after inoculation of mice directly into the brain as well as intraperitoneally, the *Dh* gene had no effect on incubation periods following intracerebral injection. The effects of the *Dh* gene did not result solely from asplenia; incubation periods were significantly longer in *Dh* asplenic mice than in surgically splenectomized genetically comparable animals with the wild-type allele at the *Dh* locus (27).

There are also nongenetic causes for familial clusters of scrapie in animals. Scrapie was more readily transmitted by natural contact to younger sheep, especially in the perinatal period, resulting in increased incidence of scrapie correlated with the length of time that young were left with infected mothers, although some lambs removed from ewes at birth still got scrapie (34); that fact probably explains the higher incidence of infected progeny born to scrapie-infected mothers compared to infected sires (34,36,64). Transmission did not appear to occur via milk (24). Similar contact spread of scrapie, especially in early life, has also been observed in goats (64) but not in mice (5,24) or monkeys (4).

Transmissible mink encephalopathy (18), which may be a form of scrapie introduced into mink (77), has been studied much less than scrapie itself. No genetic influences have been recognized in mink, all color phases of which were susceptible to the encephalopathy (76). There appeared to be no transmission of infection *in utero* (19,62), and the apparent high susceptibility of mink kits to food-borne infection may reflect inoculation through bite wounds (77).

KURU

When kuru in the Fore people of Papua New Guinea was first described, a genetic factor was suspected to play an important role in its etiology, although an additional unknown environmental factor was postulated, and infection considered (45). Early studies of the Fore people found many families with kuru in several members. A particularly severely affected kindred is shown in Fig. 1; in most genealogies, more mothers of kuru patients were affected than fathers. Bennett and co-workers reviewed about 250 such pedigrees including more than 2,000 subjects, some 200 of

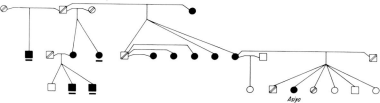

FIG. 1. Representative genealogy of a family affected with kuru (family of Asiyo). *Open box:* male; *open circle:* female; *diagonal hatch in box or circle:* died of causes other than kuru; *solid box or circle:* died of kuru; *dash:* died in childhood.

whom were thought to have had kuru (6,7), and concluded that they were consistent with the hypothesis that kuru is "controlled" by a single autosomal gene that is recessive in males and dominant in females, in whom homozygotes have onset in childhood and heterozygotes in adult life. However, the gene frequencies in the Fore population predicted on the basis of the single-gene hypothesis were implausibly high (85), especially since the disease is rapidly lethal with no evidence for compensatory heterozygote advantage (2,41), and kuru seems to have spread rapidly from a central geographic focus over a period of only 40 to 60 years (2,41,85).

Even after the simple autosomal recessive hypothesis was dismissed, and the infectious nature of kuru established (43), it still seemed likely that genetic susceptibility was required, as suggested by the absence of kuru in neighboring people who shared a common culture with the Fore (2,41). A search for genetic markers showed no obvious correlation between kuru and several blood groups (95), dermatoglyphic patterns (91), or the immunoglobulin G allotypes *Gm* and *Inv* (96). There were larger numbers of kuru patients than expected with the "aboriginal" group-specific serum protein *GcAb* gene, especially as the homozygous genotype (73). However, the same gene was also present in high frequency in villages free of kuru and absent in many patients with kuru (73); furthermore, it was recognized that analysis of interactions between genetics and infectious diseases was unreliable in small, heterogeneous, and rapidly changing population groups of the sort found in Papua New Guinea, especially since many individuals from the supposed "control" uninfected group were actually incubating kuru (100).

Subsequent changes in the incidence and epidemiology of kuru have made the concept of genetic differences in susceptibility unnecessary to explain the etiology of the disease. Kuru has become progressively less common, and the age of the youngest patients has risen each year, so that no children, adolescents, or young adults get the disease now; at the same time, the ratio of female to male cases of kuru has fallen almost to one (3,40). These changes probably resulted from the most striking social change in the lives of the Fore people—the end of cannibalism by 1960 (3). No mechanism other than cannibalism need be invoked to explain the ecology of kuru among the Fore and their neighbors, and no case has ever been confirmed without a history of participating in cannibal consumption of another kuru patient between 4 and 30 years earlier (3).

Not only can hereditary susceptibility be discarded in explaining the epidemiology of kuru, but also "vertical infection" clearly played no part. The predominance of affected mothers over fathers (7,100) was neither genetic nor a result of transmission of kuru from mother to fetus *in utero* (78) but rather was entirely a social phenomenon caused by contamination of babies with virus by their mothers during cannibalism.

Experimental studies of kuru in animals confirm that, at least in a small number of cases, no "vertical" transmission of infection *in utero* has occurred (4) and that the eating of infected tissues was sufficient to transmit disease (49), although direct intragastric inoculation was not (5). The wide range of animals experimentally infected with kuru (51) further suggests that individual differences in genetic susceptibility play no role, although the resistance of some species to kuru infection must be genetic (53). In short, although initial observations of kuru in families suggested that there was a major hereditary influence, subsequent epidemiological and experimental studies revealed no important genetic contribution to susceptibility in humans or animals.

CREUTZFELDT–JAKOB DISEASE

Soon after the earliest descriptions of the encephalopathy now called Creutzfeldt–Jakob disease (66,67,72), the condition was recognized in a family (71,72) in which new cases continued to occur as years passed (65,86,97), ultimately involving four generations (72), as shown in Fig. 2. Since then, many families with CJD have been reported (78,80).

Although the nosology of CJD remains unsettled, it seems clear that at least two somewhat different clinical–pathological syndromes of dementia are included, that both may occur in families, and that both are associated with transmissible agents. The first is a rapidly progressive dementia with myoclonic jerks or tremors dominating the clinical picture and spongiform changes in cerebral cortical gray matter, as described by Jakob (68,72), Kirschbaum (71), and many others (72), probably including Heidenhain (8,63,72) whose eponym is sometimes reserved for cases of CJD with prominent cortical blindness. The other is a similar dementia with or without myoclonus or tremors plus marked ataxia and histological changes of cerebellar degeneration, as described by Brownell and Oppenheimer (15).

In cases of the cerebellar syndrome, several types of amyloid plaques have been observed (21,79). That drew attention to several patients resembling those reported by Gerstmann and co-workers (48); in his patients, cerebellar ataxia and a peculiar reflex posturing (47) predominated over dementia, at least early in the course of illness, and both spongiform encephalopathy and amyloid plaques were found at autopsy. Creutzfeldt–Jakob disease with amyloid plaques ("Gerstmann–Sträussler syndrome") apparently differs from CJD with kuru plaques or senile plaques in that patients with amyloid plaques were younger at onset and had much longer durations of illness (79); all patients with CJD and plaques of any type had clinical durations of illness longer than those of CJD patients without plaques (79). Cases of the Gerstmann-Sträussler variant with mild dementia are remarkably similar to kuru and constitute an intriguing link with that disease. Although it is not yet clear which of the two or three clinical variants of CJD merits its own separate taxon, brain tissues from patients of each type contain infectious agents that have transmitted to animals clinically and histopathologically indistinguishable spongiform encephalopathies (37,79).

In addition to the clinical forms of CJD briefly described, there is a syndrome called "amyotrophic CJD" characterized by slowly progressive dementia, sometimes by extrapyramidal disturbances, and by signs of lower motor neuron degeneration including muscular fasciculations and wasting. The amyotrophic syndrome differs from typical forms of CJD in histopathology, lacking spongiform changes; brain

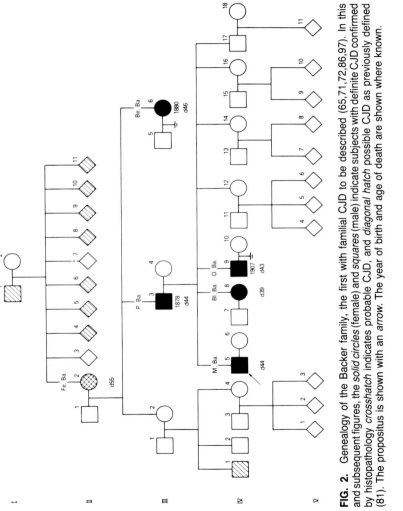

FIG. 2. Genealogy of the Backer family, the first with familial CJD to be described (65,71,72,86,97). In this and subsequent figures, the *solid circles* (female) and *squares* (male) indicate subjects with definite CJD confirmed by histopathology *crosshatch* indicates probable CJD, and *diagonal hatch* possible CJD as previously defined (81). The propositus is shown with an *arrow*. The year of birth and age of death are shown where known.

suspensions from patients with it have not transmitted disease to experimental animals (94,99). We consider the term "amyotrophic CJD" to be a misnomer for a syndrome of different etiology better classified as a motor neuron disease with dementia. Although dementia with lower-motor-neuron disease may also occur in families (23), it will not be considered further.

Familial clusters of CJD have been found among most populations studied. In the first epidemiological case-control study of CJD in the United States, one familial case was found among 38 patients identified (2.6%), and seven (18%) had family histories of some kind of neurological disease versus only 6% for controls (8). In a series of 46 patients from England and Wales, an approximate annual incidence of 0.09 cases of CJD per million population during one decade, Matthews (83) identified three (6.5%) with histories of similar illness in a sibling. In Israel, the incidence of CJD among Libyan Jews is the highest reported anywhere in the world—almost 40 cases per million people per year adjusted for age, as opposed to only about two cases per million for other Israeli ethnic groups (69). Much of that excess in CJD reflects a very high number of familial cases; three and possibly four affected families were found among 20 Libyan Jewish genealogies available for analysis (88), whereas of 11 non-Libyan patients with CJD, none had a family history of the disease (55). Of 56 cases of CJD found by Brown and Cathala in Paris (11), where there was a maximum incidence of about one case per million people per year during the decade studied, four (7%) were familial, whereas among 169 cases in all of France during the same period, with a much lower average annual incidence of 0.32 cases per million, 12 (7.1%) were familial (12), and 9.7% of 124 well-verified cases had histories of definite or possible CJD in family members (14). Overall, 6 to 9% of cases in France appeared to be familial, more than half from two families (13). In Chile, with an average annual mortality of 0.3 per million people per year for the whole country and 0.7 for Santiago over 20 years, nine of 35 patients (26%) were from five families in which cases resembling CJD had occurred (46). In our world survey of 1,435 cases of CJD, with an average annual mortality of at least 0.26 per million population per year for the United States, about 15% had family histories suggestive of CJD (80–82). However, a recent exhaustive survey over a 3-year period in Japan revealed 60 cases of CJD, none of which was familial (74), although familial CJD has been reported from Japan (1). Some of these discrepancies may result from increased recognition and reporting of familial cases in various countries, but differences in relative incidence as marked as those between Japan and Israel are probably significant.

Since the earliest description of the Backer family (Fig. 2), most genealogies of such families have been interpreted as showing autosomal dominant inheritance of the disease (20,39,61,71,72,78,80–82,84,93). This pattern was suggested by the occurrence of CJD in successive generations of families without consanguinity, affecting males and females in approximately equal numbers and approximately half of the siblings in affected sibships. No other single-gene hypothesis has been proposed.

It is beyond question that infectious agents capable of transmitting spongiform encephalopathy to experimental animals (53,54) can be found consistently in brains of patients with different clinical forms of CJD [except the "amyotrophic syndrome" discussed above (37,94,99)]. The rates of recovery of virus are quite high in brains of both familial and sporadic cases of CJD; the agent has been demonstrated in 17 of 26 patients with familial CJD adequately studied (65%), and 25 more cases have not been under investigation long enough to be considered negative. This rate is similar to that for sporadic cases of CJD—151 demonstrations of CJD virus in 220 adequately studied cases (69%) as of October, 1981. The clinical pictures and histopathological changes of familial and sporadic cases of CJD are indistinguishable, as are the syndromes transmitted from them to animals. Although the agents found in various cases of CJD cannot yet be compared in conventional ways, such as by analysis of viral ultrastructure, antigenic similarities, or molecular homologies, the absence of consistent biological differences between familial and sporadic cases of CJD suggests that they are caused by the same virus.

The reasons for familial occurrence of CJD are not known. The situation is especially difficult to understand because the modes of transmission of the CJD virus are still unknown except in a few iatrogenic cases (44,51,52,101). Possible mechanisms considered have included the following: (a) CJD may be acquired by direct or indirect contact with CJD patients or people incubating the disease; (b) CJD may be an occasional infection of humans with an agent that usually infects animals, like scrapie virus; and (c) CJD may result from the occasional activation of a latent infection that is more common than the disease itself but which remains unexpressed in most people; the latent infection may be transmitted from one generation to the next. Many variations of these three hypotheses are possible. At the moment, limited evidence favors the first explanation (9,10) because case-to-case spread of CJD has actually been documented in iatrogenic cases, and that mechanism fully accounts for transmission of kuru (40) and scrapie (64). However, none of the other modes of transmission of CJD can be completely dismissed yet (22,40).

The patterns of occurrence of CJD in families, as noted above, frequently suggest an autosomal dominant inheritance. Masters et al. (78,80) identified 73 families with CJD and selected for closer study 27 in which there were at least two histologically confirmed cases. There were no striking differences between the clinical and histopathological pictures of familial cases and a group of some 100 cases of sporadic CJD used for comparison, but there tended to be special similarities among cases in the same families. No skipped generations were found in families with CJD (78). Cases of familial and sporadic CJD were similar in ratios of males to females (1 : 1) and in durations of illness (11 and 8 months, respectively). Cases of familial CJD were, however, somewhat younger than those with sporadic CJD (mean age of about 51 years for familial and 58 years for sporadic cases of CJD, $p < 0.001$).

Table 1 summarizes the occurrence of CJD in siblings and first cousins of propositi in the study of Masters et al. (78). Thirty-six percent of siblings and 21% of the

TABLE 1. *Incidence of familial CJD and AD in siblings and first cousins*

	No. affected	Total no. of siblings or first cousins aged 40 years or more	Percentage affected
Proportional incidence in all affected sibships			
Familial CJD	116	322	36
Familial AD	154	319	48
Proportional incidence in first cousins of propositi			
Familial CJD	11	52	21
Familial AD	14	70	20

From Masters et al. (78).

small number of first cousins about whom information was available had died with neurological diseases suggesting CJD. Individuals without CJD at or after age 40 were considered to be free of disease, although that age is well below the mean and results in a lower estimated incidence of CJD than that computed for older people; lack of available data prevented using the later age. Those rates seem roughly consistent with an autosomal dominant pattern of inheritance of CJD. They do not resemble other single-gene patterns. The rather high percentage of CJD might be explained by better recollection of affected family members or by overdiagnosis of neurological disease. However, as noted by Dickinson and Mackay (30), susceptibility of hosts to infection may be affected by multiple genes superficially resembling a single gene, and the difference may be apparent only after performing appropriate back crosses that are not possible in humans. Their warning must be heeded: "The short history of genetics is littered with examples of data being forced into oversimplified genetic models which give the appearance of fitting the facts."

Nongenetic explanations for the familial occurrence of CJD must also, therefore, be seriously considered. Since CJD is indisputably an infection, the possibility of "vertical" transmission from parents to offspring during fetal life or in the perinatal period must be examined. Transmission of infection *in utero* or in infancy, by whatever mechanism, might be expected to result in a predominantly maternal pattern of "inheritance," as is the case for pedigrees of scrapie in sheep and kuru in humans. However, in our families, there was no evidence that mothers of patients with CJD were affected more frequently than fathers, nor was there any difference in age at onset of disease or death between children born to affected parents of different sexes (78). Furthermore, discordant nonidentical twins with CJD have been observed (61), mitigating against vertical transmission of infection, but no concordant twins (78). If parents acquired infection from outside the family but passed infection to their children *in utero* or in early life, then the phenomenon of anticipation, a decrease in age of affected members of succeeding generations, might be expected. In our series of patients with familial CJD, such a tendency was observed: in 33 instances the affected parent's age at onset of CJD or death

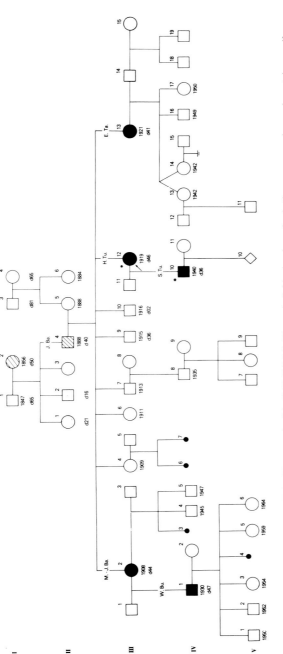

FIG. 3. Genealogy of a family (H. Tu) with transmissible CJD (39,84). Those individuals whose brain tissues caused experimental spongiform encephalopathies in animals are indicated by *asterisks*.

exceeded that of the affected child, whereas there were only 21 in which the parent had disease at a younger age, but the difference is not striking (78). So the actual limited evidence for acquisition of CJD infection *in utero* or in early childhood is unconvincing.

Creutzfeldt–Jakob disease has occurred in patients long separated (for as much as 40 years) from other members of their families. That fact can be interpreted in at least two ways: (a) transmission of infection may take place very early in life; and (b) at least some members of the family may be more susceptible than are members of the general population to a virus that is widespread in the environment. The fact that affected siblings tend to die of CJD at about the same age [mean difference of less than a year in 66 pairs analyzed by Masters et al. (78)] is not consistent with a common point exposure of all siblings in a generation to the infectious agent but does not distinguish among other possibilities: true genetic susceptibility to an infectious agent, pseudogenetic transmission of infection in early life, or even the integration into the host's genome by the genetic material of the pathogen, a phenomenon that seems unlikely in light of the fact that it clearly does not happen in scrapie, mink encephalopathy, kuru, or experimental CJD.

A limited number of experimental findings also fail to suggest "vertical" transmission of CJD from infected animals to their progeny. Manuelidis and Manuelidis (75) saw no such transmission in 24 offspring of several CJD-infected guinea pigs, nor did we in a smaller number of progeny born to CJD-infected primates (4). The possible occurrence of CJD in two subjects who married into families with the disease suggests that contact transmission to adults in a family setting must still be considered (78). The transmission of CJD to monkeys allowed to handle and eat infected tissues (49) in a fashion similar to that for kuru and scrapie suggests that, as for kuru, the portal of infection may be through lesions in skin and mucosa.

ALZHEIMER'S DEMENTIA AND OTHER CHRONIC DEGENERATIVE DISEASES OF THE CNS

Alzheimer's disease (AD) occurs not only as a sporadic dementia but also, less commonly, in families in a pattern similar to that of familial CJD and suggestive of an autosomal dominant inheritance (59,78). Alzheimer's disease occasionally has clinical and anatomic similarities to the much less common CJD, including myoclonus and amyloid plaques (78). The cell-fusion-promoting factor described by Moreau-Dubois et al. (87) in brains of many patients with CJD and familial AD but rarely in the sporadic form of AD constitutes an additional point of similarity the significance of which is unknown. The most dramatic association between the two types of dementia is suggested by the occurrence of typical CJD in four families with AD (78), although none of these cases of CJD has been proven to be transmissible yet.

Possible transmissions of disease to primates from brain specimens of two patients with familial AD have been described (40,51,92,99), and evidence for and against an infectious etiology for familial AD was reviewed recently (58). Those initial

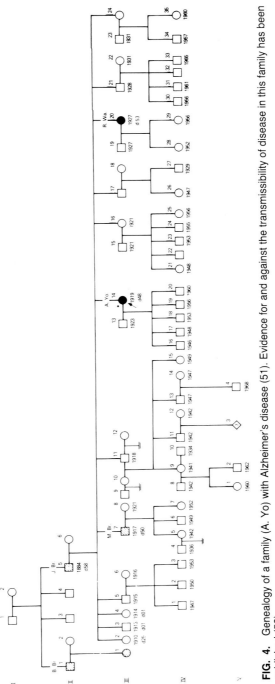

FIG. 4. Genealogy of a family (A. Yo) with Alzheimer's disease (51). Evidence for and against the transmissibility of disease in this family has been published (58).

reports of transmission of familial AD remain unconfirmed, and of 13 other cases of familial AD and 41 cases of sporadic AD adequately tested, none has transmitted disease. In view of the failure to confirm the demonstration of infectious agents from the two cases, failure to find agents by the same techniques in many other cases of AD, and, most importantly, failure to reproduce the typical histopathological changes of AD in the brains of animals to which AD had putatively been transmitted, we can only conclude that no infectious virus of the type causing the spongiform encephalopathies has been convincingly implicated in AD. No other chronic degenerative disease of the CNS, sporadic or familial, has yet been found to belong to the spongiform encephalopathy group.

SUMMARY AND CONCLUSIONS

Each of the spongiform encephalopathies, scrapie, mink encephalopathy, kuru, and Creutzfeldt–Jakob disease, may occur in relatives of affected cases. Familial clusters of scrapie in sheep probably result mainly from contact transmission of infection to lambs. There is also hereditary influence; after inoculation with small doses of virus by nonneural routes, animals of some genotypes have incubation periods so long that they seem practically resistant to infection, while animals of other genotypes have short incubation periods and appear uniformly susceptible. At least one gene in sheep and two in mice have been implicated in control of the incubation period.

The familial occurrence of kuru seems to be entirely a social phenomenon resulting from exposure of infants and children to virus by their mothers; neither differences in susceptibility nor "vertical" transmission of infection of kuru *in utero* appear to occur.

Familial clusters of CJD have been observed in many populations studied. Each of the two or three clinical forms of transmissible CJD has been found in families. Patterns of occurrence of familial CJD have generally been attributed to a single autosomal dominant gene controlling either expression of disease or perhaps susceptibility to the etiologic infectious agent. However, the single-autosomal-dominant hypothesis for familial CJD remains conjectural; the familial clustering and patterns of "inheritance" might conceivably be pseudogenetic, resulting from increased opportunities for transmission of infection in early life. "Vertical" infection with CJD *in utero* has not been documented either in humans or experimental animals.

Alzheimer's disease also occurs in families in patterns resembling those of CJD genealogies. The two dementias have also been found in the same families. However, Alzheimer's disease has not been convincingly demonstrated to be of infectious etiology.

REFERENCES

1. Akai, J., Kato, Y., and Oyanagi, S. (1979): Familial Creutzfeldt–Jakob disease. Clinico–pathological observation on cousins. *Shinkei Kenkyu No Shimpo [Adv. Neurol. Sci. (Tokyo)]*, 23:472–483.

2. Alpers, M. P. (1968): Kuru: Implications of its transmissibility for the interpretation of its changing epidemiologic pattern. In: *The Central Nervous System, Some Experimental Models of Neurological Diseases*, edited by O. T. Bailey and D. E. Smith, pp. 234–251. Williams & Wilkins, Baltimore.
3. Alpers, M. (1979): Epidemiology and ecology of kuru. In: *Slow Transmissible Diseases of the Nervous System, Vol. 1*, edited by S. B. Prusiner and W. J. Hadlow, pp. 67–90. Academic Press, New York.
4. Amyx, H. L., Gibbs, C. J., Jr., Gajdusek, D. C., and Greer, W. E. (1981): Absence of vertical transmission of subacute spongiform viral encephalopathies in experimental primates. *Proc. Soc. Exp. Biol. Med.*, 166:469–471.
5. Asher, D. M., Gibbs, C. J., Jr., and Gajdusek, D. C. (1976): Pathogenesis of subacute spongiform encephalophies. *Ann. Clin. Lab. Sci.*, 6:84–103.
6. Bennett, J. H., Gray, A. J., and Auricht, C. O. (1959): The genetical study of kuru. *Med. J. Aust.*, 2:505–508.
7. Bennett, J. H., Rhodes, F. A., and Robson, H. N. (1958): Observations on kuru. I. A possible genetic basis. *Australas. Ann. Med.*, 7:269–275.
8. Bobowick, A., Brody, J., Matthews, M., Roos, R., and Gajdusek, D. C. (1973): Creutzfeldt–Jakob disease: A case control study. *Am. J. Epidemiol.*, 98:381–394.
9. Brown, P. (1980): An epidemiologic critique of Creutzfeldt–Jakob disease. *Epidemiol. Rev.*, 2:113–135.
10. Brown, P. (1982): The author replies (Re: "An epidemiologic critique of Creutzfeldt–Jakob disease"). *Am. J. Epidemiol.*, 115:149–151.
11. Brown, P., and Cathala, F. (1979): Creutzfeldt–Jakob disease in France: I. Retrospective study of the Paris area during the ten-year period 1968–1977. *Ann. Neurol.*, 5:189–192.
12. Brown, P., and Cathala, F. (1979): Creutzfeldt–Jakob disease in France. In: *Slow Transmissible Diseases of the Nervous System, Vol. 1*, edited by S. B. Prusiner and W. J. Hadlow, pp. 213–227. Academic Press, New York.
13. Brown, P., Cathala, F., and Gajdusek, D. C. (1979): Creutzfeldt–Jakob disease in France: III. Epidemiological study of 170 patients dying during the decade 1968–1977. *Ann. Neurol.*, 6:438–446.
14. Brown, P., Cathala, F., Sadowsky, D., and Gajdusek, D. C. (1979): Creutzfeldt–Jakob disease in France: II. Clinical characteristics of 124 consecutive verified cases during the decade 1968–1977. *Ann. Neurol.*, 6:430–437.
15. Brownell, B., and Oppenheimer, D. R. (1965): An ataxic form of subacute presenile polioencephalopathy (Creutzfeldt–Jakob disease). *J. Neurol. Neurosurg. Psychiatry*, 28:350–361.
16. Bruce, M. E. (1978): Amyloid plaques in experimental scrapie: Factors influencing the occurrence of cerebral amyloid in mice. *J. Neuropathol. Exp. Neurol.*, 35:595.
17. Bruce, M. E., and Dickinson, A. G. (1979): Biological stability of different classes of scrapie agent. In: *Slow Transmissible Diseases of the Nervous System, Vol. 2*, edited by S. B. Prusiner and W. J. Hadlow, pp. 71–86. Academic Press, New York.
18. Burger, D., and Hartsough, G. R. (1965): Transmissible encephalopathy of mink. In: *Slow, Latent and Temperate Virus Infections*, edited by D. C. Gajdusek, C. J. Gibbs, Jr., and M. P. Alpers, pp. 297–305. U.S. Government Printing Office, Washington.
19. Burger, D., and Hartsough, G. R. (1965): Encephalopathy of mink. II. Experimental and natural transmission. *J. Infect. Dis.*, 115:393–399.
20. Cathala, F., Chatelain, J., Brown, P., Dumas, M., and Gajdusek, D. C. (1980): Familial Creutzfeldt–Jakob disease. Autosomal dominance in 14 members over 3 generations. *J. Neurol. Sci.*, 47:343–351.
21. Chou, S. M., and Martin, J. D. (1971): Kuru plaques in a case of Creutzfeldt–Jakob disease. *Acta Neuropathol.*, 17:150–155.
22. Davanipour, Z., Alter, M., and Kahana, E. (1982): Re: "An epidemiologic critique of Creutzfeldt–Jakob disease." *Am. J. Epidemiol.*, 115:145–149.
23. Davison, C., and Rabiner, A. M. (1940): Spastic pseudosclerosis (disseminated encephalomyelopathy; cortico–pallidospinal degeneration). Familial and nonfamilial incidence (a clinico–pathologic study). *Arch. Neurol. Psychiatry*, 44:578–598.
24. Dickinson, A. G. (1967): Ataxias and transmissible agents. *Lancet*, i:1166.
25. Dickinson, A. G. (1970): Classification of scrapie agents based on histological and incubation period criteria in mice. In: *Proceedings VIth International Congress of Neuropathology*, pp. 841–842. Masson et Cie., Paris.

26. Dickinson, A. G., and Fraser, H. (1969): Genetical control of the concentration of ME-7 scrapie agent in mouse spleen. *J. Comp. Pathol.*, 79:363–366.
27. Dickinson, A. G., and Fraser, H. (1972): Scrapie: Effect of Dh gene on incubation period of extraneurally injected agent. *Heredity*, 29:91–93.
28. Dickinson, A. G., and Fraser, H. (1979): An assessment of the genetics of scrapie in sheep and mice. In: *Slow Transmissible Diseases of the Nervous System, Vol. 1*, edited by S. B. Prusiner and W. J. Hadlow, pp. 367–385. Academic Press, New York.
29. Dickinson, A. G., Fraser, H., and Outram, G. W. (1975): Scrapie incubation time can exceed natural lifespan. *Nature*, 256:732–733.
30. Dickinson, A. G., and Mackay, J. M. K. (1967): Virus hosts and genetic studies. In: *Methods in Virology, Vol. 1*, edited by K. Maramorosch and H. Koprowski, pp. 19–61. Academic Press, New York.
31. Dickinson, A. G., and Meikle, V. M. H. (1969): A comparison of some biological characteristics of the mouse-passaged scrapie agents, 22-A and ME-7. *Genet. Res.*, 13:213–225.
32. Dickinson, A. G., Meikle, V. M. H., and Fraser, H. (1968): Identification of a gene which controls the incubation period of some strains of scrapie agent in mice. *J. Comp. Pathol.*, 78:293–299.
33. Dickinson, A. G., Meikle, V. M. H., and Fraser, H. (1969): Genetical control of the concentration of ME-7 scrapie agent in the brain of mice. *J. Comp. Pathol.*, 79:15–22.
34. Dickinson, A. G., Stamp, J. T., and Renwick, C. C. (1974): Maternal and lateral transmission of scrapie in sheep. *J. Comp. Pathol.*, 84:19–25.
35. Dickinson, A. G., Stamp, J. T., Renwick, C. C., and Rennie, J. C. (1968): Some factors controlling the incidence of scrapie in Cheviot sheep injected with a Cheviot-passaged scrapie agent. *J. Comp. Pathol.*, 78:313–321.
36. Dickinson, A. G., Young, G. B., Stamp, J. T., and Renwick, C. C. (1965): An analysis of natural scrapie in Suffolk sheep. *Heredity*, 20:485–503.
37. Ferber, R. A., Wiesenfeld, S. L., Roos, R. P., Bobowick, A. R., Gibbs, C. J., Jr., and Gajdusek, D. C. (1973): Familial Creutzfeldt–Jakob disease: Transmission of the familial disease to primates. In: *Genetic and Transmissible Dementias, Proceedings of the Tenth International Congress of Neurology, Barcelona, September 8–15, International Congress Series No. 319*, edited by A. Subirana, J. M. Espadaler, and E. H. Burrows, pp. 358–380. Excerpta Medica, Amsterdam.
38. Fraser, H., and Dickinson, A. G. (1968): The sequential development of the brain lesions of scrapie in three strains of mice. *J. Comp. Pathol.*, 78:301–311.
39. Friede, R. L., and DeJong, R. N. (1964): Neuronal enzymatic failure in Creutzfeldt–Jakob disease. A familial study. *Arch. Neurol.*, 10:181–195.
40. Gajdusek, D. C. (1977): Unconventional viruses and the origin and disappearance of kuru. *Science*, 197:943–960.
41. Gadjusek, D. C., and Alpers, M. (1972): Genetic studies in relation to kuru. I. Cultural, historical and demographic background. *Am. J. Hum. Genet.*, 24:S1–S38.
42. Gajdusek, D. C., and Gibbs, C. J., Jr. (1968): Slow, latent and temperate virus infections of the central nervous system. In: *Infections of the Nervous System*, edited by H. Zimmerman, pp. 254–280. Williams & Wilkins, Baltimore.
43. Gajdusek, D. C., Gibbs, C. J., Jr., and Alpers, M. (1966): Experimental transmission of a kuru-like syndrome to chimpanzees. *Nature*, 209:794–796.
44. Gajdusek, D. C., Gibbs, C. J., Jr., Asher, D. M., Brown, P., Diwan, A., Hoffman, P., Nemo, G., Rohwer, R., and White, L. (1977): Precautions in medical care of, and in handling tissues from, patients with transmissible virus dementia (Creutzfeldt–Jakob disease). *N. Engl. J. Med.*, 297:1253–1258.
45. Gajdusek, D. C., and Zigas, V. (1957): Degenerative disease of the central nervous system in New Guinea. The endemic occurrence of "kuru" in the native population. *N. Engl. J. Med.*, 257:974–978.
46. Galvez, A., Masters, C., and Gajdusek, D. C. (1980): Descriptive epidemiology of Creutzfeldt–Jakob disease in Chile. *Arch. Neurol.*, 37:11–14.
47. Gerstmann, J. (1928): Über ein noch nicht beschreibenes Reflexphänomen bei einer Erkrankung des zerebellar Systems. *Wien. Med. Wochenschr.*, 78:906–908.
48. Gerstmann, J., Sträussler, E., and Scheinker, I. (1936): Über eine eigenartige hereditär-familiäre Erkrankung des Zentralnervensystems. Zugleich ein Beitrag zur Frage des vorzeitigen lokalen Alterns. *Z. Ges. Neurol. Psychiatrie*, 154:736–762.
49. Gibbs, C. J., Jr., Amyx, H. L., Bacote, A., Masters, C. L., and Gajdusek, D. C. (1980): Oral transmission of kuru, Creutzfeldt–Jakob disease, and scrapie to nonhuman primates. *J. Infect. Dis.*, 142:205–207.

50. Gibbs, C. J., Jr., and Gajdusek, D. C. (1971): Transmission and characterization of the agents of spongiform virus encephalopathies: Kuru, Creutzfeldt–Jakob disease, scrapie, and mink encephalopathy. In: *Immunological Disorders of the Nervous System*, edited by L. P. Rowland, pp. 383–410. Williams & Wilkins, Baltimore.
51. Gibbs, C. J., Jr., and Gajdusek, D. C. (1978): Subacute spongiform virus encephalopathies: The transmissible virus dementias. In: *Aging, Vol. 7: Alzheimer's Disease: Senile Dementia and Related Disorders*, edited by R. Katzman, R. D. Terry, and K. L. Bick, pp. 559–567. Raven Press, New York.
52. Gibbs, C. J., Jr., and Gajdusek, D. C. (1978): Atypical viruses as the cause of sporadic, epidemic, and familial chronic disease in man: Slow viruses and human diseases. In: *Perspectives in Virology, Vol. 10*, edited by M. Pollard, pp. 161–198. Raven Press, New York.
53. Gibbs, C. J., Jr., Gajdusek, D. C., and Amyx, H. (1979): Strain variation in the viruses of Creutzfeldt–Jakob disease and kuru. In: *Slow Transmissible Diseases of the Nervous System, Vol. 2*, edited by S. B. Prusiner and W. J. Hadlow, pp. 87–110. Academic Press, New York.
54. Gibbs, C. J., Jr., Gajdusek, D. C., Asher, D. M., Alpers, M. P., Beck, E., Daniel, P. M., and Matthews, W. B. (1968): Creutzfeldt–Jakob disease (spongiform encephalopathy): Transmission to the chimpanzee. *Science*, 61:388–389.
55. Goldberg, H., Alter, M., and Kahana, E. (1979): The Libyan Jewish focus of Creutzfeldt–Jakob disease: A search for the mode of natural transmission. In: *Slow Transmissible Diseases of the Nervous System, Vol. 1*, edited by S. B. Prusiner and W. J. Hadlow, pp. 195–211. Academic Press, New York.
56. Gordon, W. S. (1959): Symposium on scrapie. In: *Proceedings of the 63rd Annual Meeting of the United States Livestock Sanitary Association*, pp. 286–294.
57. Gordon, W. S. (1966): Variation in susceptibility of sheep to scrapie and genetic implications. In: *Report of Scrapie Seminar held at Washington, D. C., January 27–30, 1964*, pp. 53–68. Agricultural Research Service, U. S. Department of Agriculture, ARS Publication No. 91–53, Washington.
58. Goudsmit, J., Morrow, C. H., Asher, D. M., Yanagihara, R. T., Masters, C. L., Gibbs, C. J., Jr., and Gajdusek, D. C. (1980): Evidence for and against the transmissibility of Alzheimer's disease. *Neurology (Minneap.)*, 30:945–950.
59. Goudsmit, J., White, B. J., Weitkamp, L. R., Keats, B. J. B., Morrow, C. H., and Gajdusek, D. C. (1981): Familial Alzheimer's disease in two kindreds of the same geographic and ethnic origin: A clinical and genetic study. *J. Neurol. Sci.*, 49:79–89.
60. Hadlow, W. J., Race, R. E., Kennedy, R. C., and Eklund, C. M. (1979): Natural infection of sheep with scrapie virus. In: *Slow Transmissible Diseases of the Nervous System, Vol. 2*, edited by S. B. Prusiner and W. J. Hadlow, pp. 3–12. Academic Press, New York.
61. Haltia, M., Kovanen, J., Van Crevel, H., Bots, T. A. M., and Stefanko, S. (1979): Familial Creutzfeldt–Jakob disease. *J. Neurol. Sci.*, 42:381–389.
62. Hartsough, G. R., and Burger, D. (1965): Encephalopathy of mink. I. Epizootologic and clinical observations. *J. Infect. Dis.*, 115:387–392.
63. Heidenhain, A. (1928/1929): Klinische und anatomische Untersuchungen über eine eigenartige organische Erkrankung des Zentralnervensystems im Praesenium. *Z. Ges. Neurol. Psychiatrie*, 118:49–114.
64. Hourrigan, J., Klingsporn, A., Clark, W. W., and DeCamp, M. (1979): Epidemiology of scrapie in the United States. In: *Slow Transmissible Diseases of the Nervous System, Vol. 1*, edited by S. B. Prusiner and W. J. Hadlow, pp. 331–356. Academic Press, New York.
65. Jacob, H., Pyrkosch, W., and Strube, H. (1950): Die erbliche Form der Creutzfeldt–Jakobschen Krankheit. (Familie Backer). *Arch. Psychiatr. Nervenk.*, 184:653–674.
66. Jakob, A. (1921): Über eigenartige Erkrankungen des Zentralnervensystems mit bemerkswertem anatomische Befunde. (Spastische Pseudosklerose—Encephalomyelopathie mit disseminierten Degenerationsherden.) *Z. Ges. Neurol. Psychiatrie*, 64:147–228.
67. Jakob, A. (1921): Über eingenartige Erkrankung des Zentralnervensystems mit bemerkenswertem anatomische Befunde (spastische Pseudosklerose—Encephalomyelopathie mit disseminierten Degenerationsherden). *Dtsch. Z. Nervenheilk.*, 70:132–146.
68. Jakob, A. (1923): Spastische Pseudosklerose. In: *Die Extrapyramidalen Erkrankungen*, pp. 215–245. Springer, Berlin.
69. Kahana, E., Alter, M., Braham, J., and Sofer, D. (1974): Creutzfeldt–Jakob disease: A focus among Libyan Jews in Israel. *Science*, 183:90–91.

70. Kimberlin, R. H. (1979): Aetiology and genetic control of natural scrapie. *Nature*, 278:303–304.
71. Kirschbaum, W. R. (1924): Zwei eigenartige Erkrankungen des Zentralnervensystems nach Art der spastischen Pseudosklerose (Jakob). *Z. Ges. Neurol. Psychiatrie*, 92:175–220.
72. Kirschbaum, W. R. (1968): *Jakob–Creutzfeldt Disease*. Elsevier, New York.
73. Kitchin, F. D., Beam, A. G., Alpers, M., and Gajdusek, D. C. (1972): Genetic studies in relation to kuru. III. Distribution of the inherited serum group-specific protein (Gc) phenotypes in New Guineans: An association of kuru and the Gc Ab phenotype. *Am. J. Hum. Genet.*, 24(Suppl.):S72–S85.
74. Kondo, K., and Kuroiwa, Y. (1982): A case control study of Creutzfeldt–Jakob disease: Association with physical injuries. *Ann. Neurol*, 11:377–381.
75. Manuelidis, E. E., and Manuelidis, L. (1979): Experiments on maternal transmission of Creutzfeldt–Jakob disease in guinea pigs. *Proc. Soc. Exp. Biol. Med.*, 160:233–236.
76. Marsh, R. F., Burger, D., and Hanson, R. P. (1969): Transmissible mink encephalopathy: Behavior of the disease in mink. *Am. J. Vet. Res.*, 30:1637–1653.
77. Marsh, R. F., and Hanson, R. P. (1979): On the origin of transmissible mink encephalopathy. In: *Slow Transmissible Diseases of the Nervous System, Vol. 1*, edited by S. B. Prusiner and W. J. Hadlow, pp. 451–460. Academic Press, New York.
78. Masters, C. L., Gajdusek, D. C., and Gibbs, C. J., Jr. (1981): The familial occurrence of Creutzfeldt–Jakob disease and Alzheimer's disease. *Brain*, 104:535–558.
79. Masters, C. L., Gajdusek, D. C., and Gibbs, C. J., Jr. (1981): Creutzfeldt–Jakob disease virus isolations from the Gerstmann–Sträussler syndrome. With an analysis of the various forms of amyloid plaque deposition in the virus-induced spongiform encephalopathies. *Brain*, 104:559–587.
80. Masters, C. L., Gajdusek, D. C., Gibbs, C. J., Jr., Bernoulli, C., and Asher, D. M. (1979): Familial Creutzfeldt–Jakob disease and other familial dementias—an inquiry into possible modes of transmission of virus-induced familial diseases. In: *Slow Transmissible Diseases of the Nervous System, Vol. 1*, edited by S. B. Prusiner and W. J. Hadlow, pp. 143–194. Academic Press, New York.
81. Masters, C. L., Harris, J. O., Gajdusek, D. C., Gibbs, C. J., Jr., Bernoulli, C., and Asher, D. M. (1979): Creutzfeldt–Jakob disease: Patterns of worldwide occurrence and the significance of familial and sporadic clustering. *Ann. Neurol.*, 5:177–188.
82. Masters, C. L., Harris, J. O., Gajdusek, D. C., Gibbs, C. J., Jr., Bernoulli, C., and Asher, D. M. (1979): Creutzfeldt–Jakob disease: Patterns of worldwide occurrence. In: *Slow Transmissible Diseases of the Nervous System, Vol. 1*, edited by S. B. Prusiner and W. J. Hadlow, pp. 113–142. Academic Press, New York.
83. Matthews, W. B. (1975): Epidemiology of Creutzfeldt–Jakob disease in England and Wales. *J. Neurol. Neurosurg. Psychiatry*, 38:210–213.
84. May, W. W., Itabashi, H. H., and DeJong, R. N. (1968): Creutzfeldt–Jakob disease. II. Clinical, pathologic, and genetic study of a family. *Arch. Neurol.*, 19:137–149.
85. McArthur, N. (1964): The age incidence of kuru. *Ann. Hum. Genet.*, 27:341–352.
86. Meggendorfer, F. (1930): Klinische und genealogische Beobachtungen bei einem Fall von spasticher Pseudosklerose Jakobs. *Z. Ges. Neuro.. Psychiatrie*, 128:337–341.
87. Moreau-Dubois, M. -C., Brown, P., Goudsmit, J., Cathala, F., and Gajdusek, D. C. (1981): Biologic distinction between sporadic and familial Alzheimer's disease by an *in vitro* cell fusion test. *Neurology (Minneap.)*, 31:323–325.
88. Neugut, R., Neugut, A., Kahana, E., Stein, Z., and Alter, M. (1979): Creutzfeldt–Jakob disease: Familial clustering among Libyan-born Israelis. *Neurology (Minneap.)*, 29:225–231.
89. Nussbaum, R. E., Henderson, W. M., Pattison, I. H., Elcock, N. V., and Davies, D. C. (1975): The establishment of sheep flocks of predictable susceptibility to experimental scrapie. *Res. Vet. Sci.*, 18:49–58.
90. Parry, H. B. (1962): Scrapie: A transmissible and hereditary disease of sheep. *Heredity*, 17:75–105.
91. Plato, C. C., and Gajdusek, D. C. (1972): Genetic studies in relation to kuru. IV. Dermatoglyphics of the Fore and Anga populations of the Eastern Highlands of New Guinea. *Am. J. Hum. Genet.*, 24 (Suppl.):S86–S94.
92. Rewcastle, N. B., Gibbs, C. J., Jr., and Gajdusek, D. C. (1978): Transmission of familial Alzheimer's disease to primates. *J. Neuropathol. Exp. Neurol.*, 37:679.
93. Rosenthal, N. P., Keesey, J., Crandall, B., and Brown, W. J. (1976): Familial neurological disease associated with spongiform encephalopathy. *Arch. Neurol.*, 33:252–259.

94. Salazar, A. M., Masters, C. L., Gajdusek, D. C., and Gibbs, C. J., Jr. (1983): Syndromes of amyotrophic lateral sclerosis and dementia. Relation to transmissible Creutzfeldt–Jakob disease. *Ann. Neurol. (in press).*
95. Simmons, R. T., Graydon, J. J., Gajdusek, D. C., Alpers, M. P., and Hornabrook, R. W. (1972): Genetic studies in relation to kuru. II. Blood group genetic patterns of kuru patients and populations of the Eastern Highlands of New Guinea. *Am. J. Hum. Genet.*, 24(Suppl.):S39–S71.
96. Steinberg, A. G., Gajdusek, D. C., and Alpers, M. (1972): Genetic studies in relation to kuru. V. Distribution of human gamma globulin allotypes in New Guinea populations. *Am. J. Hum. Genet.*, 24(Suppl.):S95–S110.
97. Stender, A. (1930): Weitere Beitrage zum Kapitel "Spastische Pseudosklerose Jakobs." *Z. Ges. Neurol. Psychiatrie*, 128:528–543.
98. Stockman, S. (1913): Scrapie: An obscure disease of sheep. *J. Comp. Pathol.*, 26:317–327.
99. Traub, R., Gajdusek, D. C., and Gibbs, C. J., Jr. (1977): Transmissible dementia. The relationship of transmissible spongiform encephalopathy to Creutzfeldt–Jakob disease. In: *Aging and Dementia*, edited by M. Kinsbourne and L. Smith, pp. 91–172. Spectrum, Flushing, New York.
100. Wiesenfeld, S. L., and Gajdusek, D. C. (1975): Genetic studies in relation to kuru. VI. Evaluation of increased liability to kuru in Gc Ab-Ab individuals. *Am. J. Hum. Genet.*, 27:498–504.
101. Will, R. G., and Matthews, W. B. (1982): Evidence for case-to-case transmission of Creutzfeldt–Jakob disease. *J. Neurol. Neurosurg. Psychiatry*, 45:235–238.

Subject Index

A areas, dopamine, inbred mice, 56–58, 60, 67–68
Absence seizures, mutations, 20, 25
Acetycholine and dopamine, inbred mice, 63
Acetylator status, 258
Acetylcholine receptor antibody, 265
Acid maltase deficiency, 240–243
Acquired traits, immunity, 262
Acute intermittent porphyria, 256–257
Adoptees
  alcoholism, 154–156
  antisocial traits, 161–162
  mental disorders, Denmark, 105–113
  obesity and thinness, 115–120
  psychoses, review, 122
Adrenergic response, affective disorders, 136
Affective disorder
  adoption study, suicide, 112
  genetics, review, 121–123, 127–136
  pathophysiology, 135–136
  transmission, 131–134
Aggression, mocha mutation, 24–26
Agoraphobia, family transmission, 122
Albinism, visual organization, 20–24
Alcoholism
  adoption studies, 154–156
  and affective disorders, families, 131
  familial transmission, 122, 148–149
  genetic heterogeneity, and sociopathy, 145–162
Alleles
  acid maltase deficiency, 242–243
  hexosaminidase deficiency, 229
  mh mutations, 25
  psychiatric illness, 123–124
  recombinant DNA techniques, 170
Alpha-galactosidase A, X chromosome, 177
Alpha-locus, hexosaminidase
  activation, 228–229
  biochemistry, mutations, 229–232
  disorders, 221–227
Altruism, natural selection, 99–100
Alzheimer's disease
  familial transmission, 284–286
  HLA links, 263–264
Amino acids
  dietary management, disorders, 257
  hexosaminidase deficiency, 230–232
Amino-levulinic acid, 257
Amphetamine
  affective disorders, 136
  inbred mice, dopamine, 64–66, 68–69

Amylo-1,6-glucosidase, 244
Amyotrophic Creutzfeldt-Jakob disease, 278, 280
Amyotrophic lateral sclerosis, HLA, 264–265
Ankylosing spondylitis, 260
Antibodies, HLA factors, 262–263
Antigens
  environmental interaction, MLC, 258–261
  t mutations, development, 47–53
Antisocial personality, see Sociopathy
Arcuate nucleus, inbred mice, dopamine, 56, 58, 60
Arylsulfatase A replacement, 184–185
Assortative mating, 108
Ataxia telangectasia, wasted mutation, 27, 29
Ataxias, dominant, 195–212
Atypical spinocerebellar ataxia, 225
Autosomal transmission
  affective disorders, 131–133
  Creutzfeldt-Jakob disease, 281–282
  enzyme deficiency criteria, 233
Axons
  neural development, 4–5, 8
  optic nerve formation, 10–11
Azores islands, Joseph disease, 199–204, 208

Backer family genealogy, 279–280
BALB/cJ mice, dopamine neurons, 56–73
Behavior, mutation effects, mice, 19–25
Beige mutation, 27, 29
Beta-locus, hexosaminidase, 222–227
  activation, 228–229
  biochemistry, mutations, 229–232
  disorders, 222–227
bg mutation, 26, 29
Bipolar illness
  in affective spectrum, 127–131
  and schizophrenia, families, 125
  transmission, 131–133
Birth injury, schizophrenia, 112
Bisector angle
  granule cell, 79, 82–83
  Reeler mouse, internodes, 84–85
Blastomeres, t mutations, 51
Blood-brain barrier, enzymes, 188–190
Blotchy mutation, 26–27
Body weight, 115–120
"Boston" pattern, 23
Brachydanio rerio, neural development, 7–16
Brain, enzyme replacement, 188–190
Brindled mouse, mottled mutation, 26–29

293

c locus, hypopigmentation, 21-22
C57BL/6J strain
　hypopigmentation, 22
　mocha mutation, 24-26
Carbohydrate, t-associated antigens, 52
Carotid artery, enzyme injections, 189-190
Cartos system, 3
Catalepsy, dopamine, inbred mice, 65-66
Catechol-O-methyltransferase, 135
Caudate size, dopamine, inbred mice, 61-63
CBA/J mice, dopamine neurons, 56-73
Cell axis, 78-79
Cell number, inbred mice, 72-83
Cell surface
　antigens, 52
　differentiation study, 47-48
Cell survival, inbred mice, dopamine, 71
Ceramide, hexosaminidase substrates, 216
Ceramide lactoside lipidosis, 182
Ceramidetrihexosidase replacement, 186
Cerebellar ataxias
　classification, 196-197
　hexosaminidase deficiency, 225-226
　and Joseph disease, 210
　single gene hypothesis, 204-205
Cerebellar cortex, mutations, 28, 37-39
Cerebellar nuclei, mutations, 39-41
Cerebellar outflow degeneration mutation, 28, 40-41
Cerebellar structure, mutation effects, 26-41
Ceroid storage disease, 215
Cheating genes, 96
Chediak-Higashi disease, 27, 29
Cherry-red spots, hexosaminidase, 221-225
Choline acetyltransferase, 62-63
Cholinergic neurons, inbred mice, dopamine, 63
Cholinergic response, affective disorders, 136
Cholinesterase, succinyl choline metabolism, 256
Choroid fissure, 10-11
Chromosome 15, Prader-Willi syndrome, 175
Chromosome masking, 7
Chromosomes
　cytogenetic techniques, 173-178
　Huntington's disease marker, 170-171
　mouse cerebellar mutants, 27-28
　and sperm function, 95-96
Class instructions, genes, 2
Clones, neuron development, fish, 7-16
Cluster analysis
　alcoholism, 151
　sociopathy, 159
cod mutation, 28, 40-41
Color blindness
　affective disorders, 134
　and G6PD, X chromosome, 177
Computer graphics, cell study, 3, 5-6
COMT (Catechol-O-methyltransferase), 135

Contralateral axons, hypopigmentation, 22
Copenhagen study, 105-113
　alcoholism, 154
　schizophrenia, 105-113
Copper therapy, mottled mutation, 26, 29
Corpus striatum, dopamine, inbred mice, 56, 60-63, 67-68, 70
Creatine phosphokinase, psychiatric disorders, 127
Creutzfeldt-Jakob disease, 278-284
　and ataxias, mechanisms, 210-211
　familial aspects, 278-284
　HLA links, 263-264
Criminality, 122, 160-162; see also Sociopathy
Curvature, dentrites, 81
CW3 status, immune response, 263
Cytogenetic techniques, 173-178

Dandy-Walker malformations, sw mutation, 38
Daphnia magna, neural development, 2-6
Darwinism, population genetics, 94
Daughter angle, dentritic trees, 78, 82
db mutation, 24
Debrancher enzyme deficiency, 243-245
Dementia; see also specific forms
　in Creutzfeldt-Jakob disease, 278, 280
　hexosaminidase deficiency, 226
Demographic patterns, natural selection, 102-103
Denmark adoption study
　mental disorders, 105-113
　obesity and thinness, 115-120
Dentate granule cell
　geometry, 82-83, 88-90
　Reeler mouse, 85
Dentritic arbor
　gene control, 91
　geometry, 77-91
　internodal correction process, 83-84
Depression
　adoption study, suicide, 112
　genetics, review, 121-123, 127-137
Dermatomyositis, HLA links, 264
Developmental noise, 6-7
Dexamethasone suppression, depression, 129
Dextroamphetamine, inbred mice, dopamine, 64-66, 68-69
Dh gene, scrapie, 275-276
Diabetes mutation, 24
Diet, genetic predisposition, 257
Differentiation
　antigens, t mutations, embryo, 47-53
　optic nerve formation, 9-12, 15-16
DNA
　chromosome effects, 97-98
　hybridization, 170
　population genetics, 96-98

SUBJECT INDEX

recombinant techniques, 167–172
  Huntington's disease, 171–172
  mental retardation, 173–178
Dominant ataxias, 195–212
Dominant hemimelia gene, scrapie, 275–276
Dopamine agonists, inbred mice, 64–65
Dopamine antagonists, inbred mice, 65–66
Dopamine-β-hydroxylase, 135
Dopamine receptor, inbred mice, 64
Dopamine system, inbred mice, 55–73
  drug response, 64–66
  neuron number, 58–60, 70, 72
  receptor density, 64
  tyrosine activity, 56–61, 66–69
DR-W2 status
  immune response, 263
  multiple sclerosis, 265–266
Drosophilia
  longevity breeding, 102
  sperm function, natural selection, 95–96
  transposons, 98
Drugs, genetic unmasking, 256–258
Duarte protein, 137

EcoRI restriction endonuclease, 168
EEG and alcoholism, 156–157
Electrocorticograph, tottering mutation, seizures, 20–21
Embryo
  clone variation, environment, 7–8
  optic nerve formation, 9–11
  t mutation, antigens, differentiation, 47–53
Encephalopathy
  hexosaminidase deficiency, 220–225
  spongiform type, familial, 273–286
Environment
  genetic predisposition, 255–268
  mental disorder, 105–113
Enzyme replacement strategies, 181–191
Epilepsy, 19–20; see also Seizures
Epinephrine response, affective disorders, 136
Equilibrium value, kin selection, 100
Evolution, population genetics, 93–103
Extroversion and sociopathy, 159
Eysenck's Personality Questionnaire, 158–159

Fabry's disease
  description, 182
  enzyme replacement, 186, 188
  organ grafts, 184
Factor analysis, alcoholism, 151–153
Fathers, alcoholism genetics, 155
Females, fragile X, retardation, 175
Finland, alcoholism, 152–154
Flecked mice, hypopigmentation mutation, 22
Flores Island, Joseph disease, 200, 205

Fragile regions, mental retardation, 174–178
Fucosidosis, 183

G_{A2}-globoside, 216–217
GALA, X chromosome, 177
Galactose, t antigen, 52
Ganglion cells
  development, 3–6, 16
  differentiation, retina, 11
Ganglioside G_{M2} and hexosaminidase, 190, 215–218, 223–224
Gangliosidosis, 183
Gap junction, 4–5
Gaucher's disease
  description, 182
  enzyme replacement, 186–191
GC 2 allele, schizophrenia, 126
GcAb gene, kuru, 277
Gel electrophoresis, DNA separation, 168–170
Gene locus, mental disorders, 123, 125, 131–133
Generalized gangliosidosis, 183
Genes; see also specific aspects
  altruism, 99–100
  natural selection, 93–103
  and relaxed selection, 102–103
Genetic compounds, hexosaminidase, 227
Genetic heterogeneity
  alcoholism and sociopathy, 145–162
  glycogen storage diseases, 239–250
  hexosaminidase deficiency diseases, 215–233
  psychiatric disorders, 123
Genetic markers
  alcoholism, 156–157
  Huntington's disease, 170–172
  recombinant DNA techniques, 167–172
Geometry, see Neuronal form
Gerstmann-Sträussler syndrome, 278
Glia
  in ataxias, mechanism, 210–211
  Joseph disease, 206–207
Gliosis, Joseph disease, 206–207
Globoid leukodystrophy, 182
Globosides
  reduction in, hexosaminidase, 185, 189
  as substrate, biochemistry, 216–217
  Tay-Sachs disease, storage, 216
Globulin, schizophrenia, 127
Glucocerebrosidase replacement, 187–188
Glucose-6-phosphate dehydrogenase, 177
α-Glucosidase, see Acid maltase
Glycogen storage diseases, 239–250
Glycoproteins, t antigen, 52
G_{M1}-gangliosidosis, 215
G_{M2}-ganglioside, hexosaminidase deficiency, 215, 217–218, 223–224
Gold salt therapy, HLA, 267
Gough's So scale, 158–159
Granule cell neuron
  geometry, hippocampus, 77–84

Granule cell neuron (*contd.*)
  mutations, 34–37
Group selection and evolution, 100–101
Growth cones, 9–10, 12
Guillain-Barre syndrome, HLA-links, 264

h proteins, Joseph disease, 206, 209
Haloperidol, inbred mice, dopamine, 65–66
HemA and G6PD, X chromosome, 177
Heminelia gene, scrapie, 275–276
Hemoglobinopathies, hexosaminidase deficiency model, 227–228, 230
Hemolytic anemia, 248
Hereditary ataxias, 195–212
Heterogeneity, *see* Genetic heterogeneity
Heterozygous clones, neuron morphology, 7
Hexosaminidase A
  deficiency of, diseases, 216–218, 222–227
  as isozyme, 217
  molecular genetics, 217–219, 229
  replacement therapy, 185, 188–189
  substrates, 216
Hexosaminidase B
  deficiency of, diseases, 217–218, 222–227
  as isozyme, 217
  molecular genetics, 217–219, 229
  substrates, 216
Hexosaminidase deficiency diseases
  genetic heterogeneity, 215–233
  replacement therapy, 185, 188–189
Hexosaminidase S
  as isozyme, 217
  molecular genetics, 217–218, 229
  mutation effects, 231
5-HIAA (5-Hydroxyindoleacetic acid), 135
Hippocampus, granule cell geometry, 77–84
HLA locus, 258–267
  affective disorders, 133–134
  ataxia genes, 198
  cirrhosis, alcoholism, 156
  and drug reactions, 266–267
  environmental interactions, 258–267
  schizophrenia, 126
Homologous cerebellar mutations, 26–29
Homozygous clones, neural development, 7
hpc mutation, 28, 39
Human leukocyte antigen, *see* HLA
Huntington's disease
  and ataxias, mechanisms, 210
  normal sib reproduction, 99
  recombinant DNA techniques, 167–172
Hybrid dysgenesis, 98
Hybridization, DNA, 170–172
5-Hydroxyindoleacetic acid, 135
Hyperspiny Purkinje cell mutation, 28, 39
Hypomania, 129–130
Hypopigmentation, visual system, 20–24

Hypothalamus
  and dopamine, inbred mice, 56, 58–60
  hormones of, diabetes mutation, 24

Immune system
  environmental interactions, 258–268
  HLA locus, 258–267
  major histocompatibility complex, 258–261
  t mutations, 52–53
Impulsivity and sociopathy, 159
Inbred mouse strains
  cerebellar mutants, 27–28
  dopamine neurons, 55–73
  experimental recommendations, 41
Inbreeding, natural selection, 101
Incest taboo, natural selection, 101
Index of Opportunity for Selection, 102
Infantile encephalopathy, hexosaminidase, 221–225
Inferior olivary neurons, 33
Internodal correction process
  granule cell, 83–84
  Reeler mouse, 84–85
Intracarotid injections, enzymes, 189–190
Ipsilateral axons, hypopigmentation, 22
Isolectric point, hexosaminidase, 231
Isoniazid acetylation, 258

J proteins, Joseph disease, 206, 209
Jakob-Creutzfeldt disease, *see* Creutzfeldt-Jakob disease
Joseph disease, 195–196, 199–212
  Azorean neuroepidemiology, 201–204
  biochemistry, 206–207, 209
  classification, 196
  genetic disease hypothesis, 204–205
  historical review, 199–201
  and OPCA, 208–210
Juvenile encephalopathy, 223–224
Juvenile G_{M2}-gangliosidosis, 224
Juvenile Sandhoff's disease, 224

Kidney transplants, Fabry's disease, 184
Kin selection, 99–100
Klebsiella antigen, 260
Krabbe's disease, 182
Kreisler mutation, 38
Kuru, 276–278

L proteins, Joseph disease, 206, 209
L subunits, phosphofructokinase, 249

SUBJECT INDEX

Laminae
  development, 4–6
  geometry, synthetic cells, 86
Landrace pigs, 256
Lateral geniculate body, albinism, 23
Lc mutation, 27, 31–32
Leaner mutation, 28, 37
Leprosy, HLA links, 263–264
Levamisole, HLA links, 267
Linkage markers
  psychiatric illness, 123–124
    affective disorders, 131–134
    schizophrenia, 126
  recombinant DNA techniques, 167–170
Lip gene, scrapie, 274
Lipid storage diseases, 181–191
Lithium
  affective disorders, 128–129, 136, 266
  HLA factors, 266
Locus ceruleus, inbred mice
  dopamine neurons, 60, 70
  tyrosine hydroxylase activity, 56, 60
Longevity, natural selection, 102
Lurcher mutation, 27, 31–33

M subunit, phosphofructokinase, 249
Machado disease, 199–200, 204–205
Major histocompatibility complex, 258–261
Males
  alcoholism genetics, 155
  fragile X, mental retardation, 174–175
  glycogen storage disease, 248
Malignant hyperthermia predisposition, 256
Maltase, *see* Acid maltase
Manic-depression, *see* Bipolar illness
Mannose enzymes, replacement, brain, 188–189
Mauthner cell morphology, 6–9
McArdle disease, 245–247
ME7 scrapie virus, 275
mea mutation, 28, 38
Meander tail mutation, 28, 38
Medulla, motor neuron morphology, 6–8
Meiosis, natural selection processes, 95–96, 98
Mendelism, population genetics, 94
Menkes' kinky hair disease
  Mo mutation, 26–27
  Purkinje cell appearance, 77, 81
Mental disorders; *see also specific disorders*
  adoptee studies, 105–113
  genetics, review, 121–137
Mental retardation
  normal sib reproduction, 99
  recombinant DNA analysis, 173–178
Menzel type ataxia, 196–197
Mesoderm cells, t mutations, 50–51
Metachromatic leukodystrophy
  description, 182
  enzyme replacement, 184–185

Mh mutation, aggression, 24–26
Mice
  cerebellar mutants, 27–28
  experimental neurogenetics, 19–41
  granule cell geometry, 77–84
  inbred strains, dopamine, 55–73
  scrapie encephalopathy, 275–276
Midbrain dopamine, inbred mice, 56, 58–60, 66–69
"Midwestern" pattern, 23
Mink encephalopathy, 273–276
Mitochondria, chromosomal control, 96
MMPI Pd scale, sociopathy, 159
Mo mutations
  mouse and man, 26–29
  and stumbler mutation, 39
Mocha mutation, aggressiveness, 24–26
Monoamine oxidase
  affective disorders, 135–136
  schizophrenia, 126–127
Monoamine oxidase inhibitors, acetylation, 258
Mood disorders, 112; *see also* Affective disorders
Morula stage, normal and t mutation, 49–51
Mosaic flecked mice, pigmentation, 22
Mothers, alcoholism genetics, 155
Motor activity, dopamine, inbred mice, 65–67, 72
Motor epilepsy, tottering mutation, 19–20
Motor neuron disease
  and Creutzfeldt-Jakob disease, 280
  hexosaminidase, 220, 226
Motor neuron morphology, 6–9
Mottled mutation, *see* Mo mutation
Mouse, *see* Mice
Mouse strains
  cerebellar mutants, 27–28
  experimental recommendations, 41
Muller cell development, 8
Multidimensional models, alcoholism, 149–154
Muscle, glycogen storage diseases, 239–250
Muscular sclerosis, HLA links, 263–266
Mutations
  antigens, 47–53
  contemporary society effects, 102
  hexosaminidase effects, 228–233
  mouse, 19–45
Myasthenia gravis, HLA links, 263–265
Myoglobinuria, 245, 247
Myotonic muscular dystrophy and ataxia, 210

Natural selection
  in contemporary society, 102–103
  population genetics, 93–103
  sperm function, 95
Nerve cells, *see* Neurons

Nerve differentiation, see Differentiation
Nerve outgrowth, 9
Nerve regeneration mechanisms, 11
Nervous mutation, 32–34
Neural development, 1–17
Neurite growth, weaver mutation, 35–36
Neuroleptics, inbred mice, dopamine, 65, 68–69
Neuronal form
   gene control, 91
   mouse granule cell, 77–84
   Reeler mouse, 84–85
   synthetic cells, 84–91
Neurons
   development, 1–17
   inbred mice, 71–73
   morphology, 5–9
Neurotransmitters, inbred mice, 61–64
Neutral maltase, glycogen storage disease, 241
Newborns, optic nerve formation, 10
Niemann-Pick disease, 182
Nigostriatal system, dopamine, 67–68, 70
Nonspecific X-linked mental retardation, 173–178
Norepinephrine system, inbred mice, 58–60
nr mutation, 32–34
Nucleus accumbens, tyrosine hydroxylase, 56

Obesity, Danish adoption register, 115–120
Olfactory bulb, tyrosine hydroxylase, 56
Olfactory tubercle, tyrosine hydroxylase, 56
Olivary neurons, 33
Olivopontocerebellar atrophy, 196–199
   classification, 196–197
   clinical manifestations, 198–199
   and Joseph disease, 208–210
   pathology, 197–198
OPCA, see Olivopontocerebellar atrophy
Ophthalmoparesis, Joseph disease, 201, 203
Optic chiasm, albinism, 23
Optic ganglion development, 3–6
Optic nerve formation, 9–15
Optic tectum
   development, Zebra fish, 9
   optic nerve connections, 11–16
Organ allografts, 181–184
Organelles, chromosomal control, 96

P subunit, phosphofructokinase, 249
Panic disorder, family transmission, 122
Parkinsonism, dopamine cell number, 73
Parthenogenesis, water flea, 2
Path analysis, affective disorders, 133
Pc 1 Duarte protein, 137
pcd mutation, 27, 32
Pedigree analysis
   affective disorder, 133
   schizophrenia, 124–125

Pewter mutation, 26–27
Phenobarbital, genetic unmasking, 257
Phenothiazines, inbred mice, dopamine, 65–66
Phenotypes, psychiatric disorders, 145–146
Phenylalanine defects, diet, 257
Phenylketonuria and diet, 257
Phobic disorders, families, 122
Phosphofructokinase deficiency, 248–249
Phosphoglycerokinase, X chromosome, 177
Phosphoribosyl pyrophosphate synthetase, 177
Phosphorylase deficiency, 245–247
Photoreceptors
   development, 3–6
   nervous mutation, 33
pI, hexosaminidase deficiency, 231
Pigment, visual system, 20–24
Planar angle
   cell geometry, 79
   granule cell geometry, 79, 82–83
   Reeler mouse, internodes, 84–85
Poecilia formosa, motor neurons, 6–9
Poliomyelitis, HLA links, 263–264
Pompe disease, 240
Population genetics, 93–103
Portuguese, Joseph disease, 199–201
Posterior ventricular nucleus, dopamine, 56, 60
Postnatal period, dopamine, inbred mice, 66–68, 71
Prader-Willi syndrome, 175–178
Preoptic region, dopamine, inbred mice, 56, 60
Programmed cell death, 33
Proteins, Joseph disease, 206, 209
Psychiatric disorders
   adoption study, Denmark, 105–113
   genetics, review, 121–137
Psychopathy, see Sociopathy
Psychoticism and sociopathy, 159
Purkinje neuron
   degeneration, 27, 32
   geometry, 77, 79–81
   mutation effects, 30–34
   weeping willow, Menke's disease, 81
Zellweger malformation, 80

Receptors
   differentiation study, 47
   dopamine, inbred mice, 64
Recombinant DNA techniques, 167–172
   Huntington disease, 167–172
   mental retardation, 173–178
Reeler mutation
   description, 27, 30, 33
   internodal correction, granule cells, 84–85
Regeneration, nerve, 11
Renpenning's syndrome, 173, 175
Residual Hex A, Tay Sachs, 222
Restriction endonucleases, 168

SUBJECT INDEX

Retina
    optic nerve formation, 9, 11, 13
    tyrosine hydroxylase, inbred mice, 56
Retinal artery, 10
Retinal ganglion cells, 22
Rheumatoid arthritis, HLA factors, 267
Ricinus II, 52
rl mutation, *see* Reeler mutation
RNA tumor virus, 97
Rolling mouse Nagoya, 28, 37

S10 antigen, X chromosome, 177
Sandhoff's disease
    enzyme replacement, 185, 189
    hexosaminidase deficiency, 217–220, 222–225
    molecular genetics, 218–220, 229
Schizoaffective disorder, 125
Schizophrenia
    adoption study, 105–113
    genetics, review, 121–127
    HLA links, 264, 266
    normal sib reproduction, 99
Schut-Haymaker type ataxia, 196–197
Scrapie encephalopathy, 273–276
Season of birth, schizophrenia, 112
Segregation analysis, affective disorders, 133
Segregation distorter genes, 95–96
Seizures
    mocha mutation, 25
    tottering mutation, mouse, 19–20
Self-reported weight, 116–118
Selfish DNA, 97–98
Sex factors
    alcoholism, 151, 155
    fragile X, mental retardation, 174–175
sg mutations, 27, 31, 33
Sheep, scrapie, 273–275
Siamese cats, albinism, vision, 23
Sibs, natural selection, 99
Sinc locus, scrapie, 275–276
Single-locus transmission, 131–134
Sip gene, scrapie, 274
Skeletonized dentritic arbor
    granule cell, 81–82, 85, 88
    synthetic cell, 88–89
Socialization Scale, Gough, 158–159
Sociobiology, population genetics, 100
Socioeconomic status, adoption, 107–108
Sociopathy
    adoption studies, 161–162
    family and twin studies, 122, 160–161
    genetic heterogeneity, and alcoholism, 145–162
Somatization disorders, 161
Sousa family, 200–204
Sperm, natural selection processes, 95–96
Spermatozoa, t mutation effects, 52

Spike-wave paroxysms, tottering mutation, 20–21, 37
Spingolipid storage diseases, 181–191
Spinocerebellar ataxia, hexosaminidase, 225–226
Spiroperidol and dopamine, inbred mice, 64–66
Spongiform encephalopathies, 273–286
Staggerer mutation, 27, 31, 33
Stereotypy, inbred mice, dopamine, 65
Stratum periventricularum, 15
Stress, genetic expression, 255–256
Striatal dopamine, inbred mice, 60–63, 67–68, 70
Stumbler mutation, 28, 38–39
Subacute sclerosing panencephalitis, 263–264
Substantia nigra dopamine, inbred mice, 56–57, 60, 67–68
Succinyl choline, 256
Suicide, adoption study, 112–113
Suppressor cell genes, 261–262
Swaying mutation, 28, 38
Synthetic cells, geometry, 84–91

T cells, major histocompatibility complex, 261
t gene mutations, 48–53, 96
t^{12} mutation, 51–53
Tarui-Layzer disease, 248–249
Tay-Sachs disease
    description, 182
    enzyme replacement, 181–185, 189–191
    hexosaminidase deficiency, 215–233
Tay-Sachs variant, *see* Sandhoff's disease
Tectum, fish, optic nerve, 11–16
Tecumseh population, 119
Testicular cells, t antigens, 52
tg mutation, *see* Tottering mutation
Thermolability, schizophrenia, 127
Thiamine deficiency
    alcoholism, 156
    neurological disorders, 257–258
Thinness, Danish adoption register, 115–120
Thomas family, 199, 204
Tor mutation, 28, 36–37
Torsion, dentrites, 81
Tortured mutation, 28, 36–37
Tottering mutation
    description, 28, 37
    and epilepsy, mouse, 19–21
Traits versus states, 124
Transketolase, thiamine deficiency, 257–258
Transposons, DNA, 97–98
Tricyclic response, mania, 129
Twins
    alcoholism, 149, 152–153
    psychiatric disorder, 122
    sociopathy and criminality, 160–161
Tyrosine hydroxylase
    diseases of aging, dopamine, 72–73

Tyrosine hydroxylase (*contd.*)
  and dopamine, inbred mice, 55–73

Ultraselfish genes, 96
Unipolar affective disorder, 127–133

Ventral tegmentum, dopamine, inbred mice, 59, 61
Vertical transmission, 282, 284
Vibrator mutation, 28, 39–40
Viruses
  ataxias, 210–211
  DNA markers, 168
  immune system, and environment, 260–267
Visual evoked potentials, albinism, 23
Visual pathways
  development, 1–17
  hypopigmentation, 20–24
Vochsyia tree branches, 77–78
Vulnerability markers, 124

Wasted mutation, 27, 29
Water flea, neural development, 1–6
Weaver mutation, 28, 34–36
"Weeping willow" configuration, 35, 81
Weight, genetic factors, 115–120
Wernicke-Korsakoff syndrome
  and alcoholism, thiamine, 156
  genetic defect, 257–258
Wilson's disease, diet, 257
"Wine glass" shape, granule cell, 77, 79, 82
wst mutation, 27, 29
wv mutation, 28, 34–36

X-chromosome
  affective disorders, 131–134
  intracellular natural selection, 94
  mental retardation, rDNA techniques, 173–178
Xg blood group, 177

Zebra fish, optic nerve development, 7–16
Zellweger malformation, 77, 80